PIMLICO

116

ON THE PSYCHOLOGY OF MILITARY INCOMPETENCE

Dr Norman F. Dixon, M.B.E., Fellow of the British Psychological Society, is Professor Emeritus of Psychology at University College London.

After ten years' commission in the Royal Engineers, during which time he was wounded ('largely through my own incompetence'), Professor Dixon left the Army in 1950 and entered university where he obtained a first-class degree in Psychology. He received the degrees of Doctor of Philosophy in 1956 and Doctor of Science in 1972, and in 1974 was awarded the University of London Carpenter Medal 'for work of exceptional distinction in Experimental Psychology'. He holds an honorary doctorate from the University of Lund. His other books include *Preconscious Processing, Subliminal Perception: the nature of a controversy*, which was described by Professor George Westby as 'one of the most substantial works of British psychology of recent years', and *Our Own Worst Enemy*, which *New Society* praised as 'an elegant play on man's chaotic nature . . . diverse and arresting'.

ON THE PSYCHOLOGY OF MILITARY INCOMPETENCE

NORMAN DIXON

With a foreword by Brigadier Shelford Bidwell

PIMLICO

PIMLICO

An imprint of Random House
20 Vauxhall Bridge Road, London SW1V 2SA

Random House Australia (Pty) Ltd
20 Alfred Street, Milsons Point, Sydney
New South Wales 2061, Australia

Random House New Zealand Ltd
18 Poland Road, Glenfield
Auckland 10, New Zealand

Random House South Africa (Pty) Ltd
PO Box 337, Bergvlei, South Africa

Random House UK Ltd Reg. No. 954009

First published by Jonathan Cape Ltd 1976
Pimlico edition 1994

5 7 9 10 8 6

Printed and bound in Great Britain by
Mackays of Chatham PLC, Chatham, Kent

ISBN 0–7126–5889–0

To Christine, Camilla and Rachel

Contents

PREFACE   9
FOREWORD BY BRIGADIER SHELFORD BIDWELL   11

1   Introduction   17

PART ONE

2   Generalship   27
3   The Crimean War   36
4   The Boer War   52
5   Indian Interlude   68
6   The First World War   80
7   Cambrai   86
8   The Siege of Kut   95
9   Between the Wars   110
10   The Second World War   123
11   Singapore   130
12   Arnhem   145

PART TWO

13   Is There a Case to Answer?   151
14   The Intellectual Ability of Senior Military Commanders   157
15   Military Organizations   169
16   'Bullshit'   176
17   Socialization and the Anal Character   189
18   Character and Honour   196

19 Anti-Effeminacy   208
20 Leaders of Men   214
21 Military Achievement   238
22 Authoritarianism   256
23 Mothers of Incompetence   280
24 Education and the Cult of Muscular Christianity   288

PART THREE

25 Individual Differences   305
26 Extremes of Authoritarianism   309
27 The Worst and the Best   318
28 Exceptions to the Rule?   354
29 Retreat   393

AFTERWORD   405
NOTES   407
BIBLIOGRAPHY   424
INDEX   440

Preface

This book is not an attack upon the armed forces nor upon the vast majority of senior military commanders, who, in time of war, succeed in tasks which would make the running of a large commercial enterprise seem child's play by comparison.

It is, however, an attempt to explain how a minority of individuals come to inflict upon their fellow men depths of misery and pain virtually unknown in other walks of life.

The book involves the putting together of contributions from a great many people—historians, sociologists, psychologists and of course soldiers and sailors. It is hoped that none of these will feel misrepresented in the final picture which their contributions make. For errors of fact, and for the opinions expressed, I alone take full responsibility.

In the writing of this book I owe a very great debt of gratitude to all those who gave generously of their time to reading and discussing earlier drafts. Their encouragement, criticisms and advice have been invaluable. In particular I would like to thank Mr Ronald Lewin, Captain Donald Macintyre, R.N., Brigadier Shelford Bidwell, Dr Penelope Dixon and Dr Hugh L'Etang for the many sorts of help they gave at every stage.

For the long hours she spent carrying out research, checking contents, and assisting with the index I owe a great debt of gratitude to Dr S. H. A. Henley.

For their generous assistance I should also like to thank Dr Halla Beloff, Mr Brian Bond and Dr Michael Dockrill of King's College, Mr Russell Braddon, Wing-Commander F. Carroll, Mr Alex Cassie, Miss Coombes of the Imperial War Museum, Professor George Drew,

Professor H. J. Eysenck, Mr Robert Farr, General Sir Richard Gale, General Sir John Hackett, Professor J. R. Hale, Professor D. O. Hebb, Mr Carl Hixon, Dr Norman Hotopf, Mr Michael Howard, Mr John James, Dr Denis Judd, Mr John Keegan and Mr Keith Simpson of the Royal Military Academy, Dr R. P. Kelvin, Sir Patrick Macrory, Lieutenant-Colonel Brian Montgomery, Lieutenant-General Sir Denis O'Connor, Professor Stanley Schachter, Mr Jack Smithers, Dr Ivor Stilitz, Dr A. J. P. Taylor and Dr Rupert Wilkinson. I would also like to express my gratitude to Miss Julie Steele for her secretarial assistance, to Miss Susannah Clapp and Mrs Jane Spender for editorial help, and to the librarians of University College, King's College, the Royal United Services Institute and Rye Public Library for their unfailing courtesy and helpfulness.

For permission to quote extracts from works in which they hold the copyright I am most grateful to: Russell Braddon, Jonathan Cape Ltd and The Viking Press, Inc., for THE SEIGE by Russell Braddon; Alan Clark, for his THE DONKEYS; and Simon Raven and *Encounter*, for 'Perish by the Sword' by Simon Raven.

Finally, I owe a debt of gratitude to that handful of people (who would probably prefer to remain nameless) whose hostility and dismay that anyone should write a book on military incompetence provided considerable, if unlooked-for, confirmation of the relationship between militarism and human psychopathology.

N.F.D.

Foreword

One day, I hope, someone will write the history of the impact of science on the conduct of warfare and also of what are loosely called 'defence studies'. When he does, I am certain that he will find this book by Dr Norman Dixon, for which I am privileged to write a foreword, to have been an important landmark. Norman Dixon is specifically concerned with the subject of leadership on the highest level, or 'generalship', which he seeks to illuminate by bringing his own branch of science, experimental psychology, to bear; but before discussing his theme from the point of view of a professional military student, it might clear the ground, perhaps, if I adumbrated, or anticipated, the history of the relationship of scientists and soldiers.

We should begin by reminding ourselves that war is only partly a rational activity directed at useful goals or benefits, such as survival, or the acquisition of desirable territory. The classical military historian sees political or religious causes playing their part as irritants; the Marxist sees purely economic factors; while others, perhaps, see the cause and conduct of war as embedded in, and the consequence of, specific cultures. The study of warfare is, perhaps, a branch of sociology. To satisfy ourselves on this last point we do not have to go very far back in history or even to leave the present. Wars are not fought solely with 'victory' as the object—victory being defined, presumably, as a net gain of benefits over costs—but for 'glory'. To achieve 'glory' the war had to be conducted according to certain rules, using only certain honourable weapons and between soldiers dressed in bizarre and often unsuitable costumes. The bayonet, the sabre and the lance were more noble than the firearm (one British cavalry regiment on being issued with carbines for the first time in the mid-nineteenth century ceremonially put the first consignment into a barrow and tipped it on to the stable dung-pile).

The leaders of such armies were chosen from corps of officers who were not recruited primarily for prowess or intelligence, but because they conformed to certain social criteria. They, for instance, had to be noble, or to profess a certain religion, or, where nobility was not a passport to rank, to belong to the appropriate class or caste. This is why successful generals when they emerge appear to be freaks or mavericks; and also, perhaps, why such a maverick as Wellington found it necessary to convert himself into a British aristocrat in the course of his ascent to fame. It also accounts for the sudden appearance of a plethora of competent generals when the mould of a society is broken, as it was by the French and Russian Revolutions, or when a new, classless and casteless society evolves, as it did in the United States in the nineteenth century. The best generals on both sides in the American Civil War could probably have beaten any comparable team from Europe, for the war made the profession of generalship a career open to talent and freed it from the rule of the authoritarians who flourish in rigid societies.

The 'scientific' breakthrough really came in the early part of this century, and I would like to dwell on this for a moment in spite of the fact that it lies in the province of applied science and engineering rather than that of behavioural sciences. 'Science' was useful, but that there could be a 'science' of war in the sense that scientific modes of thought could be used in strategic problems was incomprehensible. Navies remained rigidly authoritarian in outlook and hierarchical in structure, but at the same time our Royal Navy, for instance, was extraordinarily open-minded and imaginative in the purely technical field. The great battleships of 1914 had highly sophisticated systems of fire control, equipped, even, with rudimentary analogue computers; the importance of the submarine was grasped; and naval aviation pioneered. Unfortunately, on land, in the First World War, the tactics of Malplaquet or Borodino were combined with the killing power of modern technology, with the bloodiest of results. This tragedy did not arise solely from incompetence: the march of science so far had provided weapons to kill but not the essential apparatus for command and control. Scientists were still only asked for *tools*. No one then dreamt of asking them the question 'How shall we do it?' — to receive the teasing, or baffling, question in response, 'Why do you want to do it at all?' Not until the Second World War did we see the birth of 'operational analysis' and men of the quality of Lindemann, Tizard and Blackett and, later on, in the 1960s, Zuckermann, brought in for the purpose of pure *thinking*.

The application of the behavioural sciences followed exactly the same cycle one war later. 'Psychology' was shrouded with myth and its application blocked by subconscious fears. It was confused with psychiatry, and psychiatrists were concerned with 'mad' people, and, moreover, were soft on discipline. To allow them to participate in leader selection, asking awkward questions about *sex*, was repugnant to many officers and the resistance offered by military commanders to their use was naturally deep and obdurate. Only the insistence of one of the most enlightened men ever to occupy the post of the Adjutant-General of the British Army, General Sir Ronald Adam, overcame these obstacles. Between 1939 and 1945 army psychiatrists, and subsequently psychologists, made the most valuable contributions, quite outside their purely clinical field, to the questions of training, officer selection, 'job-satisfaction' and discipline. Both the Royal Air Force and the United States Air Force made good use of both branches of the science in the field of the effects of stress and motivation, which hitherto had been dominated by purely moral and unscientific assumptions. By the end of the Second World War we knew a great deal about the nature of leadership on the level of pilots and platoon commanders. But no one so far has had the temerity to apply the same criteria to generals, and this is why I think Norman Dixon's book is by way of being a landmark.

He is a bold man. The subject of generalship is peculiarly the province of military historians of 'classical' outlook, who are perfectly ready to fall on each other, let alone any outsider who may trespass therein, and also of the new wave of social scientists and professors of international relationships and politics whose minds are not necessarily any more open than those of their military colleagues. Norman Dixon is therefore likely to come under a hot fire from several quarters. Fortunately, he is accustomed to heat. As a former regular officer in the Royal Engineers, including nine years in bomb disposal, he was moulded in a corps where intellect habitually meets danger and he has exchanged his old discipline for a new one to become an experimental psychologist. I cannot think of anyone better qualified to attempt this synthesis.

It must be emphasized that his book is neither yet another fashionable attack on British generals, nor one of those fascinating but immature exercises in arranging the heroes of the military pantheon in order of merit, as if picking a world cricket team. Psychologists (he argues) can identify a distinct personality type in whom a fundamental conflict between the dictates of conscience and the need for aggression may seriously interfere with the open-mindedness, imagination and intellect needed to reach correct decisions. Obviously the human personality is far too complex to be represented by a simple stereotype, but Norman Dixon's approach is to use the well-documented 'authoritarian' personality as a template against which to measure some famous commanders.

In my view, at any rate, Norman Dixon's theme does not upset the 'classical' appreciation of the characteristics of a successful general. Surely, he resolves the problem of conflicting qualities: ruthlessness and consideration, relentless pursuit of the aim and flexibility of approach, which so confuse the old-fashioned historian. He speaks, in modern terms, of the 'noise' which the general must filter out from the total input of information he receives in the stress and confusion of battle. But in classical terms, this is old and familiar to us; was it not once said of Massena that 'his mental faculties redoubled amid the roar of cannon'?

I believe that this book should be required reading at all places where future officers are selected, trained or prepared for higher command. Both professional soldiers and the equally useful generation of young academic students of warfare will find new knowledge and valuable insights in this challenging study of how some men in high command may react when under the appalling stresses of war.

SHELFORD BIDWELL

Competence, then, is the free exercise of dexterity and intelligence in the completion of tasks, unimpaired by infantile inferiority.

E. H. ERIKSON, *Youth, Change and Challenge*

With 2,000 years of examples behind us we have no excuse when fighting, for not fighting well.

T. E. LAWRENCE, letter, in Liddell Hart, *Memoirs*

No general ever won a war whose conscience troubled him or who did not want 'to beat his enemy too much'.

BRIGADIER SHELFORD BIDWELL, *Modern Warfare*

I

Introduction

'. . . We only wish to represent things as they are, and to expose the error of believing that a mere bravo without intellect can make himself distinguished in war.' C. VON CLAUSEWITZ, *On War*

By now most people have become accustomed to, one might almost say blasé about, military incompetence. Like the common cold, flat feet or the British climate, it is accepted as a part of life—faintly ludicrous but quite unavoidable. Surely there can be nothing left to say about the subject.

In fact, military incompetence is a largely preventable, tragically expensive and quite absorbing segment of human behaviour. It also follows certain laws. The first intimation of this came to the writer during desultory reading about notorious military disasters. These moving, often horrific, accounts evoked a curious *déjà vu* experience. For there was something about these apparently senseless goings-on which sent one's thoughts along new channels, making contact with phenomena from quite other, hitherto unrelated, contexts; and then back again to the senseless facts, not now quite so senseless, until gradually a theme, continuous as a hairline crack, could be discerned throughout the stirring tales of derring-do.

If this pattern was real, and meant what it seemed to mean, certain predictions would follow. These were tested and found correct. Yet other pieces began falling into place, until gradually the mosaic of elements took on the semblance of a theory. This book is about that theory. It is concerned with placing aspects of military behaviour in the context of general psychological principles.

This sounds fine—a cheerful marriage of history and psychology. Unfortunately, however, such a union may not be entirely agreeable to some of the potential in-laws. Judging from the attitude of some historians, a putting together of psychology and history is, to say the least, bad form, while a putting together of psychology and *military* history is positively indecent. There are at least two reasons for this anxiety. The first is that since there are few things more annoying than having one's behaviour *explained*, there exists a natural distaste for explanations of historical figures with whom one perhaps identifies.

The second reason is a distrust of reductionism—of the idea that anything so complex as a military disaster could possibly be *reduced* to explanations in terms of the workings of the human mind, and this by a psychologist (of all people).

In answer one can only say that of course historians know more about history than do psychologists. Of course historical events are determined by a complex set of variables—political, economic, geographical, climatic and sociological. But ultimately history is made by human beings, and whatever other factors may have contributed to a military disaster, one of these was the minds of those who were there, and another the behaviour to which these minds gave rise. Now these are complex variables; hence it *has* been necessary to play down the other factors in order to focus more clearly upon possible psychological determinants. Consider the analogous case of aircraft accidents. Nobody would deny that aeroplanes crash for a number of different reasons, sometimes working independently, sometimes in unison; but this does not mean that the selecting out for particular study of a single factor, such as metal fatigue, necessitates dwelling on such other variables as bad weather, indifferent navigation, or too much alcohol in the bloodstream of the pilot.

The case for a reductionist approach, however, also rests upon another consideration: namely that the nature of military incompetence and those characteristics which distinguish competent from incompetent senior commanders have shown a significant lack of variation over the years, despite changes in the other factors which shape the course of history. Whether they are well equipped or ill equipped, whether they are in control of men who are armed with spears or men with tanks and rockets, whether they are English, Russian, German, Zulu, American or French, good commanders remain pretty much the same. Likewise, bad commanders have much in common with each other.

One rewarding by-product of writing this book has been the many enjoyable conversations I have had with people in the armed services.

Here again, however, a very small minority viewed the enterprise with dismay, as something lacking in taste if not actually bordering on the sacrilegious.*

To this understandable sensitivity I can only say that no insult is intended. In point of fact, for devotees of the military to take exception to a study of military incompetence is as unjustified as it would be for admirers of teeth to complain about a book on dental caries. In an imperfect world the activities of professional fighters are presumably as necessary to society as those of the police, prostitutes, sewage disposers and psychologists. It is just because we cannot do without these callings (except, possibly, the last) that any serious attempt to understand their peculiarities should be welcomed and, indeed, taken as a compliment. For it is a token of their importance that they should merit such attention. Moreover, it is only by contemplation of the incompetent that we can appreciate the difficulties and accomplishments of the competent. If there were no incompetent generals it might appear that the direction of armies and the waging of war were easy—tasks well within the compass of all who had the good fortune to reach the highest levels of military organizations.

However, it is not only when contrasted with the inept that great commanders look their best, but also when seen in the context of the organizations to which they belong. The thesis will be developed that the possibility of incompetence springs in large measure from the unfortunate if unavoidable side-effects of creating armies and navies. For the most part these tend to produce a levelling down of human capability, at once encouraging to the mediocre but cramping to the gifted. Viewed in this light, those who have performed brilliantly in the carrying of arms may be considered twice blessed, for they achieved success *despite* the stultifyingly bad features of the organization to which they happened to belong. This alone would seem to justify an unabashed excursion into the realms of military incompetence. But there are additional grounds, if anything more pressing. They concern the related issues of cost and probability.

While few would dispute that the *cost* grows exponentially with the growth of technology, so that the price of wrong decisions must now be reckoned in mega-deaths, the *chance* of military incompetence

---

* It is fair to add that certain common characteristics of those civilians and servicemen who took the extreme view provided a very useful clue as to the possible origins of military incompetence.

remains a matter for debate. We might hope that this would be a declining function of better education, more realistic values, greater fear of immeasurably worse consequences, and a decrease in jingoism. But there are strong grounds for taking the pessimistic view that the chance, like the cost, continues to increase with positive acceleration.

Several reasons may be advanced for this depressing hypothesis. Firstly, the gap between the capabilities of the human mind and the intellectual demands of modern warfare continues that expansion which started in the eighteenth century. It is probably opening from both sides. While modern war becomes increasingly swift and deadly, and the means by which it is waged increasingly complex, the intellectual level of those entering the armed services as officers could well be on the wane. This tentative supposition is based on the fact that fewer and fewer of the young consider the military to be a worthwhile career. One has only to look at contemporary recruiting advertisements to realize the evident difficulties of finding officer-material. They spare nothing in their efforts to convince an unresponsive youth. The services are depicted as glittering toyshops, where handsome young men enjoy themselves with tanks and missiles while basking in the respect of lower ranks hardly less godlike than themselves. In their eagerness to drum up applicants these calls to arms attempt the mental contortion of presenting the services as a classless society in which officers nevertheless remain gentlemen. The clear implication of such expensive pleading can surely be only that the market for a military career is shrinking, to say the least. To meet this fall-off in officer recruitment insufficient has been done, in the writer's opinion, to improve the real as opposed to the advertised incentive-value of a military career.

Needless to say, a perceived decline in the attractiveness of a military career may actually deter those who might otherwise have opted for one. According to Alexis de Tocqueville, this is particularly so in democratic armies during times of peace. 'When a military spirit forsakes a people, the profession of arms immediately ceases to be held in honour, and military men fall to the lowest rank of public servants; they are little esteemed and no longer understood ... Hence arises a circle of cause and consequence from which it is difficult to escape—the best part of the nation shuns the military profession because that profession is not honoured, and the profession is not honoured because the best part of the nation has ceased to follow it.'[1]

In short, possibly *less* able people are being called upon to carry out a *more* difficult task with a *heavier* price to pay for error, and at the highest levels their responsibilities are staggering.

In the Vietnam war alone, military commanders were responsible for executing policies which cost the United States 300 billion dollars. They were responsible for releasing thirteen million tons of high explosives (more than six times the weight of bombs dropped by the U.S.A. in all theatres during the whole of the Second World War). They were responsible for the delivery of 90,000 tons of gas and herbicides. And they were responsible for the deaths of between one and two million people. These are great responsibilities. Errors of generalship on this scale would be very costly.

Of course many of the arguments put forward in this book are equally applicable to other human enterprises. Indeed, there is no reason to suppose that incompetence occurs more frequently in military subcultures than it does in politics, commerce or the universities. There are, however, apart from the heavy cost of military disasters, special reasons for studying cases of military ineptitude.

The first is that military organizations may have a particular propensity for attracting a minority of individuals who might prove a menace at high levels of command, and the second is that the nature of militarism serves to accentuate those very traits which may ultimately prove disastrous. In theory, then, errors of generalship could be prevented by attention to these causes.

Thirdly, the public has, at least in the democracies, some real say as to who should make its political decisions. This control does not apply to generals. Even the *worst* government and *most* inept prime minister come up for possible dismissal every so often. This is not true of armies and navies. We may have the governments we deserve but have sometimes had military minds which we did not.

Fourthly, if one of the main differences between military and political organizations is in the degree of public control, that between the military and commerce lies in decision pay-offs.* A wrong decision by a company chairman or board of directors may cost a great deal of money and depress a sizeable population of shareholders but military errors have cost hundreds of thousands of lives and untold misery to civilians and soldiers alike.

* So relatively trivial and unimportant are most academic decisions that it would be arrogant to discuss them in the same breath. But similar principles apply.

But the case for a study of military incompetence also rests upon other issues. Not the least of these is the need to examine a view of military behaviour diametrically opposed to, though in its way no less extreme than, that of people who would vehemently defend senior commanders against even the faintest breath of criticism.

This other, hypercritical stance seems remarkably widespread. Thus, for many people with whom the author discussed the central topic of this book the notion of military incompetence struck an immediate and responsive chord. Rejoinders ranged from 'You'll have no shortage of data' to 'Surely that's the whole of military history!'.

But when pressed for details there was a tendency to become vague, and retire behind a '1066 and All That' attitude to the subject. Psychological causes were usually reduced to a single factor: low intelligence or, as one historian has put it, the 'bloody fool theory' of military history. Doubtless this view has been contributed to by such recent books on military ineptitude as Alan Clark's *The Donkeys*, an abrasive critique of generals in the First World War. Certainly its title, taken from the famous conversation between Ludendorff and Hoffman* and such captions as 'Donkey decorates Lion' (below a photograph of a general pinning a medal on a lance-corporal), seemed to suggest an equation of incompetence with mulish stupidity. The contents of the book imply, however, that while stupidity may possibly have played a part, limited intelligence was certainly not the cause of the behaviour for which the generals have been criticized. Judging from the spate of books among which *The Donkeys* appeared, it looked as if a tabu had been lifted on peering into the military woodshed. But, mixing our rural metaphors, the erstwhile sacred cows were once more being transmogrified into nothing more than very unsacred asses. Thus one historian has ascribed a series of military mishaps to 'bone-headed leadership',[2] another spoke of 'the long gallery of military imbecility',[3] while a third has said of British soldiers that 'their fate was decided for them by idiots'.[4] The view taken here is that besides being unkind, these views are probably invalid.

The hypothesis of intellectual incapacity leaves two questions quite unanswered. How, if they are so lacking in intelligence, do people

* According to the memoirs of Field-Marshal Von Falkenhayn (cited by Alan Clark), Field-Marshal von Ludendorff's comment 'the English soldiers fight like lions' was greeted by Major-General Max Hoffmann with, 'True, but don't we know they are lions led by donkeys.'

become senior military commanders? And what is it about military organizations that they should attract, promote and ultimately tolerate those whose performance at the highest levels may bring opprobrium upon the organizations which they represent?

To answer these questions, however, it is first necessary to discover what the job of generalship entails and how it could come to be done so badly or so well. This, the bare bones of good and bad generalship, is examined in the next chapter in terms of information theory.

The main part of the book is divided into two halves. The first is concerned with case histories—examples of military ineptitude over a period of some hundred or so years. Much of this material will, no doubt, be all too familiar to the reader. It is included here, and the selections made, with two main purposes in mind—to provide an aide-mémoire, and because it is believed that the common denominators of military incompetence emerge most clearly when looked at in a longitudinal study. One special virtue of this approach is that it highlights the influence (or, more often, regrettable lack of influence) of earlier upon later events.

For the most part, cases of incompetence have been taken from British military history. Far from being unpatriotic, this apparently one-sided approach springs from a sentimental regard for the forces of the Crown, whose record of valour and fighting ability is second to none, and whose ability to rise above the most intense provocation, either from a civilian population, as in Northern Ireland today, or from the lapses of their top leadership in days gone by, must surely occupy a unique position in the history of warfare. Because it is exceptionally well documented, and has been going on for rather longer than most, British military endeavour also provides a particularly useful datum for a comparative study. Finally, it is surely no more than common courtesy that a critical analysis of one's own 'beams' should take precedence over a listing of the other fellow's 'motes'.

The second half of the book is devoted to discussion and 'explanation'. It is subdivided into two parts, the first concerned with the social psychology of military organizations, and the second with the psychopathology of individual commanders.

The approach here is essentially eclectic. Drawing upon ethological, psycho-analytic and behaviourist theories, it attempts to explain military ineptitude in the light of five inescapable, if unfortunate, features of human psychology. These are:

1. Man shares with lower animals certain powerful instincts.

2. Unlike lower animals, most men learn to control, frustrate, direct and sublimate these instinctual energies.

3. While by far the largest part of this learning occurs in early childhood, its effects upon the adult personality are profound and long-lasting.

4. Residues of this early learning, and in particular unresolved conflicts between infantile desires and the demands of punitive morality, may remain wholly unconscious yet provide a canker of inexhaustible anxiety.

5. When this anxiety becomes the driving force in life's endeavours, the fragile edifices of reason and competence are placed in jeopardy.

In due course we shall examine the scientific basis for these propositions and their relevance to a theory of military incompetence.

Because this is a book about incompetence rather than competence, about disasters rather than successes, these chapters may appear to take an unnecessarily jaundiced view of the military profession and to dwell more upon what is bad rather than what is good in man's attempts to professionalize violence. But without teasing out and enlarging upon the less pleasant features of a multifaceted phenomenon there could be no theory to account for those human aberrations which have caused so much unnecessary suffering in war. As Clausewitz wrote of war, 'This is the way in which the matter must be viewed, and it is to no purpose, it is even against one's better interest, to turn away from the consideration of the real nature of the affair because the horror of its elements excites repugnance.'[5]

To the reader who recoils in disgust from these chapters I can only say that the theory they advance is based upon the emergence of a pattern, of which each small piece may in itself seem trivial, possibly ludicrous, even obnoxious, but which, when put together with other pieces, begins to make sense. This interdependence between the parts necessitates keeping an open mind, and, however much one may dislike or disbelieve the existence of individual trees, postponing judgment until the wood is seen in its entirety.

For the reader who is obsessed with trees, and thinks that history should be left to historians, ideas about soldiering to soldiers, and that psychological theorizing should never go below the belt, this is the moment to stop reading and save yourself some irritation.

PART ONE

AUTHOR'S NOTE

For a long time attempts to write this book were deterred by what seemed an insurmountable difficulty, that of knowing how to present the raw data. Should they be confined to a table of errors that appeared to recur in military disasters (backed up by an extensive bibliography) or should they be allowed to emerge gradually from long and detailed histories of the events in question? The first approach (when tried) seemed arid, and would have left the average reader with the onerous task of ploughing through a vast amount of military history. The second approach would have meant that this book would have run to several volumes. Faced with this dilemma the writer adopted the uneasy compromise of attempting to precis well-known accounts of military disasters in the pious hope that certain common denominators of these events would become apparent and, no less important, that the discerning reader would acquire a sort of feel for the psychological processes involved.

Since the object of the exercise is not the writing of another military history but rather something more analogous to the detecting of weak signals in a noisy background, these precis are deliberately selective and deliberately superficial in their treatment of surrounding context; for it is only by amplifying the signals and playing down the noise that the pattern (if there is one) comes to light.

Obviously this approach will be anathema to trained historians. They will no doubt raise scholarly eyebrows at flimsy descriptions of momentous battles and deplore the fact that the prolonged agony of the Crimean War, or, say, Operation Market-Garden should be reduced to a mere handful of ignoble pages.

To them I say skip Part 1 and turn to p. 149.

2

Generalship

'War is the province of uncertainty: three-fourths of those things upon which action in war must be calculated, are hidden more or less in the clouds of great uncertainty.' C. VON CLAUSEWITZ, *On War*

'In a situation where the consequences of wrong decisions are so awesome, where a single bit of irrationality can set a whole train of traumatic events in motion, I do not think that we can be satisfied with the assurance that "most people behave rationally most of the time".' C. E. OSGOOD

War is primarily concerned with two sorts of activity—the delivering of energy and the communication of information. Most combatants are involved with the former, a few—generals among them—with the latter.

In war, each side is kept busy turning its wealth into energy which is then delivered, free, gratis and for nothing, to the other side. Such energy may be muscular, thermal, kinetic or chemical. Wars are only possible because the recipients of this energy are ill prepared to receive it and convert it into a useful form for their own economy. If, by means of, say, impossibly large funnels and gigantic reservoirs, they could capture and store the energy flung at them by the other side, the recipients of this unsolicited gift would soon be so rich, and the other side so poor, that further warfare would be unnecessary for them and impossible for their opponents.

Unfortunately, such levels of technology have not been reached. In the Vietnam war alone, the United States delivered to Indo-China enough energy to displace 3·4 billion cubic yards of earth—ten times the amount dug out for the canals of Suez and Panama combined[1]—

and enough raw materials in the shape of fuels, metals and other chemicals to keep several major industries supplied for years. In fact, apart from a little slum-clearance this abundance of energy was wasted—consumed in the making of 26 million craters, the laying waste of 20,000 square kilometres of forest, and the destruction of enough crops to feed two million people for a year. However, while the *reception* of energy is still totally uncontrolled this is certainly not true of its direction and delivery. Indeed, these have become a matter of some sophistication and the prime concern of military and naval commanders: theirs is the job of deciding how, when and where to dispose of the energy which their side makes available. They do this by occupying nodal points on a complex communication network.

In other words, the ideal senior commander may be viewed as a device for receiving, processing and transmitting information in a way which will yield the maximum gain for the minimum cost. Whatever else he may be, he is part telephone exchange and part computer. These, the common denominators of generalship, are depicted in Figure 1. For those who don't relish flow diagrams let it suffice to say that on the basis of a vast conglomerate of facts to do with the enemy, his own side, geography, weather, etc., coupled with his own long-term store of past experience and specialist knowledge, the senior commander makes decisions that, ideally, accord with the directives with which he has been programmed.

Ideally. But these ideals are hard to meet. For this there are two main reasons. The first is that senior commanders have often to fill a number of incompatible roles. According to Morris Janowitz these include 'heroic' leader, military manager and technocrat. To these we would add politician, public relations man, father-figure and psychotherapist. The second reason for a breakdown is what communication engineers call 'noise in the system'. 'Noise' is what interferes with the smooth flow of information. Its destructive power hinges on the fact that senior commanders, like any other device for processing information, are channels of limited capacity. If they want to deal with more information, they will tend to take longer about it. If they don't take longer, they will make mistakes. Here we are using the term 'information' in a special, and perhaps its most important, sense as 'that which reduces uncertainty'. Let us expand this a little.*

* In this discussion the concepts of information theory are used descriptively and somewhat loosely.[2]

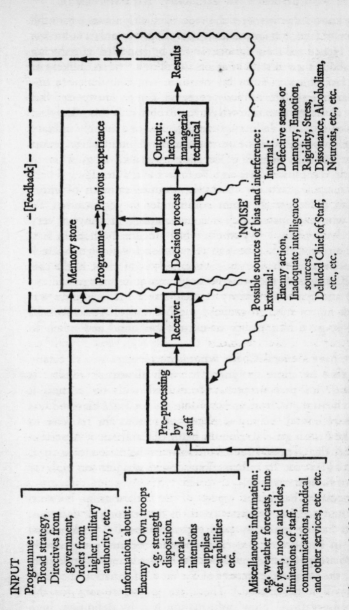

INPUT

Programme:
Broad strategy,
Directives from
government,
Orders from
higher military
authority, etc.

Information about:
Enemy   Own troops
e.g. strength
disposition
morale
intentions
supplies
capabilities
etc.

Miscellaneous information:
e.g. weather forecasts, time
of year, moon and tides,
limitations of staff,
communications, medical
and other services, etc, etc.

Pre-processing by staff → Receiver → Decision process → Memory store: Programme → Previous experience

Output: heroic managerial technical → **Results**

- - [Feedback] - -

'NOISE'

Possible sources of bias and interference:

External:
Enemy action,
Inadequate intelligence sources,
Deluded Chief of Staff, etc, etc.

Internal:
Defective senses or memory, Emotion,
Rigidity, Stress,
Dissonance, Alcoholism,
Neurosis, etc., etc.

The upper half of the above diagram provides a simplified view of the information processes before, during and after the making of military decisions. The very high information load, the great 'uncertainties' that have to be reduced or tolerated, the many stages involved, the interaction between past experience and a 'programme' of perhaps dubious validity, the dependence of decisions upon knowing the likely outcomes of possible alternatives and the necessity for possible revisions in the light of feedback of earlier outcomes constitute a large potential for error or breakdown in the smooth flow of information.

The lower part of the diagram outlines typical sources of 'noise' which could produce disturbance at one or other of the stages between Input and Output.

Figure 1. *Generalship as a communication channel of limited capacity*

Acquiring knowledge involves the reduction of ignorance through the acquisition of facts, but ignorance is rarely absolute and its reduction rarely total. Hence reducing ignorance can be regarded as reducing uncertainty about a given state of affairs. It follows that an unlikely or unexpected fact contains more information (i.e., reduces more uncertainty) than one which is already expected. But an unexpected fact is less readily absorbed than one which was expected. If this is less than crystal clear, consider the following example, cast in a suitably military context. The message in this case consists of an intelligence report which states: 'Enemy preparing for counter-attack.' It goes on to detail strength, disposition, date and likely sector for attack.

Now this message, factually so simple, contains amounts of information which differ greatly from commander to commander. To General A, who anticipated such a counter-attack, it conveys very little; it merely confirms a hypothesis which he already held. In fact, since he had already made extensive preparation for a counter-attack the intelligence report when it came was largely redundant. In the case of General B, however, the same message was quite unexpected. So little had he anticipated an enemy counter-attack that the news was charged with information. It reduced a great deal of ignorance and uncertainty. It gave him plenty to occupy his mind and much to do.

Finally we have General C, for whom the message was so totally unexpected that he chose to ignore it, with disastrous results. It conflicted with his preconceptions. It clashed with his wishes. It emanated, so he thought, from an unreliable source. Since his mind was closed to its reception, he found plenty of reasons for refusing to believe it. Like British generals after the battle of Cambrai, or American generals before the German counter-offensive in the Ardennes in 1944, he ignored it at his cost. Its information-content was just too high for his channel of limited capacity.

One particularly hazardous aspect of the relationship between information and decision processes concerns the revising of decisions. It seems that having gradually (and perhaps painfully) accumulated information in support of a decision people become progressively more loath to accept contrary evidence. As Edwards and his colleagues have shown, the greater the impact of the new information the more strenuously will it be resisted.[3] There are several reasons for this dangerous conservatism. 'New' information has, by definition, high informational content, and therefore firstly it will require greater processing capacity, secondly it threatens a return to an earlier state of gnawing uncertainty, and thirdly it confronts the decision-maker with the nasty thought that he may have been wrong. No wonder he tends to turn a blind eye!

So much for a broad description of this most vital dimension of knowledge, its prior improbability. Let us return now to the other side of the coin, the problem of 'noise'. 'Noise', as we saw, is the enemy of information. 'Noise' takes up channel space and thereby disrupts the flow of information. The more limited the channel capacity, the greater the disrupting effects of 'noise'. The more 'noise', the less information that can be handled.

A glance back at Figure 1 suggests that not only does a senior military commander receive more than his fair share of information, but the communication system of which he forms a part is peculiarly susceptible to 'noise'. This may be external in origin, ranging from static on a radio link to the delusions of a Chief of Staff. Or it may be internal, ranging from such peripheral sources as poor eyesight (a common feature of generals in the Crimean War) to such central and usually more disastrous causes as defective memory, brain disease, neurosis and alcoholism.

'Noise' from all these sources may act upon the flow of information through a general's head and eventuate in decisions varying in gravity from the mildly inept to the utterly catastrophic. But decisions hinge upon more than available information. They also depend upon 'pay-offs'—the anticipated consequences of choosing one course of action rather than another. Pay-offs may be positive or negative, beneficial or costly. They are the criteria according to which decisions are made. Obviously, if a commander gets his criteria wrong—if the possible loss of self-esteem or social approval, or fear of offending a superior authority, is given greater weighting than more rational considerations—the scene is set for calamity.

The possibility of this happening is increased by the fact that the 'fog of war', unlike the uncertainties which attach to most civilian enterprises, extends not only to the input but also to the pay-offs. Not only does the general have to make decisions on the basis of a great volume of dubious information and meet a programme of perhaps questionable validity; he may also not know the costs and benefits of what he does propose. He is like a man who places a bet without knowing the odds or where the bookie might be found when once the race is over.[4]

As well as those problems which are inherent in any communication system, the human decision-maker is the victim of another hazard— namely that attention, perception, memory and thinking are all liable to distortion or bias by emotion and motivation. The potential for this state of affairs is depicted in Figure 2 which shows the human operator represented by two interlocking feedback loops. The solid line represents an individual's interaction with his environment—perception leading to response. The broken line represents an internal loop, that between need and satiation. The latter acts upon the former. As needs arise, whether they be social or biological, neurotic or adaptive, so they act upon the way a man perceives his external world, what he attends to, the sort of memories which he conjures up and the decisions which he makes.

He is like a computer which not only has to receive, store, process and deliver information but also has to postpone sleep, cope with hunger, resist fear, control anger, sublimate sex and keep up with the Joneses. When it is considered that the capacity for perception and response, for memory and thought, presumably evolved for the satisfaction of needs, it is a remarkable achievement at the best of times to keep these informational processes of mind free from bias by the needs which they were originally designed to serve. In war such an achievement borders on the miraculous and this for one very simple reason: the effects of needs upon cognition are *maximized* when the needs are very strong and external reality ambiguous or confused. It is under such conditions that need and emotion have the greatest freedom of manœuvre, the greatest capacity for imposing themselves upon the uncertainties of thought. These are the conditions which obtain in war.

Contemplation of what is involved in generalship may well occasion surprise that incompetence is not absolutely inevitable, that anyone can do the job at all. Particularly is this so when one considers that military decisions are often made under conditions of enormous stress, when *actual* noise, fatigue, lack of sleep, poor food and grinding responsibility add their quotas to the ever-present threat of total annihilation. Indeed, the foregoing analysis of generalship prompts the thought that it might be better to scrap generals and leave decision-making aspects of war to computers.

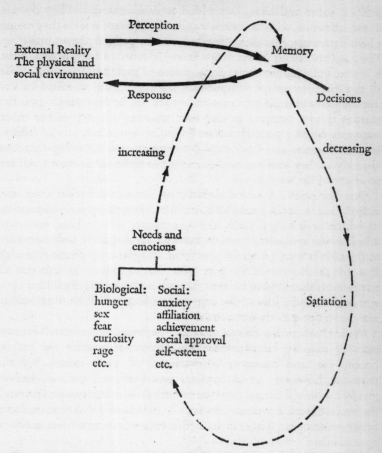

Figure 2. The way in which an individual perceives and acts towards his environment is partly determined by the quality and strength of his motives, needs, attitudes, and emotion.

A similar argument has been advanced in connection with medicine. Why leave diagnosis and therapeutic decisions to fallible human brains when a computer could make them with far less chance of error? The answer is, of course—and this no doubt contributes much to the relief of generals and doctors—that computers make poor leaders and indifferent father-figures. They may be quick and efficient, unprejudiced, sober and alert, but withal remain cold fish. They do not inspire affection, with its consequent desire to please, nor do they exude a bedside manner. Paradoxically they are also perhaps just *too* infallible. They are, moreover (as far as we know), devoid of feelings and, what is worse, quite indifferent to the outcome of their decisions. But while all this militates against computers as leaders of men, so-called leadership-qualities in military commanders are just as dependent upon the various factors outlined in our flow chart as are any of the other responses which a general makes. Prejudice, ignorance, fear of failure, over-conformity and sheer stupidity may disrupt leadership-decisions as surely as they interfere with planning or technical decisions. All are products of the same brain.

One last point. A senior military or naval commander does not, indeed cannot, act in lonely isolation but is fettered by the organization to which he belongs. He is like a computer or telephone exchange whose *modus operandi* is based on rules which may have little relevance to the tasks it is called upon to perform. Imagine a telephone exchange that, for the honour of the post office, has to follow the rule that all telephonists should have red hair, 38-inch busts and heavily lidded eyes, and one has some idea of the restricting effects which an organization may have upon its own functioning.

In the chapters that follow we shall be examining some well-known cases of military incompetence, to discover if possible the precise reasons for, and common denominators of, these events. For the moment, however, let us consider one brief and less well-known incident which illustrates how the smooth flow of information through the brains of senior commanders may be so distorted that their decisions prove catastrophic. The culprits, in this instance, are naval, not military, commanders.

The place is Samoa and the date 1889.[5] Seven warships—three American, three German and one British—are lying at anchor in the harbour of Apia. They are there as a naval and military presence to watch over the interests of their various Governments in the political upheavals that are taking place ashore. Accordingly they anchor in what has been described as one of the most dangerous anchorages in the world, for to call Apia a harbour at all is at best an unfortunate euphemism. Largely occupied by coral reefs, this saucer-shaped indentation lies wide open to the north, whence the great Pacific rollers come sweeping in. In fair weather Apia provides an uneasy resting-place for no more than four medium-sized ships. For seven large ships and numerous smaller craft, under adverse conditions, it is a death-trap.

This was the situation in which the seven men-of-war witnessed the first bleak portents of an approaching typhoon. Even to a landsman a rapidly darkening sky and falling glass, squally gusts of wind, and then a lull, would bode ill. For seven naval captains the signs were unmistakable. They knew they were in a region of the world peculiarly subject to typhoons, which, in a matter of minutes, could lash the sea into a furious hell of boiling water. They knew that such storms generate winds travelling at upwards of a hundred knots, gusts that could snap masts like carrots, reduce deck fittings to matchwood and throw ships on to their beam-ends. They knew that it was the worst month of the year and they also knew that only three years before every ship in Apia had been sunk by such a storm. In short, and in the terms of our flow chart, their stored information coupled with present input pointed to only one decision: to get up and get out. And, as if this was not enough, the urgency of weighing anchor and putting to sea was respectfully suggested by subordinate officers.

But the captains of the warships were also naval officers and so they denied the undeniable and stayed where they were. Their behaviour has been described as 'an error of judgment that will for ever remain a paradox in human psychology'.

When the typhoon struck, its effects were tragic and inevitable. Without sea room, their anchors dragging under the pressure of mountainous seas, their hulls and rigging crushed by the fury of the wind, three of the warships collided before being swept on to the jagged reefs of coral. Another sank in deep water; two more were wrecked upon the beach. Of all the ships in the harbour the only survivor was a British corvette, which, thanks to its powerful engines and superb seamanship, squeaked through to the open sea.

Why did senior naval commanders, versed in the ways of the sea and provided with ample warning, thus hazard their ships and the lives of their men? A superficial answer might be pride, or fear of appearing cowardly, or fear of criticism from their superiors. These are matters to be pursued in later chapters. For the moment the apparently incorrigible behaviour of these men illustrates how decision-processes can be thrown into disarray by noise of internal origin and how, in this instance anyway, incompetence cannot be attributed to ignorance or ordinary stupidity.

3

The Crimean War

'As far as concerns the military art, the Crimean War is usually regarded as worthy of remembrance only as perhaps the most ill-managed campaign in English History.'
Encyclopaedia Britannica, 1960 Edition

There must be few who would dispute the general tenor of the view quoted above. While Field-Marshal Montgomery concedes that there may be runners-up—'One of the most ill-managed campaigns in all recorded history'[1]—David Divine takes an even more extreme position. '... the campaign plumbed levels of incompetence never before attempted. One hesitates to challenge the *Encyclopaedia Britannica* but it is necessary to question the word "perhaps".'[2] As to the precise 'levels plumbed', the *Observer*, commenting on *Victoria's Heyday*, is even more explicit. 'The Crimean War touched the nadir of stupidity.'[3]

Nadir or not, the Crimean War certainly marked an exceedingly low point in British military history. The poor quality of the officers, most of whom had bought their commissions and for whom no standard of education was required, stood in marked contrast to the excellence of the men, described by one observer as 'the finest soldiers I ever saw in stature, physique and appearance'.

Amongst the officers there seemed to be an inverse relationship between rank and efficiency. The more senior they were, the less competent they appeared. At the apex of this pyramid of mediocrity stood (or rather sat, for he was always on his horse or in his quarters, and, being inordinately shy, rarely walked amongst his men) Lord Raglan. His qualifications for leading a British expeditionary force appear to have been his age—sixty-seven; his lineage—he was youngest of the Duke of Beaufort's eleven sons; and his experience—twenty-five years as military secretary to the Duke of Wellington, and then Master-General of the Ordnance. No one could accuse him of having a mind cluttered by any previous experience of command, for he had none—not even of a company. His appointment, however, was not wholly inappropriate, for of him it was said: 'His chief merit was that, despite his incurable habit throughout the campaign of referring to the enemy as the "French", he was admirably adapted to lessen the friction in coalition wars.'[4] In fact, Raglan seemed to agree with most French proposals: it was a characteristic of the man that he hated conflict!

At the next lower level were the members of Raglan's headquarters staff. Their role it was to edit and pass on such gems of military wisdom as might be uttered by their commander-in-chief. On paper this 'nest of noodles', as they were sometimes known, might have looked a well-knit, closely integrated group, especially since five of them were blood relations of their chief. Unfortunately, this nepotism was not based upon the possession of any military expertise. Their contribution to the smooth running of the campaign seems to have been somewhat abrasive.

If Raglan and his staff constituted the nerve-centre of the army in the Crimea, the sinews comprised a field force of five infantry and two cavalry divisions under commanders who, for the most part, did little to inspire confidence. Here too the problem was partly one of age. Apart from the thirty-five-year-old Duke of Cambridge, cousin to the Queen, all the senior commanders were between sixty and seventy, with Sir John Burgoyne, Chief Engineer, topping the list at seventy-two! Certainly it could be said of them that what they lacked in experience they made up for in years.

In the words of George Maude, a highly capable officer: 'There is an old Commander-in-Chief, an old Engineer, old Brigadiers—in fact everything old at the top. This makes everything sluggish.' In the light of events this was something of an understatement.

As has so often been the case, the next lower level of command did contain some leaders of vigour with a talent for war. Such a one was Sir Colin Campbell. His command unfortunately was no larger than a brigade.

As usual in those days the cream of the army was the cavalry, commanded in this instance by Lord Lucan, an impulsive man of moderate intellect and lacking in experience. Directly under Lucan, in charge of the Heavy and Light Brigades respectively, were James Scarlett and Lord Cardigan. The arrangement was not a happy one. To select Cardigan for a position subservient to his brother-in-law Lucan was hardly less felicitous than subordinating a mongoose to a snake. In terms of the concepts introduced in the previous chapter the mutual dislike which existed between these noble lords was such as to constitute severe 'noise' in any system of communication which involved the pair of them. It was in fact a major factor in that breakdown of communication which resulted in what some would regard as the biggest single blunder in British military history—the charge of the Light Brigade at Balaclava.

The shortcomings of those who administered the Crimean campaign were not long in making themselves felt. The first hint of trouble came with their arrival on Russian soil. Even the gross incompetence of those responsible for the transportation and disembarkation of troops for the Dardanelles expedition in 1916, or, forty years later, those who carried out comparable arrangements during the Suez crisis (on both occasions the most urgently needed stores had been thoughtfully packed at the bottom of the holds), was as nothing beside that shown by the Crimean entourage: 'No army comparable with the British has in modern history ever landed upon a foreign shore more inadequately equipped for invasion.'[5] The men were put ashore with no more than each could carry. They spent their first night without tents or blankets, soaked to the skin by the incessant rain, their sodden uniforms whipped by an icy wind.

Such conditions compared unfavourably with those enjoyed by their allies, the French and the Turks, who had excellent tents, medical supplies* and transport. Considering Britain's infinitely superior maritime resources, some tiny extra part of which would have sufficed to carry what was needed, the contrast is as remarkable as it was inexcusable.

For those who wish to exculpate the military command it might be thought that these deficiences could be laid at the door of a tight-fisted Treasury. But, as one writer has put it:

In the ultimate resort a commander-in-chief had it in his power

* One British Army surgeon, with the backing of his commandant, did in fact submit a report on the lack of medical supplies for a hospital that was full of cholera and diarrhoea. His complaint, however, was judged to be 'frivolous' by General Airey, the new Quartermaster-General.

to force a government's hands by threat of resignation. No such threat came from Lord Raglan. On the other hand it may be that such matters were deliberately concealed from him in pursuance of the policy adopted by his staff and cynically explained to Captain Adye on taking up his appointment: 'Never trouble Lord Raglan more than is absolutely necessary with details, listen carefully to his remarks, try to anticipate his wishes and at all times make as light as possible of difficulties.'[6]

This advice from a member of his staff reflects what was perhaps the prime characteristic of Lord Raglan: his almost compulsive non-participation. Aristocratic, courteous and aloof, he seemed to display many of the characteristics of the extreme introvert. So distasteful was it to have any direct contact with his fellow men that he could hardly bear to issue an order; and when he did so it was couched in such a way as to ensure a vast gulf between his wishes and the comprehension of those for whom it was intended. Combined with his never-failing courtesy, this idiosyncrasy often rendered his role upon the battlefield rather less effective than that of the regimental mascot. Thus in the first battle of the campaign, that of the Alma, he issued only one order — to advance until the Alma was crossed. The lack of further guidance might not have mattered had the generals under his command shown some spark of military ability. Unfortunately some did not, as the following excerpt from an account of the battle illustrates:

Buller — 'Gentleman George' — was not a general of distinction. The weight of his responsibilities for the left flank of the army seems to have numbed his faculties. Having led his brigade across the Alma with less difficulty than Codrington because in his path there were fewer obstacles, he seems to have been content to have complied with the command not to stop till across the river. About his next step he was so much in the dark that he actually turned for enlightenment to his youthful A.D.C., Henry Clifford, who advised an advance similar to Codrington's on the right.[7]

Another passage from the same description of the battle exemplifies only too well the sort of effects that might be expected when the commander-in-chief refuses to take part. This time it is the problem of consolidating and exploiting a position gained.

When the Light Division reached the Great Redoubt after their exhausting and perilous attack it was inevitable that tension must relax ... The moment that his victorious troops were in the Great Redoubt, the controlling and directing hand of a great commander-in-chief should have been in evidence. Now should supports be arriving to reinforce the victors and fortify and inspire them against counter-measures. But as the wearied Light Division looked back they saw nothing on the long slope leading up from the Alma but the dead and wounded 'lying like grass on the ground'. As they glanced to their flanks the prospect was even more discouraging. To the right and left were massed columns of Russian soldiers from whose throats came 'a long, sorrowful, wailing sound', the ominous precursor to some desperate venture ... It was a moment when nerves are easily lost, when perception is blunted and credulity ranges uninhibited. A frantic figure of a staff officer on horseback, perhaps misled—that at least is the most charitable explanation—by the Russian failure to fire, came galloping along the ranks shouting, 'Don't fire! Don't fire—the column's French!'

Few stopped to ask themselves how the French could have moved to the British front from having been not half an hour earlier on their right flank, nor how they came to be dressed in long grey overcoats reaching to their ankles. Obediently one bugler after another from left to right sounded the 'Cease Fire' followed by 'Retreat'. To Colonel Chester of the 23rd this was insanity. 'No, no', he shouted. 'It's a Russian column, fire!' They were his last words and they went unheeded.[8]

Despite the fact that Raglan, watching from afar, played little part in bringing it about, Alma *was* a victory for the Allies, thanks to the courage and superb fighting qualities of the soldiers and their junior officers. Through what one observer described as 'a great want of generalship', the victory was achieved with much unnecessary loss of life and, even worse, because of a total failure to follow it up, yielded few if any dividends for the campaign as a whole.

There is one final point of some relevance to the thesis of this book. It concerns the matter of initiative. Lack of direction from those at the apex of a hierarchical authoritarian organization provides a special dilemma for those at lower levels in the chain of command. Confronted with an absence of clear-cut orders, what are they to do? If they take the law into their own hands they run the risk of being accused of insubordination, particularly if their plans happen to miscarry, but if they do not show initiative then they are equally likely to suffer for *not* having done so. At Alma the field officers, for want of higher direction, used their own initiative with considerable success. In so doing they saved the day if not the campaign. Curiously, not one was mentioned in Raglan's dispatches. It is one thing to let your juniors do the thinking and take the risks but quite another to admit the fact.

Like a rudderless ship that bumps and grates from one reef to the next, the army lurched from the hazards of Alma to the far more disastrous ones of Balaclava. As touched on earlier, trouble started in dividing command between the two generals, Lord Lucan and Lord Cardigan. Individually neither was fitted to his post; together they were a disaster. As one of their fellow officers wrote in his diary: 'The more I see of Lord Lucan and Lord Cardigan the more thoroughly I despise them. Such crass ignorance and such overbearing temper.'

Raglan did not excel in dealing with these men. Instead of loyally supporting Lucan he appeared to condone even the most flagrant excesses of the incorrigible Cardigan. Not only did he allow Cardigan to bring his private yacht into Balaclava, where for weeks it took up valuable space in the congested harbour, but he also permitted him to live on board even while his brigade and divisional commander were roughing it ashore, on rations, under canvas. By forfeiting his position of authority and exacerbating the already bitter enmity between his subordinates, Raglan's laissez-faire handling of these relatively minor matters sowed the seeds of the ultimate disaster, the destruction of the Light Brigade.

As to the latter, so much has been written about their ill-fated charge that it would be superfluous to recount the details here. There are, however, several points particularly apposite to the subject-matter of this book. Firstly, the way it was treated in subsequent accounts did much to strengthen those very forms of tradition which put such an incapacitating stranglehold on military endeavour for the next eighty or so years. By the same token, this understandable consolation for those who lost so much did, by laying undue emphasis on the magnificence of the charge, tend to obscure some crucial issues. For behind the colour and the glory, behind the valour and the dash, the charge of the Light Brigade was a blunder of monumental proportions and an object-lesson in what can happen when the promotional machinery of a military organization is such as to put troops at the mercy of men like Raglan, Lucan and Cardigan.

Their ineptitude was manifested in two successive stages of the battle. The first started immediately after the successful charge of the Heavy Brigade under the Hon. James Scarlett. Again it was a matter of failing to exploit a position gained.

> To grasp how incomplete was Scarlett's triumph it is only necessary to ask one question: what was the Light Brigade doing when the Russian cavalry wheeled out of the north valley and, passing right across its front, went down to engage the Heavies? Not 500 yards separated Cardigan's men from Rykoff's. Every moment the 'gallery' expected this incomparable chance of a flank attack to be seized. To the dullest of brains there seemed no other possible course for this brilliant, dashing, eager brigade to take. But to the amazement of the spectators and the fury of the Brigade nothing of the kind was attempted.[9]

The explanation of this curious lapse hinges upon the fact that Lucan had impressed upon Cardigan that his job was to stay put and defend the position, attacking only such enemy forces as came within reach. Under the circumstances Cardigan determined that he would not give his brother-in-law the slightest grounds for making a complaint should the attack fail. If it did, then Lucan should take the blame. Lucan had ordered him to defend the position and defend it he would, even if it cost him his life.

However, this missed opportunity, which might have sealed the Russians' fate, recedes into unimportance when compared with what happened next.

It seems that the charge of the Light Brigade, from which only fifteen per cent of the original force of 673 rode back, was the end-result of faulty communication between five men: Raglan, his Quarter-master-General Airey, Lords Lucan and Cardigan, and the impetuous Captain Nolan.[10] Raglan's contribution was that he issued orders the precise meaning of which has remained a matter for debate.* The fourth and more disastrous of these orders Airey wrote out on a flimsy

* The third order, a copy of which was retained by Raglan, read 'Cavalry to advance and take advantage of any opportunity to recover the Heights. They will

piece of paper. In so doing he made no attempt to unravel the enigma posed by the words of his master. Which front? What guns? In its new written form the order was then passed to the unbalanced Captain Nolan, who loathed both Lucan and Cardigan. This glittering young officer of the 15th Hussars, who made up in arrogance what he lacked in perspicacity, delivered the order to Lord Lucan. Lucan, whose comprehension of Raglan's wishes seems to have been minimal but who was not going to demean himself by bandying words with Nolan, conveyed *his* interpretation of the order to Cardigan. Cardigan, to give him his due, realizing that he was being asked to charge the Russian guns down a valley flanked by enemy artillery, expressed considerable astonishment at what would so evidently be the *coup de grâce* for his brigade. But once again communication foundered on the rocks of mutual dislike, pride and jealousy.

Joined, and then overtaken, by the irrepressible Nolan, Cardigan led his brigade into the 'jaws of Death'.

As is usually the case after a national disaster of such proportions, the final stage in this sorry tale concerns the apportioning of blame—the means by which society obtains a modicum of revenge for the wrong it has suffered, expiates its own guilt for such responsibility as it may have had for the event in question, and finally seeks to prevent a repetition of the disaster. Of these only the last is in any way ennobling, for it is only thus that the disaster can be turned to good account.

It is a sad feature of authoritarian organizations that their nature inevitably militates against the possibility of learning from experience through the apportioning of blame. The reason is not hard to find. Since authoritarianism is itself the product of psychological defences, authoritarian organizations are past masters at deflecting blame. They do so by denial, by rationalization, by making scapegoats, or by some mixture of the three. However it is achieved, the net result is that

---

be supported by the infantry which have been ordered to advance on two fronts.' But by the time it reached Lucan the final words had been changed to 'They will be supported by the infantry which have been ordered. Advance on two fronts.'

The fourth order read 'Lord Raglan wishes the cavalry to advance rapidly to the front—follow the enemy and try to prevent the enemy carrying away the guns. Troop Horse Artillery may accompany. French Cavalry is on your left. Immediate. (Sgd.) Airey.'[11]

no real admission of failure or incompetence is ever made by those who are really responsible; hence nothing can be done about preventing a recurrence. In this instance, as in many others to be considered presently, scapegoats were found. One of these was Captain Nolan, an easy choice since he had, very considerately, allowed himself to be killed.

If Alma was remarkable for a paucity of orders, and Balaclava for their confused nature, then Inkerman, the next battle in the campaign, has the distinction of being fought in a fog without any direction at all. One diarist, an officer, wrote: 'No orders were given from first to last but to advance. No attempts to reform shattered battalions, no plan of operations.' The same writer, a certain Major Patullo, said: 'I feel gratitude to the courageous British soldier who fought all day, replenishing his ammunition from his wounded comrade's pouch without direction or hint from superior authority, only the example of his officer who was left equally without guidance, not to the generals who in my opinion have not distinguished themselves.'[12]

Clearly such opinions make a mockery of the dispatches and later the honours which succeeded the battles. To Raglan came the baton of a field-marshal, and to his staff, whether or not they had been within range of Russian guns, a step in brevet rank.

However, Crimean mismanagement reached its apogee, not in the battles so far considered, but in the winter which followed them. Despite the fact that between October 1854 and April 1855 there was no fighting whatsoever, Raglan's army suffered a thirty-five per cent decline in its active strength. This loss was due to a total disregard for the army's physical welfare, and a refusal to ameliorate the cold and wet of a Russian winter. Men died of cholera, of exposure, of malnutrition. They died of untreated wounds, of scurvy, gangrene and dysentery. As one surgeon observed of earlier losses at Balaclava: 'We [now] bury three times the number of men every week and think nothing of it!'

They died because there was no issue of fuel and stores.

Now, on logical if not humanitarian grounds, it would seem an inescapable fact that a general's prime responsibility is the welfare of his men. All his skills as leader and military planner will avail him nothing if, for one reason or another, there is no one left to do the fighting. Assuming, then, that the physical and mental welfare of troops is the *sine qua non* of successful military endeavour, how did it happen that the British Army in the Crimean campaign was virtually destroyed by something other than the enemy?

Several reasons may be given. Firstly, there was that unrealistic over-confidence in rapid victory which has characterized so many military adventures. As we shall see, it was a notable feature of the Boer War, of the First World War, of the Second World War and even, through what was by now a quite extraordinary incapacity to profit from experience, of the Suez crisis and Bay of Pigs fiasco. In the present context Lord Raglan was so confident that Sebastopol would fall before the winter of 1854 that no plans had been made to house and maintain his army on the high ground above the town. How anyone as ungifted in the waging of the three preceding battles could have clung with such childlike innocence to the myth of speedy success against the 'inferior' Russians is remarkable, to say the least.

Secondly, after he had been proved wrong, Raglan and his staff, unlike the French, who managed to improvise tolerable conditions, seemed incapable of adapting to the circumstances thrust upon them. This inertia and inflexibility were rooted partly in ignorance (thus no one seems to have had the remotest idea as to how cold it could get during a Crimean winter), partly in red tape, and partly in a refusal to admit that things were not as they should be.

The shortcomings of the high command are particularly well illustrated in connection with the fuel crisis. According to one writer: 'What killed more men than Russian bullets, what made life miserable, what sent men in the hundreds to the hospital tent or the grave—they were frequently synonymous—was the want of firewood. Without it not only were men never warm, not only could they never cook their ration of "cold grunter" but they were never dry.' One of Raglan's colonels wrote: 'They go down to the trenches wet, come back wet, go into hospital wet, die the same night and are buried in their wet blankets next morning.' And an Army surgeon wrote: 'I never thought the human subject would endure so much privation and suffering.' Even at this remove these accounts are deeply moving and the scenes they conjure up painful to contemplate. But, condemnatory though they are, these tragic statements neglect the real enormity, namely that the shortage of firewood could have been easily rectified.

Just across the Black Sea lay the forests of Anatolia. A little foresight, a breath of leadership, an iota of compassion, and two or three ships could have brought back enough firewood to last everyone the winter through. But there was none, and they did not. Instead, men soaked to the skin, and dropping with fatigue from long hours in the trenches, had to forage far and wide for meagre sticks of brushwood and, when this failed, grub up their sodden roots.

This disinclination to take even the simplest steps to mitigate the hardships of the troops or to seek on their behalf any assistance from elsewhere was no less apparent when it came to the matter of sleeping accommodation. For the most part this consisted of wet blankets and mud. By January 1855 even the wet blankets had acquired such survival value that they could no longer be spared for shrouds. A corporal wrote: 'They bury our men quite naked and throw him [sic] into the grave like a dog. I wonder what the people of England would think if they were to see it.'

The Chief Engineer, Sir John Burgoyne, suggested that the Turks, that 'nation of carpenters', should be asked to make floorboards for the tents. Any difficulty Raglan might have met with in arranging this would have been speedily resolved by our ambassador in Constantinople. But the suggestion, which might have saved hundreds of lives, was ignored. Even in the so-called hospital tents 'men lay on the bare grass and died'.

There were other areas of appalling neglect. On food and clothes Laffin has this to say:

> In the Crimea, as elsewhere before and after, the soldier suffered from poor food and poor clothing. The clothing was good enough for parades in Britain, lamentable for field service. Some units did not even possess greatcoats and suffered terribly from the intense cold. Most uniforms were in rags and tatters within a few months.
>
> Often the food was literally rotten or in other ways not fit to be eaten. The whole catering system was inept and inadequate and was even worse in the hospitals than in the camps.[13]

Far from agitating for reform and attempting to ameliorate the effects of these deficiencies, the military high command did their best to preserve the status quo. In their treatment of Florence Nightingale and her zealous helper, the London chef Alexis Soyer, the relatively well-fed generals betrayed a lack of compassion that was scarcely credible. With his army dying of malnutrition General

Eyre said of Soyer and his new cooking stove: 'Soldiers don't require such good messes as these while campaigning. You will improve the cook but spoil the soldier.'[14]

That men who had survived Alma and Inkerman, and even the charge at Balaclava, were now dying in their thousands through the gross incompetence of their own headquarters staff created something of a vicious circle, for with every death there were fewer left to construct and man the trenches. But there were other vicious circles. While Raglan and his staff enjoyed the comfort of warm beds and well-cooked food, their rapidly reducing labour force not only had to work for longer hours but even suffered more serious malnutrition than hitherto. This was because the combination of fatigue, damp and cold caused their gums to become so inflamed that they could not eat their stand-by, the biscuit. As a result, for lack of nourishment, they became even less capable of resisting the cold. By the end of January the British Army could muster only 11,000 men, its sick and wounded totalling 23,000.

Eventually, after a winter of terrible privation, Raglan's army came to the last battles of the war—those of the Redan. They involved the storming and capture of a fortress on the outskirts of Sebastopol.

It seems that little had been learned. Again there was gross under-estimation of the enemy's ability. Indeed, the forthcoming engagement was regarded with such equanimity that it attracted a large assembly of sightseers. While the band of the Rifle Brigade played light music, an audience of officers' wives, 'travelling gentlemen', and even a number of serving soldiers, took up position on the surrounding hills.

Raglan, with his mind closed to all that had gone before and an enduring over-confidence in his army, chose only 400 of his 25,000 men for the first stage of the battle: the occupation of some quarries, from which the assault on the Redan would be mounted. This proud economy in manpower was his first mistake. They occupied the quarries but at considerable and unnecessary cost—more than half the attacking force, including a large number of junior officers. In fact, the whole affair very nearly ended in disaster. Staff work had been poor, with orders confused and often contradictory. Raglan's staff had miscalculated the strength needed to occupy the quarries and to repel counter-attacks. Reserves had been inadequate, and unavailable when most needed. Little thought had been given to selecting the troops to be used. The proportion of veterans was low. Many of the officers, although unquestionably brave, were young and inexperienced.

But it was with the second stage in the battle, the main assault, that things went really wrong. Again Raglan and other high-ranking officers underestimated enemy strength and overestimated the effects of the artillery bombardment with which he preceded the attack — nor did he appreciate that between the bombardment of the Russian fort and the dawn attack, forced upon him by the French commander, the enemy would be well able to repair their defences and resite their guns.

In theory, the attack was to consist of a three-pronged charge, with soldiers from the left, right and centre of the British front converging on the enemy position. In theory, this sudden synchronized assault from three sides would sweep away all opposition. In theory, the Russian artillery would have been obliterated by the preceding bombardment. And, in theory, the British ladder parties would have no difficulty in porting their eighteen-foot loads across the 450 yards of rising ground to the walls of the Redan. In practice, things went rather differently. For a start, the general in charge of one of the three assault parties mistook the starting signal and dispatched his men half an hour too early. Even without this mistake, any possibility of a concerted simultaneous assault was ruled out because no one had arranged any means whereby the attackers could rise from their trenches in unison. There was nothing for them to stand on. At the best they came out in ones and twos. Under the circumstances it was hardly surprising that the great volume of fire from the 'obliterated' Russian guns brought the British attack to a very bloody halt. It was at this moment, just when it was most needed, that Raglan's artillery received an order to cease firing. It was this last blunder which transformed an aborted attack into a massacre. No longer intimidated, enemy muskets poured a hail of lead into Raglan's stricken army. The ladder parties moving like snails beneath their loads were mown down as they struggled up the slope.

Thus ended the first battle for the Redan — until then the most disastrous of the war. Raglan's army had no illusions as to the incompetence of their general and his staff. A staff officer wrote: 'We had been told from Headquarters and other high authority that success was certain; that the arrangements for the plan of attack were so perfect that they must succeed; when put to the test they turned out to be so execrably bad that failure was inevitable.' Others described the battle as 'mismanaged', 'botched', 'bungled', 'feeble and ill-conducted', 'bad business', 'a bungling disgracefully childish failure'.

Later Lord Wolseley wrote: 'Upon this occasion what we asked from them was beyond the power of men to give. Our plan for the attack was simply idiotic and was bound to fail.' Another writer has this to say: 'Not only was it a question of defective tactics. At Headquarters there was not merely ignorance but an entire lack of perspicacity. How was it possible that Raglan and those about him, knowing as they ought by this time the remarkable Russian ability to repair damage overnight, could believe that 2,000 soldiers would be able to advance over a shell-swept glacis 250 yards in length, thread their way through an undestroyed abattis, cross a ditch 20 foot wide and then assail an escarpment all without preliminary bombardment?'[15] How indeed!

For the student of psychosomatic disease the aftermath of the battle is not without significance. Immediately following the defeat Raglan was seen 'to age visibly'. Within a few days he had contracted cholera and before ten days had passed he was dead. Two of his generals were similarly stricken. Raglan's demise added little to the depression of the army. Had they known that his replacement would be sixty-three-year-old General Simpson their grief might have been rather more acute.

It was not that Simpson was a harsh taskmaster. On the contrary, he was a gentle old man but of very mediocre ability. He was as devoid of useful experience as had been his predecessor. His methods were rather simpler than those of Raglan's. Presumably to avoid giving a wrong order he gave *no* orders at all and he devised *no* plan. In the words of one observer: 'He did not command the army.' On the day of his promotion he is credited with saying: 'They must indeed be hard up when they appointed an old man like me.'[16]

In fact the Government was not so hard up that they had to entrust the army to this gout-ridden old general. A far better choice would have been the energetic and outspoken Sir Colin Campbell, a man of considerable ability and wide experience. But Campbell was a maverick and as such unpopular with the military establishment. He also came from a relatively humble background.*

Under Simpson's quavering and ineffectual hand the second battle

* It was another two years before Campbell became Commander-in-Chief in Bengal, during the Indian Mutiny.

for the Redan, the last battle of the war, proved even more disastrous than the first. Once again a massive bombardment was followed by a frontal assault across a heavily defended triangle of ground flanked by Russian guns. But this time the troops were younger and greener, and despite all their training on the parade-grounds of Aldershot less inclined to valour than discretion. Having sustained 2,447 casualties in two hours of fighting, they turned tail and fled, thus adding humiliation to defeat. But the fault lay with their commander, whose planning, in the words of Lord Wolseley, was 'as faulty in every detail as it was puerile in conception'.

As an example of protracted military incompetence at high levels of command the Crimean War is not, unfortunately, unique. It was, however, the prototype for subsequent ineptitude. Though small in uumber in comparison with those of later wars, the 18,000 who died owed their untimely demise to an admixture of poor planning, unclear orders, lack of intelligence (in both senses of the word) and fatal acquiescence to social pressures on the part of their commander. They died because they were mismanaged by men whose positions in the military hierarchy owed less to their ability than to their wealth, their place in society, or their reputation for 'fitting in'. They died because soldiers were too readily regarded as expendable objects.

The Crimean War was fought at a time of the greatest prosperity this country had ever known, when British efficiency, inventiveness and sheer entrepreneurial vigour knew no bounds. Why then was it fought so badly? ... so badly that the casual observer might have been forgiven for thinking that, at some level, we did not really want to win? Of course there are some obvious and immediate reasons. Governmental stinginess clearly played a part, as did the deliberate policy of entrusting military matters to an aristocratic, rich, but essentially amateur, elite: this on the grounds that such a class would have neither the motivation, nor indeed the skill, to turn upon the State.

But this is only to touch the surface of the problem. Such reasons do not explain the passivity and non-participation, the monumental errors of judgment, the ludicrous appointments, the paralytic inability to improvise or innovate. They do not explain the staggering and ultimately self-destroying wastage of manpower, which seemed to have its origin in a curiously detached attitude towards human suffering.

Finally they do not explain the even greater depths of incompetence shown on this occasion by the enemy, of whom it has been said: 'The Russians, with more men in the field and immense potential reserves, were even bigger muddlers than their invaders, and seemed to move in a vague dream of battle.'[17]

And what of the man who led the Crimean army—the aloof yet courteous Raglan? Christopher Hibbert, who has written the most sympathetic of biographies in defence of Raglan, makes three points which as we shall see have considerable significance for the theory advanced later in this book. Firstly, Raglan had 'an emotional antipathy' towards the use of spies. Secondly, he nursed an intense dislike of the press and shunned publicity. Thirdly, he lacked enthusiasm for intellectual pursuits.

He cared little for the changing world outside it [the Army]. Science and mechanics, which were beginning already to change the whole life of Europe, meant nothing to him. Nor did painting, nor music; nor did books. In fact in the great mass of his personal correspondence only once does he mention having read one. It was *The Count of Monte Cristo*. 'So far as I have got in it,' he confessed, 'I find it is tiresome—very poisonous.'[18]

4

The Boer War

'It was in sum the second failure of a military machine.'
DAVID DIVINE, *The Blunted Sword*

The most extraordinary thing about the events of the Boer War was that they could have occurred not only after those of the Crimean War, but also after those of the First Boer War of 1880–81.

In terms of psychological jargon the Boer War showed very little evidence of positive transfer ('positive transfer' being merely what happens when something learned in one situation is utilized successfully in some subsequent, similar situation) and this was odd because there were many striking similarities between it and the Crimean War. Unlike most actions of the Victorian Army, both were against white races. Both were waged far from home, in what were, to say the least, trying climates—too cold in the one case, too hot in the other. There were, however, two great differences, and it was these which were largely responsible for depths of military incompetence on the British side that stretch credulity to breaking-point.

The first difference was one of the available technology. Thanks to Maxim's invention of an efficient machine-gun in the 1880s and Armstrong's production of breech-loading rifled field guns, soldiers at the time of the Boer War could (in theory anyway) kill each other in larger numbers with greater accuracy and over much longer distances than ever before. The transition from the old smooth-bore 'Brown Besses' (later replaced by Minié rifles) of the Crimean War to the high-powered, rapid-firing rifles, using smokeless powder, that were available by 1890 constituted what is probably the single greatest advance in fire power since the invention of gunpowder.

The second difference was in the nature of the adversary. While the

Russians at Sebastopol had been resolute and courageous fighters, they, like the British troops, were regimented and disciplined to act in unison without opportunity for thought or personal initiative. The Boers, however, were a very different proposition—in their motivation, style of fighting and expertise. As one observer remarked, the Boer Army consisted of 35,000 generals, each combatant his own master defending his homeland. They were also good marksmen, agile horsemen, and determined members of a flexible, knowledgeable guerrilla force.

These two differences between the wars would not have mattered had those who ran the British Army managed to keep up with the times in their thinking, and shaken off habits acquired over the preceding five hundred years. This they seemed unable to do. For a start, officers were still so busy being gentlemen, in or out of gorgeous uniforms, that they had little time for their men and a total absence of concern for the latter's welfare. Even as late as 1902, according to Sir Evelyn Wood, commanding officers in Southern Command did not even know the whereabouts of the soldiers' latrines, much less their condition. It was this attitude of mind, in conjunction with climatic conditions, that accounted for the fact that out of 22,000 British dead 16,000 died of disease.

Progress in adopting new military techniques was also conspicuous by its absence. Right up to the outbreak of war, training manœuvres were characterized by a disregard of new weapons. 'The accent was on solid line formations, mechanical precision, rigid dependence on order, firing strictly in volleys at a word of command.'[1]

The position as regards artillery was little better. According to one writer: 'The artillery doctrine of the time was older than the guns, older almost than the Crimea. The artillery galloped smartly into action, unlimbered in the open (for it had no notion of indirect fire), and opened fire over primitive sights with no vestige of a gun-shield to protect its crews.'[2]

Considering its inadequate handling of the Crimean War, one might have expected that the high command would try to rehabilitate itself with a rigorous training programme during the ensuing years of peace. Evidently it did not. According to Kruger: 'Only two months a year were spent training. For the rest a man was parading.'[3]

As for new inventions, the War Office was sporting rather than pragmatic. Machine-guns like the Maxim, which in its modified form became the Vickers of the First World War, were written off as suitable only for the destruction of savages and hardly suitable for use against white men.

It was in decisions of this sort that the generals showed a curious illogicality. On the one hand the colour of the Boer soldiers elevated them from the level of savages, thereby saving their white skins from exposure to machine-guns, but on the other they were regarded, in terms of their believed military expertise, as no better than savages. Of all the factors which contributed to the succession of disasters which marked the war this underestimation of the enemy was perhaps the most important. Largely because they eschewed any form of sartorial elegance and preferred the wearing of civilian attire, dark cloaks and floppy hats to the sorts of uniforms affected by the British, the Boers were dubbed a rabble of illiterate peasants and their army utterly ludicrous. In reality, as events were to prove, it was the British not the Boers who despite their smart appearance showed up in a far from satisfactory light.

The soldiers themselves were not to blame; they were victims of the system. As Pemberton points out: 'So far as British politicians were concerned, it might almost be said that the greater the inefficiency of the Army (except on military parades and on annual military tournaments) the greater their contentment.'[4] According to the same writer: 'Ours was an army of amateurs; for officers a congenial club.' Any sign of keenness or desire for self-improvement in military matters was frowned on as not 'good form'. Those who signified a wish to attend staff college were likely to be told: 'Isn't the regiment good enough?' Polo and the highly symbolic sport of pig-sticking were the applauded occupations and any sort of book-work quite beyond the pale.

So absorbed were the generals by the equation which they made between military appearance and fighting efficiency that their contempt for the Boers was matched only by their neglect of most ordinary training exercises for their own men. As Lord Roberts was to remark, the men were not trained to use their eyes or estimate distance; in his opinion far too much attention was given to order and regularity and far too little to developing individuality. One further and related factor which plagued British military thinking then, as at other times, was the tendency to equate war with sport. The notion that certain acts were 'not cricket' was carried to such absurd lengths that the trooper was given no training in the 'cowardly' art of building defensive positions or head cover.

To this mixture of potentially disastrous attitudes was added one other—a self-righteous indignation towards the other side. This came to the fore when it was discovered that the British methods of running the war fell short of those practised by the Boers. As Lord Kitchener said: 'The Boers are not like the Sudanese who stood up to a fair fight. They are always running away on their little ponies ... there are a good many foreigners among the Boers but they are easily shot as they do not slink about like the Boers themselves.' 'A fair fight' was evidently one in which primitive tribesmen (from a poor country) by standing their ground obligingly presented an easy target for the soldiers of what was then the richest nation on earth. By the same token, disengagement in an action against superior numbers, in order to fight again another day, was deemed cowardly and a poor show.

This then was the background of attitudes and expertise which the British Army brought to the Boer War. Any residual doubts about its unfittedness for the expedition tend to dissipate when one considers the behaviour of the generals put in charge.

The leading character was the commander-in-chief, General Sir Redvers Buller. According to a contemporary description there could be no finer choice for our South African adventure: 'There is no stronger commander in the British Army than this remote, almost grimly resolute, completely independent, utterly fearless, steadfast and vigorous man. Big-boned, square-jawed, strong-minded, strong-headed ... Smartness ... sagacity ... administrative capacity ... He was born to be a soldier of the very best English type, needless to say the best type of all.'[5]

Unfortunately this assessment was at variance with the facts in all but two particulars. Firstly, he was indeed big. Secondly, though sadly lacking in moral courage, he was undoubtedly brave when it came to physical danger. In this respect, as in many others, he was not unlike Raglan of the Crimean War, and indeed some other commanders of subsequent years.

Of Sir 'Reverse' Buller, as he came to be known by his troops, Rayne Kruger writes: 'At the risk of marring [the] contemporary description ... it should be mentioned that his big bones were particularly well covered, especially in the region of the stomach, and that his square jaw was not especially apparent above a double chin. He had entered the army with no disadvantage, his mother being a Howard and niece of the Duke of Norfolk, and he was very wealthy, which was fortunate in view of his preference for a diet of ample good food and champagne.'[6]

Kruger attests to his bravery. 'His record was something to conjure with. Since the age of twenty-one he had fought in five campaigns, including the Zululand wars when he was the dashing leader of irregular cavalry. A winner of the V.C., he had such a reputation that many people thought he and not Wolseley should have been head of the British Army.'[7]

Never has a nation been more wrong-headed in its selection of a general. Never has a general been more disastrous in the execution of his duties. Like Raglan, Buller had no experience of commanding a large body of men. For the previous decade he had held a number of different posts in the War Office. According to contemporary accounts he was bereft of creative imagination and totally lacking in discrimination. He was also without that gift of intuition which impels a good general to choose the right course of action.

His first step towards disaster was to shelve the official British strategy. This was to capture Bloemfontein and then Pretoria, thus knocking out those centres of the Boer movement, the Orange Free State and Natal. Instead, distracted by the sieges of Ladysmith and Kimberley, he split his army into three unequal parts, for the simultaneous relief of these beleaguered towns and the capture of Stormberg Junction. While his motives for this move are not entirely clear, it was certainly consistent with three characteristics of the man. Firstly, he was a kind and gentle creature, and therefore undoubtedly distressed at what he imagined was going on behind the sieges. Secondly, he lacked confidence, particularly for the role of active Supreme Commander. As he remarked to Lord Lansdowne: 'I have always considered that I was better as second in a complex military affair than as an officer in chief command ... I had never been in a position where the whole load of responsibility fell on me.' Lastly, he lacked singleness of mind. Contrary to popular belief at the time, which was based on two grossly misleading clues—his appearance and his bravery—he was in fact the antithesis of the steadfast son of Mars: irresolute, indecisive and without faith in his own plans.

As a consequence of these traits Buller lost no time in trying to rid himself of any direct responsibility for the conduct of the war, by handing over the reins to subordinate commanders to whom he gave no further directives. This abdication, as opposed to delegation, might have passed unnoticed had the subordinates been men of great ability. Unfortunately they were not. As a consequence, within five days three battles had been lost.

Since the principal reasons for this 'impossible' state of affairs— wherein a large army of professional soldiers could be defeated three times in quick succession by a handful of 'untutored' peasants—bear on the central theme of this book, we must outline the nature of these engagements. However, to appreciate the factors which led to these defeats we should first consider an encounter which immediately preceded them. This was a battle in which Buller's subordinate, General Methuen, with 8,000 men was very nearly defeated by 3,000 Boers.

Methuen's objective was the Modder River, a natural defence-line for the Boers. Accordingly, without any reconnaissance he ordered his troops to make a frontal attack. Since he could not see the enemy he wrongly assumed that no enemy was there. Led by their officers the men advanced across the flat and open veld towards the river. All went well until they were within easy range of the Boers, who had concealed themselves, with what was subsequently described as 'fiendish cunning', below the deep banks of the river.

Those of Methuen's army who were not killed outright by the sudden blast of fire from the invisible Boers spent the day lying prostrate under a scorching sun. In a temperature of 110°, unable to move forward or back, they, including the seventy wounded, suffered extreme discomfort from thirst and slowly blistering skin. Methuen's remedy was to direct heavy artillery fire on to the Boer positions. Thanks to the latter's use of cover this barrage had very little influence on the course of events, apart from killing a number of his own troops through faulty range-finding.

It was only under the cover of darkness that the British eventually withdrew, leaving behind 500 dead and wounded. Because the Boers also used the night to pull back their line, to the Magersfontein range behind the river, Methuen claimed the action as a victory. Others were not so sure. According to one observer, Methuen 'failed to display a much higher degree of generalship than a promising young subaltern straight from Sandhurst'. According to another: 'Everyone here is furious with Methuen for his bad generalship. He goes slap bang at the position with a frontal attack and never thinks of turning the flanks.' Yet others noted that Methuen had studiously ignored two sound, if obvious, observations by Lord Roberts, one of the great generals of that time. The first was that modern weapons make frontal attacks over open ground impossible. The second was that the first duty of a commander is reconnaissance.

While all this may seem obvious, it evidently did not occur to Methuen. Within a few days of his performance at Modder River he confirmed the worst fears of his critics in the even more disastrous battle of Magersfontein.

On the quite unjustified assumption that the Boers were occupying the Magersfontein hills he subjected their deserted slopes to an intense artillery barrage at considerable expense to the British taxpayer. Total enemy casualties from this enterprise were three Boers killed. With a confidence born of ignorance he then ordered a frontal assault by the Highland Brigade. This was made at dawn, following a night march.

Unknown to Methuen, the Boers were concealed in narrow trenches some distance in front of his objective. As at Modder River they waited patiently until the British came within easy range. Surprise was complete. When they opened fire a hail of lead swept through the ranks of the Highland Brigade. Within minutes the ground was carpeted with dead soldiers, including the Highland commander, General Wauchope. It was too much for the remainder; despite their training and their discipline, despite the honour of the regiment, despite all those factors which the high command fondly believed would induce uneducated soldiers to sacrifice themselves for the shortcomings of their generals, they broke ranks, turned tail and fled. As they did so they were further pounded and demoralized by hitherto undetected batteries of Boer artillery.

From his position well behind the firing line, it should have been clear to Methuen that the only way to save the day was to outflank the Boer positions. This he refused to do. He could only order 'Forward, forward the Highlanders', and then, like some latter-day Canute, watch the rising tide of defeat as the Highlanders fell back. The battle ended with his army, less a thousand dead and wounded, back at the starting line.

It is worth comparing the cerebrations of the British command with those of the Boers. While Methuen, despite his very recent experience at Modder River, could still respond only with frontal assaults, the Boers had applied some *thought* to their defences. They reasoned that much was to be gained by surprise and concealment. They also reasoned that a high-powered rifle fired at ground level has more chance of finding a target than one fired down from above.

But it would be uncharitable to condemn Methuen as typical of his time, for he was not. That he failed to anticipate the patently obvious was ascribed not so much to his limitations as to the 'extraordinary ingenuity' of the Boers! Thus *The Times History* refers to the Boer trenches as 'one of the boldest and most original conceptions in the history of war.'[8]

Moreover, Methuen was by no means the most foolhardy of the generals. There was General Featherstonehaugh who at the battle of Belmont insisted on riding up and down in front of his men in full regalia, thereby announcing his importance to the enemy and effectively hampering the fire of his own men. It was not long before the Boers rectified his error by shooting him off his horse. And there was General Hart who, at the battle of Colenso, inflicted thirty minutes' parade-ground drill on his brigade before marching them shoulder to shoulder, in barrack-square precision, across the open veld against the Boer position. Since it was broad daylight his densely packed column provided an irresistible target for every Boer gun and rifle within range. In this battle the British were defeated with the loss of 1,139 casualties and 10 guns against Boer losses of 6 dead and 21 wounded. In the words of a German staff historian: 'The general and not his gallant force was defeated.'

And there was General Gatacre, whose performance at the battle of Stormberg Junction was singularly lacking in panache. To capture *his* objective, Gatacre settled for a night march followed by a dawn attack. Not only did he not know the route but he succeeded in forgetting to bring along the one man who did, a certain Captain of Intelligence. As a result of the appointment of two 'guides' (who knew no more than he did), dawn found him and his army behind the hills he was supposed to be in front of. After some moments of consternation, during which he lost all sense of direction, the general resolutely faced his army the wrong way with their backs to the enemy.

Having recovered from the novel experience of being attacked by an army which appeared to be moving in reverse, the Boers opened fire with such devastating results that within half an hour Gatacre's force was in full retreat. When they finally reached their original starting-line they were delighted to find they had suffered only 90 casualties. Their euphoria was shortlived, for a second count revealed that 'by a mere oversight' 600 British soldiers had been left behind on the enemy-held hills. Since nobody had told them to retreat, they became prisoners of the Boers. As for other blunders by senior commanders in this and other wars, a cover story was soon put out to preserve the general's reputation. On this occasion Gatacre had been 'treacherously led into an ambush'.

In mitigation it might be claimed that since attack is more difficult than defence men like Methuen and Gatacre were disadvantaged in their conflict with an enemy who were, after all, defending their own terrain against an invading army. Two points, however, deserve to be made. Firstly, British Army training up to that time had always laid great stress on attack, with an almost total neglect of defensive tactics. As we shall see, there are good psychological reasons for this one-sided preparation for war, a bias in training for which this country paid dearly in subsequent conflicts. In the Boer War the Army was doing what it had been trained for. The other point is this. While in the battles so far described British troops were on the offensive, there *were* other military events in which they occupied a defensive role, as for example in the sieges of Ladysmith and Mafeking.

This makes it possible to examine the suggestion that military incompetence was confined to the handling of offensive rather than defensive actions. When we do so, however, the hypothesis is found untenable. Even in defence, incompetence still reigned. The best example is that afforded by Sir George White, V.C., Commander-in-Chief Natal, who, in trying to resist the enemy, failed to carry out the most elementary precautions. Like some deranged householder who refuses to lock his door when he knows burglars are about, White omitted to carry out any measures to deny the Boers use of their most valuable mode of access—the railway. He failed to mine passes, block tunnels, blow up bridges, or in any way destroy their prime means of transportation. Of this serious dereliction, *The Times History* comments: 'The least damaging explanation is that Sir George White never realized fully that the Boers were civilized opponents who could make use of a railway for military purposes.'*

* Sir George White, whose statue can be seen near Broadcasting House in London, was subsequently made a field-marshal and Companion of Honour (against the advice of A. J. Balfour).

As we shall see, there are remarkable parallels between this lapse and the events which preceded the fall of Singapore in 1942. In both cases the generals in charge were responsible for the safety of an English civilian population. In both cases they seriously underestimated the ability of the enemy, and this in the face of overwhelming evidence. And in both cases they lacked the imagination to carry out the most elementary and obvious of precautions.* In short, the argument that they failed to be efficient only because of the difficulties attending *attack* scarcely holds water.

It is at this point that it becomes necessary to introduce another concept which is relevant to the conduct of the South African War. It is that of the effects of psychological stress upon decision-making. It is perhaps in their resistance to stress, in their ability to carry on when things go wrong, that good generals are most easily distinguished from poor ones.

By this standard General Buller, physically so huge, failed dismally. Irresolute from the outset, the three defeats at Magersfontein, Stormberg Junction and Colenso sapped whatever confidence he ever had. From being weak and fearful he became a veritable jelly of indecision. His plans became vague and indefinite, and his specific orders scarcely more enlightening. His lack of moral courage in the face of adversity revealed itself most clearly in his propensity for making scapegoats of his unfortunate subordinates, those admittedly incompetent generals who had blundered on without direction or assistance from above, while taking none of the blame himself. 'Nothing in his despatches at the time, or later in his evidence before the Royal Commission [convened to investigate the reasons for the series of defeats] suggests that, even in the most roundabout way, he who planned the whole [campaign] was in any manner responsible for its failure. The nearest he came to such an admission was a reference to "bad luck".'[9]

Bad luck it may have been, but worse luck was to follow in the shape of that 1,400-foot monument to military ineptitude, Spion Kop. The totally unnecessary storming of this mini-mountain was to the Boer War what the charge of the Light Brigade had been to the Crimean War. The details are as follows. While still numbed by the series of

---

* The fact that in the case of Singapore even greater errors of judgment had been made by politicians and Army leaders *before* the war is immaterial to this argument.

defeats just recounted, Buller's army of 29,000 infantry, 2,600 mounted men, 8 field batteries and 10 naval guns was enriched, if that is the word, by the arrival of a fresh division commanded by Sir Charles Warren, R.E. Together these two forces, under the supreme command of Buller, were employed to try and lift the siege of Ladysmith. Unfortunately, and despite their immense superiority in men and equipment, they failed to do so.

So far as the British were concerned, the operation involved crossing the river Tugela and then closing on Ladysmith via a complex of small hills and ranges of high ground. There were two places at which the river could be forded. Buller ordered Warren to lead the assault across one of these 'drifts', while another force, under another general, Lyttleton, created a diversionary crossing at the other. For success, the plan depended upon speed of movement, surprise and synchronizing the two crossings. Ideally, Warren and his force, on the left flank, should have been over the river and well on their way to Ladysmith while Lyttleton was still occupying the Boers' attention on the right flank.

The plan went wrong for several reasons. In the first place Warren's division was far too small for the main attack. As to why Buller should have used an attacking force whose size, in comparison with the total number of troops he had available, constituted a mere pinprick, it suffices to note that it accorded with his general policy of avoiding any direct responsibility for whatever might transpire. If the worst happened it would be Warren's army, not his, that would carry the blame. In due course we shall examine two deeper reasons for this particular form of military incompetence.

The second reason for disaster lay in the character of Warren, who has been described as 'dilatory yet fidgety, over-cautious yet irresolute and totally ignorant regarding the use of cavalry'. He was also obsessive, obstinate, self-opinionated and excessively bad-tempered.

While Lyttleton crossed the Tugela with his diversionary force and successfully convinced the Boers that this was Buller's line of advance, Warren failed to exploit the situation. Instead of crossing the river with all speed he seemed 'to give way to certain fads and fancies'. These included an obsession with his enormous baggage train* and the fear

* One of the factors which slowed up Buller's military movement in the Boer War was the quantity of the baggage with which officers went on active service. According to Kruger this might well include pianos, long-horned gramophones,

that it might be destroyed by non-existent enemy guns on the small mountain Spion Kop. So concerned was he with his baggage that he spent twenty-six hours personally supervising its transfer across the river. The delay was invaluable to the Boers.

It was at this point that mere tardiness and inefficiency gave way to something more approaching madness. Under the mounting strain of inactivity a curious *folie à deux* seemed to descend upon Buller and his subordinate. In chronological order the events were as follows:

1. A cavalry reconnoitre by Lord Dundonald of the territory beyond the river revealed an obvious line of advance for Warren's army.

2. Warren was furious that Dundonald should have used his cavalry to make the recce.

3. Partly through his obsession with the baggage train and partly because of the unsolicited and unwelcome information from Dundonald, Warren rejected the projected movement and opted instead for a direct advance across the Tabanyama range, directly to his front. Unfortunately no recce had been made of this area.

4. It was at this point that Buller began describing Warren's behaviour as 'aimless and irresolute'. Nevertheless, he still refused to assume command.

5. Warren's assault on the Tabanyama range was hardly a success. This was because he found the Boers well dug in on a second crest of whose existence he had been ignorant. He still refused to outflank the Boer positions.

6. Buller, who was becoming increasingly restless, rode over to proffer criticism and advice. He still refrained from giving any orders to Warren.

7. Warren's eye now lit upon the cone-shaped eminence of Spion Kop. He knew instantly that it must be captured. Buller readily agreed, and this though neither general had previously considered such a course of action, let alone worked out what it would entail.

8. The job of attacking what has been called 'an unknown mountain on a dark night against a determined enemy of unknown strength' was given to General Talbot-Coke. His 'qualifications' for the venture

---

chests of drawers, polo sticks, and in Buller's case an iron bathroom and well-equipped kitchen.

were that he had only just arrived and was seriously affected by a game leg. At least he was no more ignorant of Spion Kop than were any of his colleagues, for they knew nothing about its summit—its extent or suitability for defence. No one wondered why the Boers had no guns up there, nor did it occur to anyone that the Boers might resent its occupation by the British. Hence no diversionary tactics were employed.

And so, while the generals stayed below, the men were ordered up the steeply sloping mountainside, into a fog hardly less dense than that which clouded the minds of their commanders. When, in almost zero visibility, they thought they had reached the summit, the assault force halted, congratulated themselves on the total absence of opposition, raised the Union Jack and tried to entrench. The operative word is 'tried', for the top was much like the rest of the mountain, solid rock. Nobody had warned them of this. They decided to use sandbags, only to find that no one had remembered to bring them. While the mists cleared they did the best they could with pieces of rock and clods of earth, only too well aware that this flimsy protection provided no head cover whatsoever.

If this gave them food for thought there was more to follow, for with a further improvement in visibility they made a second disquieting discovery. They were not where they thought they were. Instead of the summit they found themselves on a small plateau some way below the mountain top: 1,700 men on a piece of ground 400 x 500 yards, and above them, on three sides, the Boers. The enemy opened fire. Within minutes the ground was littered with corpses, many with bullet-holes in the side of the head or body. Owing to the lack of head cover the losses from shrapnel were even greater. Trapped in this seemingly hopeless position without any guidance or directives from their general, two hundred Lancashire Fusiliers laid down their arms and surrendered to the Boers. Their place was taken by reinforcements sent up from below.

Meanwhile Warren and Buller did nothing to help the hard-pressed troops. No doubt appalled by what was happening to his army on the heights above, Warren, supine at the best of times, went into a state that has been described as paralytic. Only once did he try to interfere with the course of events. This was to stop his battery of naval guns from shelling Boer positions on a neighbouring peak. He did so in the mistaken belief that the troops they were shelling were British.

Although possessing the necessary equipment, he had failed to establish telegraphic communication with his troops on Spion Kop. Had he done so this particular costly error would never have occurred.

As to why he, the commanding general, should deliberately cut himself off from the main source of intelligence, his own front-line troops, one can only surmise that, at some level, he just did not want to know. This hypothesis, that Warren was using what is technically known as the mechanism of denial, receives support from another curious incident. A war correspondent who had witnessed the dire events on the top of the mountain hurried down to the commanding general. But instead of receiving this, admittedly unsolicited, information with gratitude, Warren flew into a rage and demanded that the journalist should be arrested for insolence. The war correspondent in question was Winston Churchill.

But Warren's behaviour, as we have said, was only part of a *folie à deux*. No less extraordinary was that of his commander-in-chief. Buller's contribution was violently to resist the pleas of his subordinate commanders for an attack upon those positions from which the Boers were so assiduously shelling his troops. He even went so far as to recall such units as had managed to reach peaks held by the enemy. Had they been allowed to remain, the massacre of British troops would have been substantially reduced.

When night came, those who had survived the constant shelling and rifle fire decided to seek permission to withdraw. Unfortunately their lines of communication were again disrupted, this time because they had not been given sufficient oil for their signalling lamps. Maintaining communications within his army was not Warren's strongest suit. He did, however, order General Talbot-Coke to go up the mountain and bring back news. But once again he took great pains to avoid hearing the worst. For a start, he selected as his messenger a lame man who did not know the country; then, just in case he did succeed in struggling up and down the mountain, Warren took the ultimate precaution of shifting his H.Q. to a new location. Since he did this in Talbot-Coke's absence, and without a word to anyone, he managed to sustain his ignorance.

So ended the battle. Having lost 243 dead and 100 wounded, the army withdrew. The following day found 20,000 sullen men marching back the way they had come. For all their superiority in numbers, for all their training on the drill squares of Aldershot, they had achieved nothing. Once again the rigid Goliath had been ousted by the astute but nonconformist David.

This, the last of their disastrous battles, showed up the weaknesses of the high command with blazing clarity. As one captured Boer artillery officer remarked: 'If *your* men had *our* generals where should we Boers be?' Though slight compared to what lay in store for them years later the cost was immense by contemporary standards: 22,000 British dead in only 31 months and a bill to the nation of £22,000,000.

As with other generals in other wars, Buller's hitherto inconspicuous energies found their outlet not on the battlefield but afterwards in making scapegoats of his subordinates. In his case it was the unfortunate Warren who bore the brunt.*

Before moving on to the next example it is worth placing the Crimean and Boer Wars in the same perspective. Both present a picture of what appears to be unrelieved stupidity, but more interesting is the psychological pattern of these events. Here was a rich and powerful nation anxious to assert its rights, first in Russia and then in South Africa. What did it do but send out highly regimented armies which endeavoured to make up in courage, discipline and visual splendour what they lacked in relevant training, technology and adequate leadership? As to the latter, in each case a commander-in-chief was selected who despite his deficiencies remained inordinately popular with his troops for far longer than he deserved. Both men were genial, courteous and kind. Both were inexperienced, irresolute and lacking moral courage. Both were rich and well connected, but both, when the occasion demanded, were only too ready to divest themselves of all responsibility for the errors which they had made. And the one seemed quite unable to learn from the mistakes of the other.

So strange are these phenomena that one is forced to consider the hypothesis that at some level in the minds of those who direct national aggression there lurks a contrary motive, a need to pull their punches. Since man is the only species which through his ability to kill at a distance cannot avail himself of those automatic inhibitors of intra-species aggression which are commonplace with other animals, it is possible that he quite unconsciously uses other means to achieve the

* Buller was dismissed from the Army in 1901. The following year the Government published his dispatches. These proved him to have been 'incompetent, blundering, defeatist'.

same purpose. The appointment of such men as Raglan and Buller would certainly constitute such other means.

A fuller development of this thesis is reserved for later chapters. For the moment it might prove helpful to keep in mind certain characteristics of the incompetence just described. They include:

1. An underestimation, sometimes bordering on the arrogant, of the enemy.

2. An equating of war with sport.

3. An inability to profit from past experience.

4. A resistance to adopting and exploiting available technology and novel tactics.

5. An aversion to reconnaissance, coupled with a dislike of intelligence (in both senses of the word).

6. Great physical bravery but little moral courage.

7. An apparent imperviousness by commanders to loss of life and human suffering amongst their rank and file, or (its converse) an irrational and incapacitating state of compassion.

8. Passivity and indecisiveness in senior commanders.

9. A tendency to lay the blame on others.

10. A love of the frontal assault.

11. A love of 'bull', smartness, precision and strict preservation of 'the military pecking order'.

12. A high regard for tradition and other aspects of conservatism.

13. A lack of creativity, improvisation, inventiveness and open-mindedness.

14. A tendency to eschew moderate risks for tasks so difficult that failure might seem excusable.

15. Procrastination.

5

Indian Interlude

From the data considered so far it might be thought that military incompetence is confined to intra-racial conflicts—white against white. Unfortunately, as suggested by the following account of a minor incident at the time of the Indian Mutiny, this particular prediction is not borne out. When it comes to *inter*-racial conflict the pattern of incompetence is little changed.

Here is the story of Fort Rooyah as recounted by P. Scott O'Connor:

General Walpole, who, it appears, had never before held independent command, was ordered to lead an expedition up the left bank of the Ganges from Lucknow to Rohilkhand, to clear the rebels out of that part of the country. The brigade set out from Lucknow on the 7th of April, 1858, and on the morning of the 15th found itself in the vicinity of Fort Rooyah. The troops had marched nine miles that morning; but Walpole, anxious to win his spurs with the least possible delay, sent his force immediately to the assault.

The fort was the residence of a rebel landholder named Narpat Sing. He had but three hundred followers at his command; but, taking advantage of the troubles which beset the British in India in the dark days of 1857, he unfurled the flag of rebellion at Rooyah and bade the Government defiance.

His stronghold was nothing very formidable. On its northern and eastern faces it was strongly defended by a high mud wall and a broad and deep ditch, and covered by a dense jungle; but from the west and south it was open to attack, as the wall on those sides was but a few feet high, the defenders relying mainly on the *jheel*, the waters of which lapped the fort, to protect them from their enemies coming from that direction. There were two gates to the fort, and these opened on the sides just mentioned; and there is no doubt that had General Walpole delivered the assault from that direction the fort must have been quickly reduced, with but a fraction of the casualties which actually occurred. It was the month of April, and the water of the *jheel* was everywhere very shallow, and in many places dried up, so that the only obstacle to an assaulting party from that side was lacking.

But General Walpole took no trouble to reconnoitre; and, without even a cursory examination of the position, launched his men in a blundering and haphazard manner against the strongest face of the fort.

The rebels, it was reported, were prepared to evacuate the place after firing a few rounds; but when they saw the British advancing against the face which could be defended, they changed their minds and determined to show fight.

Now Walpole, under the mistaken impression that there was a gate on the east side of the fort, directed Captain Ross Grove to advance with a company of the 42nd Highlanders through the wood in that direction, and to hold the gate and prevent the enemy from escaping. The company advanced in skirmishing order through the jungle before them; and dashing across the open space of ground which lay between the forest and the fort, found their progress impeded by the ditch, which had up till then been invisible. There was no alternative but to lie down on the edge of the counterscarp; and as there were only a few paces between them and the enemy, and no shelter whatever, they were exposed to a galling fire and suffered severely. They held on to their position, however, in a most heroic manner, awaiting the development of the attack in other directions; but finding, after a time, that no other attack was being made, Grove sent word to the general to tell him that there was no gate, and requested scaling-ladders for an escalade. Meanwhile Captain Cafe, wholly unaware of the ditch which had checked Grove in his advance came up with his Sikhs and dashed into it. With no ladders to help them out again, they were shot down without mercy by the enemy ...

No orders had as yet reached Grove, nor were the scaling-ladders forthcoming, so a second messenger was despatched to the general, asking for reinforcements. The general, apparently now alarmed at the consequences of his own rashness, hastily sent the heavy guns round to the west, and ordered a bombardment of the fort from that side.

A very natural result followed. Some of the balls from the guns, going over the fort, fell among our men on the other side, for they had not yet been withdrawn. A report to this effect was carried to Adrian Hope, who at once rode off to inform Walpole, but from what followed it appears the latter doubted the accuracy of the statement, for Hope immediately returned to see for himself.

'Good God! General,' exclaimed Grove, on seeing him, 'this is no place for you. You must lie down.' But the kindly warning came too late, for even at that moment Hope fell back into the speaker's arms, shot through the chest. Soon after came the order to retire and General Walpole rode back to camp.

Under cover of the darkness that night the rebels slipped out of the fort and made good their escape.

The loss the country sustained by the death of Willoughby, of Douglas, of Bramley, of Harrington, and of the hundred and odd men uselessly sacrificed before Rooyah was great; but the loss of Adrian Hope was a cause for national sorrow. His death was mourned on the spot by every man in the camp. Loud and deep were the invectives against the obstinate stupidity which had caused it.[1]

The tale of Fort Rooyah speaks for itself. There is little to add beyond pointing out that the traits and behaviour of the unfortunate General Walpole do not depart significantly from those of commanders in the Russian and South African campaigns.

But General Walpole's unhappy expedition was not the first disaster to befall the British Army in India. Sixteen years previously, in 1842, a catastrophe occurred beside which the events at Fort Rooyah seem scarcely worth a mention.

The road was strewn with the mangled corpses of their comrades and the stench of death was in the air—All along the route they had been passing little groups of camp-followers, starving, frost-bitten, and many of them in a state of gibbering idiocy. The

Afghans, not troubling to kill these stragglers, had simply stripped them and left the cold to do its work and now the poor wretches were huddling together naked in the snow, striving hopelessly to keep warm by the heat of their own bodies. There were women and little children among them, who piteously stretched out their hands for succour ... Later the Afghans were to report with relish that the unhappy fugitives, in their blind instinct to preserve life a little longer, had been reduced to eating the corpses of their fellows. But they all died in the end.[2]

The British retreat from Kabul in the First Afghan War has been described by Field-Marshal Sir Gerald Templer as 'the most disgraceful and humiliating episode in our history of war against an Asian enemy up to that time'. Judging from the details of how a British army of 4,500 men was wiped out by what was, in comparison with the British strength, a handful of Afghan tribesmen, the field-marshal's words are nothing of an overstatement.

For events leading up to the disastrous retreat the reader is referred to *Signal Catastrophe* by Patrick Macrory. Suffice it to say that in 1842 a British army was stationed in Kabul, the capital of Afghanistan, for the purpose of supporting the puppet ruler Shah Soojah. This unwise move was motivated by the belief that without a pro-British ruler in Afghanistan that country and then India might be lost to Russia.

From the outset the situation in Kabul did not bode well, nor were conditions such as to inspire confidence. Cut off from India by some of the worst country in the world—towering crags interspersed with deep ravines and narrow passes—the army's lines of communication could hardly have been more vulnerable. And if the terrain was hostile to peace of mind, the climate was even more so. Temperatures, depending on the place and time of year, ranged from 120° to 40° below freezing. Death from heat-stroke vied with death from exposure for men unlucky enough to soldier in such a place. As if this was not enough, there were the Ghilzyes, the Kuzzilbashes and the Uzbegh, described by one witness as 'savages from the remotest recesses of the mountainous districts ... many of them giants in form and strength, promiscuously armed with sword and shield, bows and arrows, matchlocks, rifles, spears and blunderbusses ... prepared to slay, plunder and destroy, for the sake of God and his Prophet, the unenlightened infidels of the Punjab'.[3]

These amiable creatures, and in particular the Ghilzyes, apparently impervious to the rigours of nature, swarmed above the British lines of communication like killer wasps above a rivulet of honey.

Surveying the scene from a stronghold in Kabul and conscious of the fact that the vast majority of Afghans did not want Shah Soojah, disliked the British and resented the army of occupation, any prescient military commander might have been forgiven for taking every precaution against a native uprising.

But the British, by their own choice, were not in a strongly fortified position, from which we may draw the not unreasonable conclusion that they were short on prescience. For reasons that defy any simple explanation they chose to site themselves in a hutted camp on a stretch of low-lying swampy ground a mile outside the town. As Lieutenant (later General Sir Vincent) Eyre remarked: 'It must always remain a wonder that any Government, or any officer or set of officers, who had either science or experience in the field, should in a half-conquered country fix their forces in so extraordinary and injudicious a military position.'4

In this worst possible site the British laid out a camp to the worst possible design. Not only was the two-mile perimeter, a purely nominal obstacle consisting of a low wall and a narrow ditch, far too long to be defended by the numbers it enclosed, but the whole was open at its northern end to a compound of dwellings for the British envoy and his staff. This hotchpotch of houses positively invited infiltration by even the least intrepid of enemies. To complete this incorrigible behaviour there had been one final act of such unbelievable stupidity that its repercussions were to lead to the death of an army. By the orders of the commanding officer, Willoughby Cotton, the army's commissariat stores were constructed a quarter of a mile *outside* the cantonment. The consequences of this decision were tragic and inevitable. When the Afghans finally rose against the British, the army were promptly cut off from their supplies. Thus it was that under the threat of starvation they ultimately capitulated to Akbar Kahn, the Afghan leader, and began the retreat which cost them all their lives.

There was thus good cause to feel uneasy about the situation in Kabul. It still might have been saved, however, had the army at this time been blessed with competent leadership.

Unfortunately it was not. Thanks to pressure from none other than the future Lord Raglan, the Government of India chose this moment to appoint Major-General William George Keith Elphinstone as Commander-in-Chief Afghanistan. He was, to say the least, an unfortunate candidate, described at the time as 'the most incompetent soldier that was to be found among the officers of the requisite rank'. Even if necessary, his qualifications were certainly not sufficient. They were that he was 'of good repute, gentlemanly manners and aristocratic connexions'. He had last seen active service at Waterloo, twenty-five years previously, and had since been on half-pay. He was elderly and so stricken with gout that he could scarcely move.

Like General Sir Redvers Buller half a century later, Elphinstone had no illusions about his unfitness for the job, and pleaded that his health made him quite unsuited to the demands that would be made upon him. But Lord Auckland, the Governor-General, was adamant, and so the gentle, courteous Elphinstone was shipped off to Kabul.

Once there, whatever shreds of self-confidence he may have had were speedily removed: firstly by the ludicrous nature of the army's cantonment and secondly by encountering for the first time his new second-in-command, Brigadier Shelton, a rough brute of uncertain temper. So appalled was Elphinstone by the army's location that he offered to buy up surrounding land so that he could then clear suitable fields of fire. His generous offer was refused. About Shelton he could do nothing.

After this events accelerated towards the final catastrophe. Aware no doubt that the British were weakly led, Afghan resentment blossomed into revolt. Sporadic attacks on British personnel culminated in the assassination of the British Resident and the sacking of the Residency. Faced with these unpleasant facts, General Elphinstone sank into a state of numbing indecision. Consumed by doubts, needled by his cantankerous second-in-command, he cast about him for advice from everyone within reach, even down to the most junior subalterns.

Finally, it was McNaghten, the British envoy, a civilian, who came up with a plan. He suggested that a force under Shelton should withdraw at once into the neighbouring fortress of Balla Hissar. Clutching at this brainwave, Elphinstone ordered Shelton to march on the fortress. No sooner had this order been received, though, than it was countermanded. Shelton, unimpressed by this stop/go policy, retorted sharply that 'if there was an insurrection in the city it was not a moment for indecision, and recommended him [Elphinstone] at once to decide upon what measure he would adopt'.

Elphinstone then countermanded his countermand and once more ordered Shelton to march at once to Balla Hissar. But barely had Shelton started before he was overtaken by another order to the effect that he should halt and remain where he was. But no sooner had this order been received, reducing the second-in-command to a state approaching apoplexy than it was followed by the inevitable counter-order. It seemed that he was, after all, to proceed with his men to the fort. And this, surprisingly, he did.

Meanwhile, Elphinstone was canvassing opinions as to what to do next. Should he enter Kabul in force to crush the insurrection or would it be more prudent to remain where he was? Should he reinforce the commissariat which contained all the army's supplies or should he withdraw its small garrison into the main cantonment? A day was wasted in futile debate—a day in which the insurrection, encouraged by British paralysis, grew apace. When eventually Elphinstone did act, it was a case of too little and too late.

With Afghans rallying to the cause in ever-increasing numbers, it soon became impossible even to reach Kabul. Similarly, through Elphinstone's procrastination, any question as to whether or not he should reinforce the commissariat became purely academic; it fell lock, stock and barrel into the hands of the insurgents.

Determined leadership might still, however, have won the day; but this quality of leadership was not to be found in Elphinstone. He considered launching a full-scale attack upon the Afghans, but just as quickly dropped the idea. His state of mind is reflected in a letter which he then wrote to the envoy: '... it behoves us to look to the consequences of failure, in this case I know not how we are to subsist or from want of provisions, to retreat. You should therefore consider what chance there is of making terms, if we are driven to this extremity.' Thus already, only three days since the unavenged murder of the British Resident, the commander-in-chief was ready to accept defeat.

Having settled for the necessity of capitulation, he backed it up with the unfounded delusion that his army was running short of ammunition. Thenceforth things went from bad to worse. Having satisfied himself that the army could not, or rather would not, fight to defend itself McNaghten, the British envoy, urged on by Elphinstone, entered into negotiations. The enemy's terms were quite uncompromising. They demanded a speedy and total withdrawal of the British from

Afghanistan. Bereft of any military backing, the envoy had to accept, and a draft treaty was drawn up. But McNaghten, a braver man than Elphinstone, then tried to doublecross the Afghans and was murdered for his pains. Apparently unmoved by this second killing of a British Government official, and wholly averse to initiating any reprisals, Elphinstone became more ingratiating than ever towards his tormentors.

While rage, and a thirst for revenge, consumed the lower ranks of the army, those at the top became increasingly indecisive and anxious to appease. Inevitably the Afghan surrender terms stiffened, until finally Elphinstone, in response to empty promises of safe conduct, found himself agreeing that his army *without* its ordnance but encumbered by twelve thousand non-combatants, including many women and children, would go back the way they had come.

Having decided upon the disastrous plan of trying to reach Jalalabad in the depths of winter, across mountain ranges infested with hostile tribesmen, Elphinstone proceeded to make matters worse by further procrastination. Right up to January 6th, 1842, he remained in an agony of mind as to whether or not he should commit his army to the march, and when, on that fateful day, they eventually set off he changed his mind when half the force were already on their way. He tried to stop them but now his order to halt was disobeyed; for good or ill the die was cast. It was for ill. With deep snow on the ground, night temperatures that fell to well below freezing, and blood-thirsty Afghans preparing to fall upon them as they traversed the narrow passes, the only hope of successfully reaching Jalalabad lay in speed of movement. To hang about on this fearful journey could only mean death from exposure for those who managed to avoid death at the hands of the marauding Ghilzyes.

To achieve their purpose the British had to move, and move fast, preferably by night when in the narrow confines of the passes. But speed was denied them. For a start, no one had bothered to reconnoitre a suitable route. Secondly, Elphinstone had refused to cancel the construction of a bridge across the Kabul River despite the fact, as was pointed out to him, that this waterway was fordable in several places. Since no one wanted to get their feet wet, they all converged on the bridge, to produce such a monumental bottleneck that it delayed the marching columns by many hours. Finally, Elphinstone, fearful of moving by night, took to calling a halt at the end of each day.

Without food, firewood or any shelter beyond that provided by holes scraped in the snow, many died each night. By day, as they traversed the grim passes of Khoord-Kabul, Jugdullok and Gandamack, thousands more died at the hands of the murderous Ghilzyes. At the end of four days, with seventy miles still to go, only 850 remained of the original 4,500 soldiers. By the end of the tenth day their number had been reduced to 450.

Throughout this pitiful venture, Elphinstone, despite the trail of corpses which lay behind him, retained a pathetic and wholly unjustified faith in the Afghan leader's promise of safe conduct.

By the end of the fifth day the total losses of soldiers and civilians had risen to 12,000. As one officer described it: 'There was literally a continuous lane of poor wretches, men, women, and children dead or dying from the cold and wounds who, unable to move, entreated their comrades to kill them and put an end to their misery.'

While this painfully prolonged disaster could be attributed to many factors—including national greed and the anxiety which had resulted in the invasion of Afghanistan in the first place, political ineptitude in the choice of military leaders, and governmental stinginess in denying sufficient funds for an extension of British fortifications in Kabul—the sheer enormity of the catastrophe which was now unfolding must be laid at Elphinstone's door.

This refined and gentle creature manifested what at first sight may appear to be some curiously inconsistent characteristics. By his own admission, he sought 'the bubble reputation' in India and yet, when given an important command, shrank from the responsibilities which it entailed. He was hopelessly indecisive, lacking in moral courage and suggestible, yet could, on occasions, manifest irrational pigheadedness. He wobbled when he should have been firm, yet was rigid when he should have been flexible. Finally, he was courteous and kind, retaining the affection of many of his followers right up to the end, yet could be totally lacking in compassion for many of those who had suffered at his hands.

His inflexibility is highlighted by his refusal to enter the fort of Balla Hissar, even though his suffering columns of soldiers and civilians passed close to this edifice on their way to Jalalabad. Of this episode Macrory writes:

Pottinger, Lawrence [officers of Elphinstone's staff] still hoped against hope that at the eleventh hour Elphinstone would come to his senses and order the army to march straight in and occupy that formidable stronghold before the Afghans could rally to prevent them. *But Elphinstone was not the man to be capable of such an audacious change of plan* (italics mine). The crossroads were reached, the advance guard turned left towards Jalalabad and the Balla Hissar died away in the winter dusk behind them.[5]

His total lack of compassion is shown in the following incident:

To the misery of hunger was added the misery of cold, for the bitter Afghan winter had descended upon the wretched inhabitants of the cantonments. Before the end of November, sleet and snow became a daily occurrence, with the thermometer at freezing point, and from mid-December onwards the ground was inches deep in snow. The Indian troops suffered particularly from the cold, but although there was a complete winter stock of firing fires were not allowed. Sturt pressed Elphinstone and Shelton that at least fires might be permitted at night, so that men coming off duty from the ramparts might warm themselves and dry their frost-encrusted clothes, but nothing was done and the miserable troops sank deeper into apathy and numbed despair.[6]

Of some significance for matters to be dealt with later, there is the following incident on the subject of dress:

Pottinger, who had noticed that as soon as the first snows fell every Afghan appeared with his legs 'swathed in rags', now urged that old horse-blankets should be cut into strips which the troops could roll puttee-fashion round their feet and legs. This sensible suggestion presumably seemed to the high command both slovenly and unsoldierly, for nothing was done, and the troops were left to the misery of their hard leather boots. Within a few hours of the start of the march the frost had done its work and hundreds were suffering the agonies of frost-bitten feet.[7]

Finally there is the case of the hostages. When the Afghan chief Akbar Kahn offered protection to hostages Elphinstone took the opportunity of saving the lives not only of women and children but also of their menfolk. However: 'No one supposed for a moment that he was referring to any but the *British* wives and children, nor was any plea put in for the far more numerous wives and children of the sepoys and camp-followers ... these were native and expendable.'[8]

Elphinstone's concern that officers wounded should also benefit from protection did not extend to wounded other ranks, who were 'apparently of no account'. These attitudes, which in the present instance saved the lives of several British women and their officer husbands but cost the lives of thousands of lesser mortals, is, as we shall see, not without significance for theories of military incompetence.

To conclude this account of the total dissolution of an army: On January 13th, 1842, soldiers on guard at the British fort in Jalalabad saw a single horseman riding towards them, with all the speed that his maimed and worn-out horse could muster. It was the surgeon Dr Brydon, the only man, it seemed, to survive the fearful journey from Kabul.*

When news of the disaster reached India and London, much mental energy was devoted to the discovery of scapegoats. The two favourites for this role were Shah Soojah, accused by his critics of treachery, and Elphinstone's Indian sepoys. In neither case were the accusations justified. Soojah had in fact remained loyal to the last. As for the sepoys, though dragged from the warmth of their native India to fight another man's war in the freezing climate of Afghanistan, they, if anything, fought more bravely and endured what were for them particularly adverse conditions more stoically than any other unit of Elphinstone's army. But they were convenient scapegoats, because they were dead.

For a fitting epitaph to these men there is the following description from a subsequent relief force:

Pollock's force was marching back along the line of Elphinstone's disastrous retreat ... at every point they came upon ghastly evidence of the fate of the Kabul force. Rotting corpses and skeletons picked clean by carrion met them at every turn. At Tezeen they found a pile of fifteen hundred corpses of Elphinstone's sepoys and camp followers, who had been stripped naked by the Afghans and left to die in the snow. In the Khoord-Kabul pass, wrote Captain Backhouse, 'the sight of the remains of the

* Brydon was the only *European* to arrive at Jalalabad, but in the days after his arrival a few Indian soldiers and a number of followers also completed the journey. Elphinstone himself died of dysentery after being made captive by Akbar Khan.

unfortunate Caubul force was fearfully heartrending. They lay in heaps of fifties and hundreds, our gun wheels passing over and crushing the skulls and other bones of our late comrades at almost every yard.'[9]

A rather more accurate levelling of blame than that applied by armchair critics came from a man who had been there, Lieutenant Lawrence. 'Our Caubul army perished, sacrificed to the incompetence, feebleness, and want of skill and resolution of their military leaders.'[10]

To conclude this sorry tale there is one last point of some significance. When he heard of the troubles in Kabul Lord Auckland chose as commander of a relief force a certain Major-General Lumley. Like Elphinstone, Lumley made up in gentle manners and courtesy what he lacked in drive or physical stamina. Fortunately he was *so* ill and *so* decrepit that his doctors ruled that he could not possibly assume the role that Auckland had wished upon him. Hence another was chosen in his place.

6

The First World War

'The opposing lines congealed, grew solid. The generals on both sides stared at these impotently and without understanding. They went on staring for nearly four years.'

A. J. P. TAYLOR, *The First World War*

Only the most blinkered could deny that the First World War exemplified every aspect of high-level military incompetence. For sheer lack of imaginative leadership, inept decisions, ignoring of military intelligence, underestimation of the enemy, delusional optimism and monumental wastage of human resources it has surely never had its equal.

In an age in which it has become fashionable to question authority, it may well seem strange that a bare sixty years ago millions of ordinary men, living in indescribable conditions, could, with a courage, fortitude and cheerfulness past human comprehension, meekly carry out the lethal decisions of well-fed generals comfortably housed many miles behind the place where their orders were being translated into several kinds of pointless death.

Apologists for this period have found good things to say of some of the generals who took part. We are told that Haig did the best he could, given the conditions of the Western Front, that he was rock-like and tenacious. Joffre's saving grace, so it has been said, was that he was 'a skilled politician' and the only man with enough prestige to dominate France's allies. And, to quote A. J. P. Taylor: 'Even Sir John French was supposed for some time to be a great military leader.'

Other views have been less charitable: 'stupid, obstinate blimps', 'butchers', 'ossified brains' and 'donkeys' are just a few of the unkind epithets which have been applied to those who bore upon their immaculate shoulders the responsibility of committing a generation of young men to various forms of mutilation on the battlefield. A contemporary expression of this point of view puts it thus: 'It is hard for a connoisseur of bad generalship, surveying the grey wastes of World War I, to single out any one commander as especially awful. There were dozens of them on both sides.'[1]

Certainly, of some generals (and admirals), such as those engaged in the Gallipoli fiasco, who showed a paralysis of leadership which approached in severity that displayed by Buller and before him Elphinstone, little eulogizing is possible.

Notwithstanding the apologists, the First World War highlighted as never before the contrast between the 'muscles' and the 'brains' of the armed services. The 'muscles' were superb, the 'brains' with a few notable exceptions—such as Plumer, Smith-Dorrien, Allenby and Monash—were not so good. As a result, the armies resembled the saurians of a bygone age, huge in strength, massive in body, but controlled by a nervous system so sluggish and extended that the organism could suffer fearful damage before the tiny distant brain could think of, let alone initiate, an adequate response.

Incompetence took several forms. These included:

1. The implementation of a plan for the disposition of the British Expeditionary Force that had been devised three years before the outbreak of hostilities and remained unmodified in the light of subsequent events.

2. A tenacious clinging to the age-old practice of frontal assaults, usually against the enemy's strongest point. The following lines from a war correspondent *in the Boer War* suggests that in this respect learning from past experience was not the forte of the high command: 'The bayonet charge of a few years back is as dead as the Grecian phalanx— the quick-firing rifle has changed the face of war . . . For nineteen dreary months the great English people has been held in check by a handful of farmers, simply because English folk cling to old traditions as sand crabs cling to seaweed in storm time ... to me it was simply incomprehensible that they did not evolve a new process of attack which would nullify the natural advantage and native astuteness of the Boers.'[2] There is one respect in which one might quarrel with this report: its implied generalization to 'the great English people'. In fact there has never been a nation so inventive and so fertile in its technical

innovations as the English at that time. But the military were, *and still were twelve years later,* sand crabs indeed.

3. An under-use and misuse of available technology. Haig's opinion that two machine-guns per battalion would be quite sufficient and the attitude of some reactionary elements to development of the tank are cases in point.*

4. A growing belief in the value of a prolonged bombardment before launching an attack. Besides being enormously expensive, such bombardments necessarily sacrificed the vital element of surprise, made the intervening ground almost impassable to the subsequent assaulting infantry, and provided numerous convenient craters to which enemy machine-gunners might betake themselves, from their deep dug-outs, after the holocaust was over, there to await the slowly moving ranks of attacking infantry.

5. A tendency on the part of the high command to ignore evidence which did not fit in with their wishes or preconceptions.

6. A terrible crippling obedience. There was at even the highest levels of command an attitude of mind so pathological and unrealistic that, on occasions, even army commanders dared not express their doubts about the viability of a particular order or venture, preferring to conceal evidence from their superiors rather than be thought wanting in courage or loyalty. As Liddell Hart wrote of the Third Battle of Ypres: 'It would seem that none of the army commanders ventured to press contrary views with the strength that the facts demanded. One of the lessons of the war exemplified at Passchendaele is certainly the need of allowing more latitude in the military system for intellectual honesty and moral courage.'[3]

7. A readiness to accept enormous casualties. In terms of the number of lives lost, relative to the ground gained, the actions of the First World War make dismal reading. In the first two hours of the battle of Loos we lost more men than were lost by all services together in the whole of D-Day 1944. On the first day of the Somme offensive the British Army suffered 57,000 casualties—the biggest loss ever suffered by any army in a single day. And yet, as one historian has put it, to see the ground gained one needs a magnifying glass and large-scale map.

Post-war critics have tended to ascribe these sad facts to blundering

* These often-quoted examples reflect a state of military conservatism that was, in fact, far more prevalent in the armies of France and Germany.

stupidity on the part of the controlling generals. At best this can only be a partial truth, for behind the apparent stupidity lay some rather more important factors — personal ambition, jealousy and the relationship in men's minds between ground (and material), lives and reputations. One aspect of this relationship has been described as 'that service tendency of mind which sentimentally values things more than lives, a tendency which may have its foundation in totemism, but is also accentuated by the peace-time shortage of material, and the penalties attached to any loss of it. The artillery man's love of his guns, and readiness to sacrifice his life to avert the disgrace of losing them, is paralleled by the sailor's adoration of his ship ... it hinders [them] from adopting the commonsense view that a ship, like a shell, is merely a weapon to be expended profitably.'[4] Another aspect is implied in A. J. P. Taylor's comment, that 'those British generals who prolonged the slaughter kept their posts and won promotion'.[5]

As for ground, even a few yards of blood-soaked mud must never be yielded, particularly when it bore a name, like Ypres or Verdun, of almost mystical significance.

Together such attitudes account for much of the carnage in the First World War.

However, since the purpose of this book is the analysis of certain psychological tendencies associated with warring behaviour and not the reconstruction of military history, only two specific incidents will be touched upon, and these albeit briefly.

The first concerns a case of jealousy and the price paid for disobedience — that of General Sir Horace Smith-Dorrien.

The story begins in August 1914 with Smith-Dorrien, commanding 2nd Corps of the B.E.F., in contact with the enemy at Le Cateau, and his commander-in-chief Sir John French installed at G.H.Q., thirty-five miles behind the lines.

Sir Horace had been ordered to retreat but, realizing that to do so would jeopardize the whole of the B.E.F., he ignored G.H.Q. and, acting on his own initiative, engaged the enemy.

His delaying action paid off. Instead of being enveloped, the army was able to withdraw, weary but intact, to fight another day, thus rescuing Sir John French from what would have been the results of his own incompetence.

Now for reasons which went deeper, and further back in time, Sir John French, weak, irascible, touchy and inefficient, nursed a jealous dislike of his competent corps commander. Under the circumstances it is hardly surprising that Smith-Dorrien's stand at Le Cateau was scarcely calculated to evoke warm feelings of gratitude in his chief. But worse was to follow.

It seems that, totally out of touch with what was going on at the front (having withdrawn his headquarters a further thirty-five miles from the scene of enemy activity), and envisaging quite incorrectly that Smith-Dorrien's troops were fleeing from an enemy in hot pursuit, French ordered that all spare ammunition and officers' kits should be abandoned.

To Sir Horace, the man on the spot, the order, based as it was on a false premise, could serve no possible advantage. Indeed, its effect upon morale (in the days when officers had to buy all their own kit) would be wholly bad. He chose to ignore it.

Only too well aware that his corps commander had, for the second time, saved him from his own ineptitude, Sir John was none too pleased. With understandable ingratitude he addressed himself to the task of removing his troublesome subordinate. His chance came nine months later when the Germans attacked at Ypres with a new weapon — chlorine gas.

Despite considerable evidence from many sources that the enemy were about to use this new weapon, neither the French nor the British high commands were prepared to meet it. No warnings were given, no precautions taken. In the event, casualties were small by later standards but in their nature singularly unpleasant. More seriously, the falling back of troops on either flank left the British Second Army surrounded on three sides in that sentimentalized death-trap, the Ypres salient. Under the circumstances, Smith-Dorrien, now in command of the Second Army, wrote an appreciation of the situation in which he stressed the high cost in life of further counter-attacks and advocated a withdrawal to a new defence line west of Ypres.

This was all that French needed. On the grounds that Smith-Dorrien had disobeyed his orders and was now a source of dangerous pessimism, he forced a resignation from one of his ablest and most valuable of generals — one whom the King himself had congratulated on his saving of the B.E.F. at Le Cateau nine months earlier.

A few days after Sir Horace had been relieved of his command and sent home (on the pretext of ill-health), Sir John authorized the very withdrawal which his subordinate had advocated. In the words of a contemporary writer: 'There is no accounting for how a man in so high a position could behave thus, or how a man capable of such behaviour could have been placed in so high a position.'[6] Fortunately Sir John's record of ineptitude caught up with him. At the Battle of Loos his failure to position reserves where they could be of any possible assistance, and the discrepancy between what he said he had done and what in fact he did, cost the British Army 60,000 casualties and himself his job.

7

Cambrai

*'Accusing as I do without exception all the great Allied offensives of
1914, 1916 and 1917, as needless and wrongly conceived operations
of infinite cost, I am bound to reply to the question—what else
could be done? And I answer it, pointing to the Battle of Cambrai,
"this could have been done". This in many variants, this in larger
and better forms ought to have been done, and would have been done
if only the generals had not been content to fight machine-gun
bullets with the breasts of gallant men, and think that that was
waging war.'* WINSTON CHURCHILL, *The World Crisis*

*'In the case of the tanks a constant war had to be waged against the
apathy, incredulity and shortsightedness of G.H.Q.'*
STEPHEN FOOT, *Three Lives*

In 1912 a private civilian inventor, E. L. de Mole of Adelaide, presented
the War Office with a design for a tracked vehicle, which, to put it at
its simplest, would help to solve the major tactical problem of the
First World War: how to get soldiers across no man's land, barbed
wire and enemy trenches without being shot. The War Office looked at
de Mole's design, and laid it on one side.

In 1915, through a total lack of personal protection, British soldiers
on the Western Front were dying at the rate of thousands a day. De
Mole was moved to resubmit his invention; again it was ignored.

In 1919, after the war was over, the Royal Commission on Awards
to Inventors said of de Mole: 'He is entitled to the greatest credit for
having made and reduced to practical shape, as far back as the year
1912, a very brilliant invention which anticipated, and in some respects
surpassed, that actually put into use in the year 1916. It was this

claimant's misfortune and not his fault that his invention was in advance of his time, and failed to be appreciated and was put aside because the occasion for its use had not then arisen.'[1]

While one must congratulate de Mole on receiving, if not financial reward, at least some belated credit for his ingenuity, there are one or two inconsistencies in the Royal Commission's statement. In the first place, the idea of an armoured fighting vehicle was neither new nor ahead of its time. Forerunners of the tank can be traced back to Caesar's invasion of Britain. Leonardo da Vinci had designed an armoured fighting vehicle in the sixteenth century, and the concept was advanced by H. G. Wells in his book *The Land Ironclads*, published in 1903. Secondly, there had already been every justification for using tanks both in the Boer War and in the First World War *prior* to de Mole's second submission of his plans.

It might be concluded, therefore, that his invention was put aside not just because it was a new idea, which it was not, nor because it was not needed, which it was, but because it conflicted with a mystical belief in the virtues of horsed cavalry and in the power of a prolonged artillery barrage. Any residual doubts as to the reactionary nature of the motives underlying the non-acceptance of armoured fighting vehicles may be dispelled by the fact that anything approaching wholehearted adoption of tanks by the British high command did not occur until well into the Second World War.

Perhaps de Mole's greatest misfortune was that he was only a civilian in 1912 and only a private in 1915. But, fortunately for those who had to do the fighting, he was not alone in his enthusiasm. Notwithstanding the resistance of such senior functionaries as the Director of Artillery and the Assistant Director of Transport, there was a handful of visionaries who took up the cause and, in the face of steady opposition, agitated for the adoption and construction of tanks. For those who seek to excuse military incompetence on the grounds that generals are only the helpless tools of their political masters, it should be pointed out that in this instance it was senior professional soldiers, *not* the politicians, who were against the use of armoured fighting vehicles. While Churchill and Lloyd George were enthusiastic supporters of the tanks, Master-General of the Ordnance General Von Donop remained implacably opposed to any such development. In the services the major proponents of tank development included, ironically, a small group of *naval* officers. The fact that the Admiralty felt less 'threatened' by tanks than did the War Office was strikingly illustrated at one of those demonstrations wherein proponents of a new idea strive to convert sceptics by confrontation with the evidence of their senses. After an impeccable display, in which prototype tanks cut through barbed wire, crossed trenches, slithered through mud and clawed their way out of craters, a naval officer was heard to remark: 'We ought to order three thousand now!' But the War Office contingent remained cool, one senior general retorting: 'Who is this damned naval man saying we will want three thousand tanks? He talks like Napoleon.'

So much for the first stage in the adoption of tanks. In the second a rather different class of error came to the fore—that of premature application. The occasion was the third phase of Haig's Somme offensive, an operation so dismal that, as one writer put it: 'Even the commanders most eager for this kind of warfare were shocked.'[2] Haig himself, to his eternal credit, had always been enthusiastic rather than obstructionist in his attitude towards tank development, but now, carried away by his own enthusiasm, decided to throw in the first few tanks then available. Opinions are divided, but some consider this to have been an unwise decision. As Churchill said in his war memoirs: 'The ruthless desire for a decision at all costs led in September to a most improvident disclosure of the caterpillar vehicles.' Likewise, Lloyd George considered their use in such small numbers premature. But Haig was nothing if not obstinate. The few tanks then available were thrown in—not en masse, but piecemeal.

Even so, compared with what had gone before, they achieved a small but spectacular success. But because they were too few in number, and their breakthrough not adequately exploited, the tanks were unable to prevent the offensive from being, in the end, a costly defeat. Cooper writes:

> The small part played by the tanks, however successful on the local scale, was overlooked in the general sense of failure ... *Doubts which many Staff Officers had previously expressed as to the value of tanks turned to scorn. Instead of trying to plan an intelligent use of the superior weapon that had been put in their hands, the military leaders could only make criticisms of minor details.* They conveniently forgot that it was they who had ordered so few tanks to be built in the first place, and that it was Haig's own decision, against the

advice of those who were beginning to understand the nature of tank warfare, to order them into battle before their crews had been properly trained and before they were available in sufficient numbers to make a worthwhile contribution.[3] (Italics mine)

Frustrated by failure, and unable to admit their own contribution to defeat, they did what the highly prejudiced do in such a circumstance, vented their feelings upon the original object of their prejudice, and in so doing precluded any chance of learning from the exercise. As Liddell Hart put it: 'Criticism fastened on faulty details and particular failures, with little sense of proportion, and still less imagination.' A year later the price of prejudice was paid in full.

If anything, the premature use of tanks, on the Somme and later at Third Ypres, sharpened the conflict between those progressives who had now seen with their own eyes what tanks could do (at St Julien alone it was estimated that they had saved over a thousand casualties) and the reactionaries, including Haig's own Chief of Staff, who did everything they could to curtail their use. Thus a thousand tanks were ordered, but then the Army Council cancelled the order. Fortunately Lloyd George stepped in and cancelled the cancellation. Then Fuller produced a pamphlet on tank tactics which, because it stressed the advantages of surprise and a short bombardment, was promptly withdrawn by G.H.Q. Since the Third Army Commander of Artillery was sympathetic towards Fuller, he too was withdrawn! Finally Swinton, who had probably done more for developing the tank than any other single man, was removed from his post as Tank Commander, largely because G.H.Q. deplored the lack of discipline in tank crews— they looked too dirty! Swinton was replaced by a General Anley, whose job it was to 'inject discipline into the Tank Corps'. Anley went on record with the reassuring comment that he 'was not interested in tanks'.

Meanwhile the futile Third Battle of Ypres continued to consume the lives of infantrymen at the rate of more than two thousand a day. Never at a loss, G.H.Q. blamed this waste of life upon the few tanks that *had* been used. It seems they disappeared into the mud along with everything else.

But the forces of progress were still at work. Despite the gloomy resistance of senior staff officers at G.H.Q., Haig was persuaded to let the Tank Corps try again, on ground of its own choosing.

The Cambrai tank offensive on November 20th, 1917, occurred in three stages. The first was eminently successful. Three hundred and eighty tanks operating on ground suited to caterpillar tracks achieved a spectacular success, overrunning three strongly held lines of enemy trenches. Whereas previous offensives had been measured in yards gained for tens of thousands of lives lost, the Cambrai advance was four and a half miles on a six-mile front with negligible casualties.

But if the first stage was an unprecedented victory, the second showed a beginning of the rot which was to turn victory into disaster. There were various contributory factors.

The first was General Harper, whose 51st Highland Division had been given the task of capturing a key objective in the centre of the attack—the village of Flesquières. Unfortunately Harper, who has been described as 'a narrow-minded soldier of the old school', was one of those who disapproved of tanks. Consequently not only were his troops given little training in working with the new weapon, but they were instructed in tactics contrary to those recommended by the Tank Corps. Even worse, Harper delayed his assault by one hour because he did not believe that the first objective, the Hindenburg Main Line, would be captured so quickly. In the event, the Hindenburg Line was crossed at 8.30 a.m. but Harper's timetable for the next stage had been fixed for 9.30 a.m. and this, despite the evidence of his senses, he resolutely refused to change. The unnecessary delay allowed the Germans an hour in which to bring up and site field guns on the Flesquières ridge. Here is one description of what followed:

... the tanks continued blithely on to the crest of the ridge, in line abreast as instructed. They came to the top, huge dark shapes silhouetted against the skyline. And there before them were the German field guns. Had the infantry been close behind the tanks as Fuller had planned, they could easily have dealt with these guns in a matter of minutes, but the infantry were far behind, not only held up by having to find their way through the wire but because of the machine-gun fire which was causing heavy casualties. The tanks were on their own.

With such perfect targets the German gunners opened fire. One by one the tanks were hit, while the crews worked desperately at the cumbersome gears to drive a zig-zag course and the gunners tried to return the fire. But taking accurate aim in all the pitching and tossing was virtually impossible. It was some minutes before the German guns had been put out of action ... but by this time sixteen tanks had been destroyed, with huge gaping holes in their sides. Most were on fire, and those crew members who had not been killed outright by the blasting shells were burned to death. There were no survivors.[4]

Woollcombe, whose grandfather was a corps commander in the Cambrai offensive, has, in his account of the tank battle, presented General Harper in a rather more charitable light. His strongest argument against criticism of Harper is that since the corps commander was renowned for being a strict disciplinarian, it is inconceivable that he would have allowed Harper to deviate in training or tactics from what had been laid down.

Be that as it may, the following facts suggest that the hold-up at Flesquières had its origins in the prejudices of a reactionary general. Firstly, Harper's assault was one hour later than it need have been. Secondly, it was only in his sector of the front that the infantry failed to follow closely upon the mobile armour. Thirdly, Harper had already gained a reputation for obstinacy and the possession of a closed mind, by his unbelievable opposition, well into the war, to development of the machine-gun. Finally, Harper and his division enthusiastically supported the legend of a mythical German artillery officer who, single-handed, destroyed all the tanks on the ridge. This legend, which served to exonerate Harper by finding another reason for the hold-up, gained significant impetus from those who still managed to sustain the belief that artillery and cavalry would always prove superior to tanks. Anyway, whatever its origin, the hold-up played a significant part in the next stage of the battle.

This was to have been an exploitation of the favourable situation created by the tanks. The force destined for this task was three divisions of cavalry. For what one observer described as 'our medieval soldiers' it was the opportunity for which they had been waiting since the outbreak of war. They failed for three reasons. Firstly, the delay at Flesquières robbed the offensive of its momentum, thereby losing the possibility of a German rout. Secondly, because they were under the control of a headquarters far behind the front line, the local cavalry commanders were unable to act promptly in the changing situation. While they hung about waiting for orders, the Germans brought up reserves and regrouped. Thirdly, it was proved, if proof was necessary, that when slowed down by wire or difficult ground there are few easier and more vulnerable targets for enemy machine-guns than a horse. Add to this the fact that, owing to the enormous losses in the Ypres offensive, there were no infantry reinforcements for Cambrai and it is not surprising that the battle ground to a halt. Through a pious and mistaken belief in the value of horsed cavalry, and a paralysis of thought occasioned by years of trench warfare, the brilliant break-through by the tanks was thrown away.

Some ten days later the Germans counter-attacked. In a matter of hours they recovered much of the ground originally lost. The British Third Army, commanded by General Sir Julian Byng, lost 6,000 men taken prisoner, some thousands killed or wounded, and a vast quantity of guns and other equipment.

The magnitude of this disaster was directly attributable to a feature of high-level military incompetence seen all too often: the ignoring of intelligence reports which did not fit in with preconceived ideas. Before the German attack, Byng had received evidence from local commanders that the Germans were massing reinforcements for a counter-offensive, but this information was ignored. No attempts were made to strengthen British positions. Requests from local commanders for artillery support to disrupt German preparations were refused. British tanks were withdrawn and prepared for enrailment to rear areas.

That the attack, when it came, was not more disastrous can be attributed to the initiative of some local commanders who, despite a total lack of encouragement from G.H.Q., took what precautions they could to stem the threatened onslaught. It was also thanks to the resourcefulness of certain young Tank Corps commanders who, on their own initiative, when the German onslaught started, halted the entrainment of tanks and made them ready for battle. This resourceful-ness, combined with the irreproachable bravery and superb fighting qualities of N.C.O.s and men, turned what might have been a rout into a costly and serious setback.

We now come to the last, perhaps saddest, stage of the Cambrai affair, the discovery of scapegoats. This process, to be efficient, must white-wash the true culprits (and their friends) while effectively muzzling those who might be in a position to question this action. This muzzling is a subtle process, the main inducement to silence being the unspoken threat that any attempt to undo the 'scapegoating' might put the undoers in jeopardy. Secondly, it must 'discover' scapegoats who are are not only plausible 'causes' but also unable to answer back. Thirdly, it must impute to the scapegoats undesirable behaviour *different* from that which actually brought about the necessity of finding a scapegoat. By so doing it distracts attention from the real reason for the disaster and therefore from the real culprits.

Using these criteria as a yardstick, the apportioning of blame which followed the Cambrai débâcle makes Raglan's treatment of Nolan, and Buller's of Warren, amateurish to say the least. When news of the disaster reached Britain, it was naturally assumed that the generals had failed again. Haig's reputation, already low, sank to vanishing-point. The War Cabinet demanded an immediate explanation.

Haig's response was to endorse a report from General Byng that the Third Army had not been taken by surprise and that the failure to stem the German breakthrough was due to shortcomings of those junior officers, N.C.O.s and men who had been involved in the fighting. In the face of so much contrary evidence, these views did not impress the critics. Byng was asked to explain why no reinforcements had been sent up to that part of the line which the enemy had chosen for his breakthrough. He replied that none had been asked for, and that he and his commanders had considered that no further troops were needed. Again, Haig supported these palpable untruths and opined that no criticism should attach to the senior commanders.

To stifle further debate, the War Cabinet called in General Smuts. Not very surprisingly, this 'great operator of fraudulent idealism' came down on the side of the generals. After studying all the reports from divisional commanders and above, while studiously ignoring the fact that the VII Corps commander had warned G.H.Q. of the impending attack and had received no response to this or to his request for reinforcements, Smuts stated: 'Higher Command Army or Corps Command were not to blame—everything had been done to meet such an attack.' He went on to say that the fault lay either with local commanders who might have lost their heads or with those lower down —junior officers, N.C.O.s and men. Of these two alternatives he preferred the latter explanation. And so Smuts, in the fashion of the day, blamed those least able to answer back—the youthful, the junior and the dead.

All in all, this black episode raises several matters of great relevance to the theory of military incompetence presented later in this book. Stupidity does not explain the behaviour of these generals. So great was their fear of loss of self-esteem, and so imperative their need for social approval, that they could resort to tactics beyond the reach of any self-respecting 'donkey'. From their shameless self-interest, lack of loyalty to their subordinates and apparent indifference to the verdict of posterity, a picture emerges of personalities deficient in something other than intellectual acumen.

As to how they look to a contemporary chronicler, there is the following passage:

> And so the whitewashing went on, to protect arm-chair generals who in the main had little conception of what the front line was like— and had no intention of going there to find out. One of those infantrymen so blamed was J. H. Everest. During the two days when he and his fellow soldiers were being pushed back by the Germans, they had no water to drink and no food to eat. At the end of the second day, while waiting in a trench for a renewed attack, Everest went up to his company commander and asked for permission to search for water. 'My request was refused,' Everest wrote later. 'Nevertheless, I went over the top and found some water in a mud-hole, thus ending two days of torture.' Shortly afterwards Everest was wounded and found himself in the Australian General Hospital at Abbeville ... But the most bitter pill of all on top of all this was to be blamed for their commanders' own mistakes.[5]

One of the consequences of these and other comparable events in the First World War was that they almost certainly terminated for all time the hitherto reverential and blind faith which troops had in their generals. In an organization renowned for striking loyalties between men in junior ranks, and in a war whose frightfulness was relieved only by comradeship and altruism in dangers shared, this betrayal by senior commanders cannot but have produced a lasting cynicism.

It could, of course, be argued that this was the one good which came out of the Cambrai affair. The same might also be said of the next example from the First World War, except that so few survived to tell the tale.

8

The Siege of Kut

If the degree of military incompetence is indicated by the ratio of achievement to cost, then the activities of 'Expeditionary Force D' under the command of Major-General Sir Charles Townshend merit examination. Firstly, there was a 250-mile discrepancy between what it was designed to do and what it tried to do. Secondly, the cost of this discrepancy was large. To reach Kut cost Townshend 7,000 casualties; during the ensuing siege a further 1,600 died; attempts to relieve his force accounted for another 23,000 casualties; when he eventually surrendered to the Turks, 13,000 of his troops went into captivity and of these 7,000 died while still prisoners of war. All this went for nothing, not one inch of ground or any political advantage, nothing, that is, beyond corpses, suffering and ruined reputations.

The story starts in 1914, with the Indian Government, under pressure from Whitehall, sending a small force to protect British oil interests in Mesopotamia. By 1915, with Turkey's entry into the war, the threat of an attack on the Ahwaz–Abadan pipeline had increased to such an extent that, again under pressure from London, the Viceroy, Lord Hardinge, and the Commander-in-Chief India, Sir Beauchamp Duff, increased the Mesopotamia force to divisional strength. Thanks to the machinations of four men—Hardinge; Duff; the Mesopotamian Army commander, Nixon; and the leader of the freshly constituted expeditionary force, Major-General Townshend—this modest venture led to a British military disaster so total yet unnecessary, so futile yet expensive, that its like did not occur again until the fall of Singapore in 1942.

Because we are primarily concerned with the more human aspects of these events, what follows has been based very largely upon Russell Braddon's book, *The Siege*, a work which has the unique advantage of being based upon eyewitness accounts by the survivors of Kut and the writings, orders, communiqués and telegrams of their commander. As such, it provides the sort of detail essential to a psychological analysis.[1]

As intimated above, the story starts with a fatal discrepancy, between the object of the campaign as laid down by the British Government and what the army actually did. Whereas Whitehall's purpose was to protect the oil refinery at Abadan with its pipeline to the coast, the army was soon busily engaged in trying to capture Baghdad. In terms of difficulty, distances involved and strength required, this discrepancy between its instructions and its endeavours was comparable to that between having a bath and trying to swim the Channel. From their point of disembarkation at Basra to Abadan is about thirty miles; from Basra to Baghdad is close on three hundred miles. The force provided to protect the oil installations comprised one division— 10,000 men—and that required to capture Baghdad was at least two corps—upwards of 30,000 men. While the lines of communication for the intended task were compatible with the supplies required and transport available, those entailed by an attempt on Baghdad were totally inadequate. This inadequacy resided in the fact that there were no roads between Basra and Baghdad, only the Tigris, a tortuous and uncharted river of reefs and sandbanks flanked by marshes and inhospitable desert. It resided in the fact that as an army penetrates into enemy territory its needs increase exponentially. More and more has to be carried farther and farther. It resided in the fact that as they stretch, lines of communication become increasingly vulnerable to enemy attack. In Mesopotamia there were four enemies: the Turkish Army, marauding Arabs, the terrain and the climate. All four played their part in hazarding the lines of communication and bringing about a defeat which cost much and gained nothing. But the real instigators of this tragedy were neither climate nor geography, neither Turks nor Arabs, but three generals: Sir Beauchamp Duff, Commander-in-Chief India, General Nixon, army commander, Basra, and Major-General Townshend, commander of 6th Division. Through an admixture of self-interest, personal ambition, ignorance, obstinacy and sheer crass stupidity this trio sealed the fate of some thousands of British and Indian soldiers.

It was in part a case of *'l'appétit vient en mangeant'*. Nixon, who made up in ambition for what he lacked in intelligence, ordered Townshend to capture Amarah, a township on the Tigris some hundred miles north of Basra. Townshend, equally ambitious but by no means stupid, did as he was bid. In so doing he and Nixon were already exceeding the directive of the British Government.

As well as occupying Amarah, Townshend struck westwards and took Nasaryeh. Nixon's appetite for glory was whetted by these easy victories; with no thought to the risks involved, he pressed Townshend to continue his advance a further ninety miles to Kut. In this he was backed by Duff, who had never visited Mesopotamia and had no idea of the conditions prevailing there. But Townshend had. He wrote to General Sir James Wolf Murray in England:

> I believe I am to advance from Amarah to Kut el Amarah ... The question is, where are we going to stop in Mesopotamia? ... We have certainly not got good enough troops to make certain of taking Baghdad ... Of our two divisions, mine, the 6th, is complete: the 12th (Gorringe) has no guns! Or divisional troops! And Nixon takes them from me and lends them to Gorringe when he has to go anywhere.
>
> I consider we ought to hold what we have got ... as long as we are held up, as we are, in the Dardanelles. All these offensive operations in secondary theatres are dreadful errors in strategy: the Dardanelles, Egypt, Mesopotamia, East Africa—I wonder and wonder at such expeditions being permitted in violation of all the great fundamental principles of war, especially that of Economy of Force. Such violation is always punished in history.
>
> I am afraid we are out in the cold out here. The Mesopotamian operations are little noticed, though we are fighting the same enemy as you have in the Dardanelles, plus an appalling heat ... The hardships in France are nothing to that.
>
> I have received great praise ... and have established a record in the way of pursuits ...

In the light of subsequent events this letter by Townshend is of interest. Of it Braddon writes: 'The letter was completely in character. It revealed a gift for strategic appreciation amounting almost to prescience. It revealed Townshend's chronic tendency to criticize his superiors, and his obsession with his own affairs to the exclusion of all others. It revealed his habitual lack of generosity to colleagues—whom he praised only if they were of inferior rank to himself—his tendency to whine and his almost embarrassing immodesty.'[2]

But the most extraordinary feature of the letter was that for all its strategic prescience it bore little relationship to Townshend's subsequent behaviour. Though he clearly realized that he was being asked to undertake a major campaign with the logistics of a subsidiary defensive operation, *he said nothing of this to his superiors*.

Seventeen days after writing to Murray, Townshend not only enthusiastically accepted Nixon's order that he should advance a further ninety miles to Kut but also, entirely off his own bat, talked of pursuing the enemy another 190 miles to Ctesiphon, and possibly beyond that to Baghdad. 'As to why he did so, there is no evidence at all—except his character. Indisputably, he was a man ambitious to the point of egomania: a man whom the lure of promotion had goaded throughout his career to such incessant intriguing and importunate letter-writing that he had incurred constant snubbing and rebuke, yet had persisted. To such a man, the smallest hint of condonation seems enthusiastic approval.'[3]

Closing his mind to his own forebodings, Townshend and his unsuspecting troops pressed on. Once again the Turks were defeated, and the British occupied Kut. But this time, though a remarkable achievement, Townshend's victory was not entirely free of blemishes. Two features in particular cast an ominous shadow over future events. Though suffering many casualties, the Turkish Army was not destroyed and escaped to fight another day. Then there were the British wounded. Townshend had estimated for six per cent but had suffered twelve per cent. The differential showed up and underlined those inadequacies of his lines of communication which were to prove so costly. What this meant in human terms is described as follows:

> The wounded suffered frightfully. Untended, they lay freezing all night—some to be stripped and murdered by Arabs—and, when daylight came, were placed on supply carts, unsprung, iron-slatted, and drawn across a cruelly uneven surface to the river bank. There, in fierce sun, they languished until they could be crammed on to the decks of iron barges and towed very slowly downstream to Amarah. What little water they were given was impure. What little treatment they could be given was ineffective.

Their wounds went gangrenous ... and they lay in a morass of their own blood and excreta, assailed by millions of flies. Quite unnecessarily, many of them died.

Sir John Nixon and Sir Beauchamp Duff had more important things with which to concern themselves than the plight of men wounded in an action it would have been wiser never to have fought. 'Their obsession was Baghdad.'[4]

It was at this stage in the campaign that Townshend's earlier pessimism returned with renewed force. Though pathologically ambitious and irretrievably egocentric, he was neither stupid nor ignorant. It now became obvious to him that to advance beyond Kut would be foolhardy and quite unjustified in view of the smallness of his force and their hopelessly inadequate lines of communication. But, like the sorcerer's apprentice, he seemed incapable of halting the flow of events that would so soon destroy not only his reputation but also the lives of his men. For one thing, he was powerless to quench the desire for glory which his earlier talk had kindled in the mind of the equally ambitious but far less talented Nixon. For another, he was, despite his appraisal of realities, loath to relinquish his own dream of becoming Lieutenant-General Sir Charles Townshend, Lord of Baghdad. And so, grossly underequipped, he marched his men beyond the point of no return towards Baghdad. He never reached that fabled city. For at Ctesiphon an army of 13,000 Turks lay across his path.

Meanwhile, while there still might have been time to turn back and abandon this suicidal mission, Nixon received intelligence that a second Turkish army 30,000 strong and led by the redoubtable Khalil Pasha was also converging upon Ctesiphon. But because this news did not accord with his desires, Nixon chose to ignore the report as untrue.

The battle of Ctesiphon marked the end of Townshend's luck. Though his conduct of the fight was exemplary if not brilliant, he sustained 4,000 casualties and, again, did not succeed in routing or destroying the enemy. This turn of events was in large part due to a Turkish counter-attack by the very reinforcements which Nixon had dismissed as non-existent—but Nixon and his entourage had now returned to the safety of Basra and so were spared confrontation with the results of their unwisdom.

As for Townshend, this reversal of his fortune had a predictable effect upon a mind already preoccupied with delusions of grandeur. He withdrew his force to Kut: Kut, which he knew to be without defence; Kut, which he had described to Murray as a position undesirably remote from Basra; Kut, which he now described as 'a strategical point we are bound to hold'. According to Braddon, Townshend's new-found delusion regarding the virtues of Kut may well have had its origins in a much earlier event, the siege of Chitral. This is a highly plausible hypothesis. When intractable desires are thwarted by reality there is a tendency to hark back to the memory of earlier gratifications, and Chitral epitomized for Townshend just such a gratification. Here, as a young officer in the Indian Army, he had withdrawn into a fort and captained his small force throughout forty-six days of siege. When eventually he did emerge, it was to find himself a hero beloved by Queen and country.

For a man of Townshend's temperament this had been a wish fulfilment not easily forgotten in time of stress, and so it was that now his eye fell upon Kut—the nearest thing to Chitral. Little wonder that he could now overlook the shortcomings of Kut and see in this smelly collection of mud huts the key to ultimate success. Kut became the strong point from which his four weak brigades, more than a match for the entire Turkish Army, would once again emerge victorious and, with the help of mythical reinforcements from England, fulfil his dream of taking Baghdad.

Another feature of delusions powered by insatiable needs is that they yield neither to reason nor to knowledge acquired in calmer times. Apprised by Brigadier-General Rimington, G.O.C. at Kut, that it would be as difficult to entrench the northern approach to Kut as it would be to keep the way open for a relief force from the south, Townshend retorted that it was Kut or nothing. His troops, he said, were too exhausted to retreat one step farther. This, of course, was nonsense, for they were evidently not too exhausted to dig six miles of trenches and then engage a determined enemy who outnumbered them by three to one.

The other inconsistency in Townshend's behaviour is that he had always prided himself upon the fact that he drew upon the lessons of history. Identifying himself, as the occasion demanded, with such great captains as Hannibal, Napoleon and Wellington, there was nothing he liked better than to quote the precepts of famous military commanders. Two such precepts were 'To make war is to attack' and 'Movement is the law of strategy'. But here was Townshend as heedless of Frederick the Great as he was deaf to the counsel of Marshal Foch. For to bottle himself up in Kut was to assume a posture of defence as stationary as it was passive. And it was unnecessary, for there was still time to fall back on the safety of Amarah, where reinforcements from Basra might in due course reach him. To have marched his force back to Amarah would have shortened his lines of communication and lengthened those of the Turks. That he did not do so cannot be ascribed to stupidity or to ignorance of the principles of war, for Townshend was neither stupid nor ignorant.

His behaviour during the next 147 days was that of a man who, while sliding inexorably towards a precipice of his own making, assumes that someone will not only step forward to break his fall but hand him a prize for having done so.

His first move towards hastening his rescue was so to manipulate his would-be rescuers that they felt compelled to try and relieve the siege before they were ready. Thus he persuaded his Army commander at Basra that since he had only a month's supply of food for his *British* troops, an early relief was essential. To sustain this lie and force Nixon's hand, he deliberately refrained from rationing either his British or his Indian troops, nor did he make any attempt to unearth the stocks of Arab grain concealed within the town.

Misled as usual by Townshend's 'inaccuracies' and fearful for his own reputation as the man partly responsible for the present débâcle, Nixon ordered the unfortunate Lieutenant-General Aylmer to break through the Turkish defences and relieve Kut.

Thus began a series of costly and futile attempts to defeat those Turkish forces which, having by-passed Kut, had taken up positions to the north of Amarah. Aylmer was handicapped by two factors. Firstly, since the Turkish lines ran from the Tigris on their right flank to an impassable marsh on their left, they could be taken only by a frontal assault, but Aylmer had neither the strength nor sufficient supplies to mount a successful attack of this kind. Secondly, he received no help from Townshend.

The lack of necessary supplies was directly attributable to Nixon, whose administration of docking arrangements at Basra had been so abysmally inefficient that the Indian Government sent him a harbour expert, Sir George Buchanan, to get supplies moving. But Nixon resented help from experts. Preferring that ships should be kept waiting three weeks before being unloaded, he 'argued so bitterly with Sir George, and defined for him so few duties that the latter returned in disgust to India—there to report that Basra's dockside arrangements were "of the most primitive order", situated in "a huge quagmire", and looking as if Force D [Townshend's force] had arrived "only last week rather than a year ago".'[5]

Starved of material and goaded by Nixon into hopelessly premature attacks, Aylmer sought help from Townshend. His eminently reasonable request was that the Kut force should create diversionary sorties to coincide with his, Aylmer's, attacks. But this Townshend steadfastly refused to do, despite the fact that in preaching the arts of war he had always emphasized the value of feints. As to why he now vetoed sorties by his troops, his explanation was that since every sortie would have to terminate in a withdrawal, this would look like failure and lower morale. And so, not very surprisingly, Aylmer failed time and again to achieve the impossible. Thanks to the combined efforts of the man he was trying to rescue and those of Nixon, the man largely responsible for the rescue being necessary, the relief force suffered 23,000 casualties, nearly twice the number of those invested.

Those not fortunate enough to be killed outright, or, less happily, to die slowly of their wounds and exposure during days and nights spent lying out in the battlefield in the rain and cold of a Mesopotamian winter, succumbed through the shortcomings of army medical services. Lacking the ruthless humanitarianism of a Florence Nightingale, they were in some respects rather worse off than the wounded of the Crimean War. Again Hardinge, Duff and Nixon were the culprits: Hardinge because he lied when answering inquiries by Whitehall as to the state of those medical services which the Indian Government was supposed to provide. His claims that all was well ignored the fact that Force D was seventeen medical officers and fifty sub-assistant surgeons under strength. As for Nixon, the following account says all that is needed. It starts with an exchange between the Secretary of State for India and Army commander, Basra. Joseph Chamberlain cabled: 'ON ARRIVAL WOUNDED BASRA PLEASE TELEGRAPH URGENTLY PARTICULARS AND PROGRESS'. Nixon replied: 'WOUNDED SATISFACTORILY DISPOSED OF MANY LIKELY TO RECOVER ... MEDICAL SERVICES UNDER CIRCUMSTANCES OF CONSIDERABLE DIFFICULTY WORKED SPLENDIDLY'.[6]

But Nixon, too, had lied, for he had just witnessed the arrival of 4,000 wounded from Ctesiphon:

The *Mejidieh*, with six hundred casualties on board and two crammed lighters in tow, had reached Basra festooned with stalactites of excreta, and exuding a stench that was offensive from a distance of a hundred yards. She had laboured downstream for thirteen days and nights. On her decks, and on the exposed decks of her lighters, men lay huddled in pools of blood, urine and faeces, their bodies slimed with excrement, their wounds crawling with maggots, their shattered bones splinted in wood from whisky crates and the handles of trenching tools, and their thighs, backs and buttocks leprous with sores.[7]

For those who like to find excuses for the behaviour of bad generals it may afford some pleasure to discover that lifting the siege was rendered doubly difficult by another factor—the weather. Once again Nixon managed to make a bad situation worse. For his policy was 'to send each new batch of reinforcements on a fourteen-day march upstream, dispatching their equipment after them, the first-line transport after their equipment, and their second-line transport (which included their blankets and medical supplies) after that. Sleeping cold in a Mesopotamian winter for fourteen successive nights, many of the troops who should have strengthened Aylmer's Relief Force were soon in hospital instead!'[8]

But it is time to get back to Townshend, safely locked up in Kut. Over the period of the siege he evinced several characteristics of some significance. First, there was his lying. In his cables to Aylmer and to Nixon, he continued to lie about his food supplies. From the outset, as we saw earlier, he maintained that he had food for only a few days; but as the days became weeks and then months, this initial falsification became something of an embarrassment, particularly when it was mooted that without food he would have to break out. This he did not in the least want to do. It was one thing to pressurize Aylmer into a premature and costly rescue but quite another for him to risk a break-out. Hence it was not surprising when he 'suddenly discovered' that his supplies would stretch to fifty-six days, nor was it wholly unexpected when he later raised the limits to eighty-four. In all, the figures seemed to suggest that the more his men ate the more food remained!

In his efforts to manipulate Aylmer, Townshend also falsified his estimates of Turkish strength, thereby encouraging his rescuers to throw themselves upon an enemy very much stronger than they had been told to expect.

Townshend's communications were not, however, confined to those outside Kut. During the siege he devoted much attention to the issuing of communiqués to his troops. These were remarkable for three features: a flagrant disloyalty towards and criticism of his superiors, a thinly veiled contempt for the valiant but unsuccessful relief force, and a total absence of gratitude towards those who were losing their lives in trying to rescue him.

Even less attractive was the hypocrisy of his behaviour towards his troops. Ostensibly he was the devoted, jolly father-figure of his 'beloved' 6th Division, and this is how they saw him, but, in small things as in big, his deeds belied this image. Although he would work his signallers to death tapping out an endless stream of trivial messages to his friends in London, not one of his other ranks was ever permitted to send a message through to his family in England, and this despite the fact that they received no mail whatsoever during the entire period of the siege. When military aircraft did drop supplies to the beleaguered force, such 'essentials' as pull-throughs, which could have been improvised, were given priority over letters from home, which could not.

As for big things, the worst was the way he abandoned his troops when the end eventually came—but this is a matter to which we shall return presently.

Further insight into Townshend's mind comes from contemplation of his more personally orientated behaviour.

Collating all Townshend's communiqués and messages from Kut, it can legitimately be deduced that after February 7th (when he had first asked Nixon to recommend his promotion to lieutenant-general), Townshend was always prepared to abandon his beloved command in the interests of either his own release or his own advancement. On March 5th he had again requested promotion. On April 9th, for the second time, he had suggested that he should attempt to escape from Kut and leave his division to its fate. Three times he had suggested negotiation to exchange Kut and its guns for the release of himself and his men, though he must have known that only he would be allowed to go. Twice he had sent ingratiating letters to the enemy commander in the field: and once he had insisted that no attempt be made on the life of an enemy field-marshal.[9]

Any doubts as to the correct interpretation of these unedifying facts are dispelled by three subsequent events. The first is a minor one, but none the less revealing. When Townshend learned that Aylmer's successor Gorringe had been promoted to lieutenant-general he burst into tears and wept upon the shoulder of a 'shrinking' subaltern, because he knew that Gorringe's promotion meant none for him. The second is that he did, in fact, leave his division to die as prisoners of the Turks. And the third is that neither then nor later did he so much as lift a finger to ameliorate their plight.

For our present purposes little remains to be said. After 147 days, Townshend's food supplies, which he had originally stated would only last a month, ran out. Confident from his exchange with the Turkish commander that he would be treated generously, he capitulated on April 19th, 1916, and handed his weak and starving men over to the not so tender mercy of the Turks. Then it was their paths diverged. While he was transported in the greatest comfort to Baghdad and thence to Constantinople, his 13,000 men began their 1,200-mile march across the arid wastes and freezing heights of Asia Minor. And while he was wined and dined, honoured and entertained as the personal guest of the Turkish commander-in-chief, his men died in their thousands of starvation, dysentery, cholera and typhus, and from the whips of their bad-tempered Kurdistan guards. They died of the heat by day and of the cold by night. They died because they wearied of staying alive—dropping out of the column, to be set upon by marauding Arabs who, having robbed them, filled their mouths with sand and stones. In all, seventy per cent of the British and fifty per cent of the Indian troops perished in captivity. But Townshend was spared these sordid details for he

travelled by train and arrived at Constantinople on June 3rd, to be met by the G.O.C. of the Turkish Army, his Staff, members of the War Office and a crowd of respectful locals ... he felt very flattered: and was even more flattered to be entertained later at Constantinople's best restaurant, then escorted by a detachment of cavalry ... to the water-front, where a Naval pinnace awaited him. His baggage, Staff and servants aboard, he sailed ten miles down the Sea of Marmara to the fashionable island of Halki, where, high on a cliff, he took up residence in a comfortable villa ... That same day, in the building the Turks called a hospital, those [of Townshend's troops] still too ill to march from Samarrah were being allowed by their captors to die in agony. There was no treatment for them and very little food, and those who fouled their beds were given an injection of brandy-coloured fluid after which they stopped fouling their beds because they were dead ... By that same day, more than a third of the British troops to whom Townshend had vowed that he was leaving them only to procure their repatriation had died.[10]

As the person most responsible for the disaster of Kut and for the misery inflicted upon his troops, Townshend might well have experienced and tried to expiate at least some modicum of guilt. That he did not raises several interesting issues, not the least of which being the suggestion that membership of a hierarchical authoritarian organization in some way absolves the individual from being hampered in his actions by this tiresome emotion. In the present instance, for example, Townshend was by no means unique in being apparently devoid of a sentiment which most people experience. Nixon, too, seemed quite unmoved at what *his* bid for glory had cost the soldiers under him. Nor was Nixon out of fashion. *His* superior, Sir Beauchamp Duff, Commander-in-Chief India, was similarly minded. This unbecoming trait showed itself most clearly when he forbade British exchanged prisoners* from Kut to publicize the suffering of those they had left behind. In this connection Braddon makes the interesting point that 'by making public the fearful conditions suffered by Japanese prisoners of war in Thailand, the British Government [in 1943] procured for them an almost immediate amelioration: in 1916, by saying nothing, and by muzzling those who wanted to speak, Townshend and Duff condemned ten thousand of their troops to months of agony and death.'[11]

In considering these data one is forced to the conclusion that the behaviour of these generals had something in common with that of Eichmann and his henchmen, who, as we know, were able to carry out their job without apparently experiencing guilt or compassion.

* Of the thousands of prisoners, 345 were exchanged for an equal number of Turkish prisoners.

As to what that 'something' might be, a suggestive clue is provided by another facet of the Kut affair to which Braddon draws attention. It concerns the extraordinary fact that of all the senior officers in Townshend's force only one chose to share the fate of the men as they marched into captivity, and this was Major-General Mellis—a general as different from Townshend, Nixon and Duff as Christ was different from the Pharisees. (Not that there was anything *overtly* Christ-like about Mellis. On the contrary, his reputation for swearing and blood-thirsty, reckless courage was second to none and higher than most. Though rough-tongued and blisteringly outspoken, Mellis nursed a great compassion for his men and used his not inconsiderable powers of invective to bully the Turks into improving conditions for their captives.)

Not so Townshend and his other senior officers, who, no doubt regretfully in some cases, allowed their loyal troops to go one way—to death—while they went another—to a life of comparative ease and comfort. Why? One reason is that King's Regulations did not stipulate that officers should go into captivity with their men. Had Townshend's officers been ordered to stay with their troops, doubtless they would have done so. But they were not and so they did not, and King's Regulations condoned their flouting of the old precept: 'No privilege without responsibility.'*

In conclusion, one point demands particular emphasis. In the mismanagement of the Mesopotamian campaign sheer stupidity played a relatively minor role. Certainly Duff was no genius and Nixon was unintelligent, but Townshend was not. Men's fates were decided for them not so much by 'idiots' as by commanders with marked psychopathic traits. Stupidity and ignorance there may have been, but it was the ambitious striving of disturbed personalities which accounted for the loss of Townshend's force.

In such matters as vanity, personal ambition, dishonesty and lack of compassion, Townshend was not unique. Where he differed from others was in possessing charm, intelligence and professional expertise. In a world of the 'square', the pompous and the desperately un-funny, Townshend had a refreshingly light touch and could radiate that bonhomie which earned for him the soubriquet 'a lovely man'. It was

* As Milgram has shown in his studies of obedience (see p. 269), even the most inhumane of acts may be carried through by the nicest of people without restraint from guilt *provided* they are sanctioned by the trappings of authority.

the possession of these qualities which so endeared him to his men that they were prepared to forgive him all his faults. The evidence suggests that he was not so popular with his fellow officers, who thought him 'frenchified'. It is possible that some awareness of this veiled criticism only served to sharpen his appetite for advancement. For underneath the agreeable veneer there lay a fatal flaw which showed itself in a ravenous, self-destructive hunger for popular acclaim. Though its origins remain obscure, Townshend gave the impression of a man who at some time had suffered traumatic damage to his self-esteem which resulted in an everlasting need to be loved. This hypothesis gains strength from an incident recounted by Braddon which, though trivial in itself, is curiously revealing. It concerns the general's dog, Spot.

> That night, on deck, attempting to sleep, Boggis [Townshend's batman] shivered more than usual: and Townshend's dog was so cold that he crept up to Boggis and snuggled against him. Each warming the other, they fell asleep.
>
> Boggis awoke to a fearful yelping and found Townshend thrashing his dog. Struggling upright, he demanded, 'What are you doing that for, sir?'
>
> 'He was sleeping with you!' Townshend snarled, still thrashing. 'He's *my* dog and he's got to learn.'
>
> 'He's a harsh bastard,' Boggis decided. But he was puzzled nevertheless. Townshend was devoted to Spot, as he was to his horse.[12]

Boggis was right in crediting his general with devotion for Spot. When Kut capitulated, Townshend's concern for the welfare of his dog was considerably more in evidence than that for his troops. He even made a successful appeal to the Turkish commander that the animal should be spared the rigours of captivity and returned to Basra.

But Boggis need hardly have been puzzled. The pathological jealousy which flared up in an assault upon the beloved Spot and overrode any feelings of compassion he might otherwise have had for the two frozen creatures lying outside his door was quite consistent with Townshend's other characteristics.

Later we shall illustrate certain qualities of personality by reference to particular top-ranking Nazis. While not for a moment suggesting an equation between the people concerned and the military incompetents who have graced these pages, the fact remains that some of the Nazi leadership exhibited, albeit in an extreme and grotesquely terrible form, some of the personality traits of our more inept military commanders. Using this yardstick for a measure of Townshend, his personality approximates most nearly to that of Göring. Like the Reichsmarschall, he exuded bonhomie, was sybaritic in his tastes and universally popular with his compatriots. But underneath the 'hail fellow well met' exterior lurked that same preoccupation with the fruits of power that consumed the stout German. Like Göring Townshend was professional, brave and narcissistic, and like Göring his goals were selfish rather than ideological. Göring had contemptuously referred to 'this ideological nonsense' and was unashamedly in it for what he could get out of it. Likewise, Townshend betrayed by his deeds and the countless letters which he wrote to those who might pull strings on his behalf that he too was 'in it' for the pickings. But Townshend also possessed one trait quite foreign to Göring, though present in large measure in other members of the Nazi elite: a totally unrealistic appraisal of the effect his actions might have on the opinions of others.* He simply could not grasp that people would fail to be amused by his abandoning the troops who had served him so loyally. He could not understand that the speech which he made when he was repatriated, a speech in which he referred to himself as having been 'the honoured guest of the Turks', would hardly endear him to the friends and relatives of the 7,000 who had died in captivity. For one so anxious for love and social approval and personal esteem, he was curiously unrealistic. And afterwards, when subject to the cool breath of official disapproval, he still persisted in writing to those whom he thought would help him to a new position of power. He never, never could take no for an answer.

---

* A common characteristic of psychopathy.

9

Between the Wars

'The British soldier can stand up to anything except the British War Office.'   GEORGE BERNARD SHAW, *The Devil's Disciple*

In theory, a major war should confer benefits on the armed forces of the victor. New lessons have been learned, new technologies developed and new confidence found. Thus equipped, they should have a head start on preparations for the next war. In practice, the reverse seems to be the case, and this was never more so than after the First World War.

During this period, preparations for future conflict seemed to spring from a nostalgic urge to refight the Boer War. It was not a happy period for the armed forces of the Crown and a strange malaise settled upon their chiefs. There were several reasons for this.

As we noted earlier, military stock is never lower than at the end of a costly war. With a million dead, society's appetite for aggression had been assuaged. People were weary of war and tired of soldiering. For the military, the truth was rubbed in by swingeing cuts in men and *matériel*. From being the most important members of the community they were now relegated to a very minor role.

This thinly veiled ingratitude had three effects upon the military. With the hoarse yet self-consoling cry: 'Now we can get back to some real soldiering,' they withdrew into cocoons of professional impotence. In accordance with the principle that the more florid aspects of militarism are defences against threats to self-esteem, there was a falling-back upon the rites of the barrack square. Renewed attention to spit and polish helped to expunge the last traces of the mud of Flanders.

At higher levels of the military hierarchy, service thinking was embodied in an extract from a paper on Imperial Defence dated June

22nd, 1926: 'The size of the forces of the Crown maintained by Great Britain is governed by various conditions peculiar to each service, and is not arrived at by any calculations of the requirements of foreign policy, nor is it possible that they should ever be so calculated.'

Of this statement, which he describes as 'a peerless gem for the connoisseur's collection', a contemporary critic has written:

> The British Chiefs of Staff advanced a proposition which, in spite of its inspired lunacy, has remained to this day at the heart of much of what passes for military thinking in this country ... We are, in other words, here because we're here because we're here.
>
> Those who subscribe to the theory that armed forces should be designed to implement the nation's chosen external policies should therefore rid their minds of such childish delusions; it is the size and shape of the armed forces, their recruiting rate, their equipment and their conditions of service which matter, and those charged with the formulation of foreign and defence policy had better order their affairs accordingly.[1]

In the period between the wars the shape and the equipment, if not the size, of the armed forces were partly determined by a number of curious military attitudes. These centred particularly around three instruments of warfare: tanks, planes and horses.

Describing a tank attack which he had witnessed in 1916, General Sir Richard Gale tells how the British command tried to exploit it with cavalry. Apparently they failed, as was borne out by the grim sight of riderless horses returning whence they had come. Of this experience he writes: 'I was as impressed by the potential of the tank as I was unimpressed by the employment of horsed cavalry in modern warfare conditions. Yet after all our experience in that war it took us a further twenty years to mechanize our cavalry. The lesson was as clear in 1916 as in 1936.' In truth it was not 1936 but 1941 before the British began to implement the lessons of 1916.

What happened between the wars shows the alarming extent to which reactionary elements can draw the wrong conclusion from what to most people would seem quite unambiguous facts. Rather than recognize the potential of the tank, they drew the conclusion that innovation and progress are inherently dangerous and therefore to be eschewed. The symptom is not without precedent, nor confined to the Army. While on naval manœuvres in 1893 Admiral Tryon wished to about-face two parallel columns of battleships. From his flagship he ordered that the two columns should reverse course by turning inwards. Unfortunately, the combined turning circles of the ships was greater than the distance between them. With mathematical inevitability, H.M.S. *Victoria* was rammed by H.M.S. *Camperdown* and sank with great loss of life. Other officers had seen what was going to happen but dared not question orders. The lesson from this disaster seemed fairly clear. Admirals should base their decisions upon information supplied by their staff, and junior officers should not be afraid of speaking up when their knowledge (e.g., of the turning circles of naval craft) and their special abilities (e.g., superior eyesight and greater capacity for mental arithmetic) led them to believe that a given order would end in calamity. The argument seems sound enough. Indeed, even the most junior charlady at the Admiralty, had she pondered the facts, could hardly have failed to draw the same conclusion. But this was not the conclusion reached by her lords and masters. For them, Tryon's lapse just went to show that it never pays to try anything new!

To return to the tank, successive Chiefs of the Imperial General Staff between 1918 and 1939, with the support of other senior officers, did not exert themselves to mechanize the Army. Some were actively obstructionist. Against these reactionary elements stood a handful of progressive Army officers and a few like-minded civilians. The progressives, who had assimilated the incontrovertible evidence from the preceding war with Germany and were only too well aware of Hitler's preparations for the next, made their views known through books, essays and lectures, and by word of mouth. These moves were countered by the military establishment in two ways. Firstly, they resisted the dissemination of progressive literature; secondly, they did their best to curtail the careers of those who questioned their own obsolete ideas. For example, when Fuller, an early protagonist of mechanization, won the R.U.S.I. gold medal for his essay on tanks, and later produced a book on the same topic, he was castigated by successive Chiefs of Staff, remained unemployed in the rank of major-general for three years, and was then forcibly retired, in 1933.

In the course of these events the C.I.G.S., Lord Cavan, whose ideas, according to Fuller, were 'about eight hundred years out of date', opined that no officer should be allowed to write a book. Not to be outdone, his successor, Field-Marshal Montgomery-Massingberd, delivered himself of a diatribe against Fuller's books, while admitting that he had never read them because it would make him so angry if he did!

Equally unambiguous was the treatment meted out to Liddell Hart, a man described by the press as 'the most important military thinker of the age of mechanization in any country'. Over the years Liddell Hart produced a number of articles and books on mechanization, on new infantry tactics, and on the strategic and tactical use of armour. His efforts encountered extreme hostility and resistance from the British General Staff. When he submitted his essay on 'Mechanization of the Army' for a military competition, it was rejected in favour of an entry on 'Limitations of the Tank'. The judges were a field-marshal, a general and a colonel.

Unfortunately Liddell Hart's entry was not entirely lost to view. Along with other products of his pen, it was enthusiastically studied by Hitler's Panzer General, Guderian, and became required reading of the German General Staff.

Like those of his fellow protagonists, Liddell Hart's Army career was prematurely cut short by the military establishment. The case is germane to the thesis of this book. Here was a man who was cultured, fluent, lucid, highly intelligent and, that rare combination, a soldier who was also a first-class military historian, one whose advice on military matters was frequently sought by such civilian leaders as Hore-Belisha and Winston Churchill, who in due course became military correspondent of the *Daily Telegraph* and subsequently *The Times*, chosen by these papers in preference to a number of retired generals who had applied for the same job. Here was a man whose views and writings were eagerly studied and acted upon by many foreign powers including Germany, Russia, France and Israel, whose prophecies in the military sphere were borne out time and again, and who lived to see his ideas on mechanization and tank tactics used against us by Germany in 1940. But here was the man so deplored by the British military establishment that Lord Gort, Chief of the Imperial General Staff at the outbreak of war, felt moved to say during a lecture to 400 officers of the Territorial Army: 'Kindly remember that Liddell Hart does not occupy a room at the War Office.'

It was this same Lord Gort, the Army's top man at the outbreak of war, whom Hore-Belisha described as 'utterly brainless and unable to grasp the simplest problem', and of whom he said, upon another occasion: 'I never could have believed that people could be so dishonest.' Clearly there was something wrong somewhere.

In their suppression of Fuller and Liddell Hart, the military leaders of the inter-war years did the country, and themselves, a grave disservice. As for the other proponents of the tank, some not immediately obvious reasons for curtailing their military careers were conveniently discovered: Broad was too quarrelsome, Pile too dashing, and Hobart had been involved in a divorce! Liddell Hart remarked: 'If a soldier advocates any new idea of real importance he builds up such a wall of obstruction—compounded of resentment, suspicion and inertia—that the idea only succeeds at the sacrifice of himself: as the wall finally yields to the pressure on the new idea it falls and crushes him.'[2]

In mitigation of military shortcomings it has been customary to blame the politicians, this on the grounds that soldiers, sailors and airmen, however senior they may be, are ultimately subservient to civil government. In theory, prime ministers and war ministers see to it that the armed forces are not run by incompetents, for, as Clemenceau put it: 'War is far too serious a business to be left to the generals.' But in practice, because they profess specialist knowledge, and because, in times of national emergency, there is an understandable dependency upon them, some military leaders, even in the democracies, have become adept at manipulating their civilian bosses. Such was the case over the issue of War Minister Hore-Belisha. It seems that he was not appreciated by the military establishment. Five reasons, relevant to our general theory of military incompetence, may be advanced for this antipathy. Firstly, he was probably brighter than some of the senior officers with whom he had to deal. Secondly, his ideas for the Army were progressive. Thirdly, he made no bones about using Liddell Hart as his military adviser. Fourthly, he was, with every justification, critical of the generals whose job it was to prepare the British Army in France against the German assault on the West in 1940. Fifthly, he was a Jew.

It was for a mixture of these reasons that the General Staff persuaded Chamberlain to sack the man who had probably done more for the Army and defence than any other single person during Hitler's rise to power. At a time when his energy and ability were most sorely needed, Hore-Belisha found himself moved from the War Ministry to the Board of Trade.

Another progressive civilian, Geoffrey Pyke, described by *The Times* as one of the most original and unrecognized figures of the past century, hung on until 1948, then committed suicide 'from despair of officialdom's imperviousness to new ideas'.

To understand the psychology of these reactionary elements in the military establishment, of men who choose to make the Army their career, painstakingly work their way up the hierarchy to the highest positions, but then behave in such a manner as to ensure that if they are remembered at all it will be only for their conservatism, we needs must have recourse to ego-psychology. Thus it seems that, in the present instance, military leaders like Deverell, Montgomery-Massingberd, Milne, Ironside and Gort displayed behaviour symptomatic of extremely weak egos. In this light, their behaviour typifies the neurotic paradox in which the individual's need to be loved breeds, on the one hand, an insatiable desire for admiration with avoidance of criticism, and, on the other, an equally devouring urge for power and positions of dominance. The paradox is that these needs inevitably result in behaviour so unrealistic as to earn for the victim the very criticism which he has been striving so hard to avoid.

Consider a few concrete examples of the syndrome. For those who had despaired of anyone ever learning anything from the events of the First World War, 1933 brought a belated gleam of hope with the publication of the Kirke Committee Report, which was not uncritical of the high command. It could hardly have been otherwise. But there were those for whom preservation of personal reputations counted for more than the need to avoid a repetition of the senseless slaughter to which their direction had given rise. One such was Field-Marshal Montgomery-Massingberd, whose immediate response to the report was to block its dissemination throughout the Army.* While one can wonder at a system which would make it possible for one man to operate such censorship, the precise reason for his behaviour is by no means obscure. Montgomery as he then was (the hyphen Massingberd was adopted later) happened to be Chief of Staff of the 4th Division during the Battle of the Somme.

Our second example is rather more complex, concerning as it does that major obstacle to military development, the horse. As a noble if

* An abridged version was subsequently issued to H.Q.s of companies, squadrons and batteries.[3]

115

uncomprehending factor in military incompetence, this animal was much in evidence between the wars.

Upon reflection, it is hardly surprising that the horse became the *sine qua non* of the military life. For a thousand years man had found in it enormous advantages. There was nothing better for transportation and load-hauling. Horses raised morale and enhanced egos. Horses took the weight off feet and enabled people to go to war sitting down. When they lay down you could hide behind them. When it was cold you could borrow their warmth, and when they died you could eat them!

Because of the traditionally rural origins of so many Army officers and military families, horsemanship in the context of sports like hunting became one of their preferred leisure activities. Since such sports as polo, pig-sticking and, in an earlier age, jousting not only act out symbolic aspects of real warfare but are also associated with a higher social class, there is little wonder that they should find so much favour with those who choose the Army as a career. All in all, it is not surprising that the cavalry became that branch of the Army with the highest status. Nor is it surprising that they should have become the most vehement in denunciation of the tank, which was seen as an 'intrusive junior rather than an heir apparent'.

Nor is it surprising that the desire of the War Office to placate the cavalry was stronger than logic. Not only did they veto any expansion of the Tank Corps but, under the direction of Montgomery-Massingberd, ruled that the new Tank Brigade should never be reassembled, and this in the mid 1930s with Hitler arming to the teeth. Such resistance to progress, in the face of gathering evidence as to German intentions, was not confined to serving soldiers. During a Commons debate in 1934, the Labour M.P. for Leigh, Mr Tinker, had the temerity to question the value of horsed cavalry. Hardly had he finished speaking than a Conservative M.P., Brigadier Making, spurred to the attack. Having cut down the unfortunate Mr Tinker, the brigadier concluded with the immortal words: 'There must be no tinkering with the cavalry!' From all accounts it seems unlikely that his wit was even deliberate.

Horses also reared their heads in the Army Estimates. By an unhappy coincidence, British Army needs for 1935–6 were published on the same day that Hitler announced that his 'peacetime' army would comprise thirty-six divisions. To meet this threat Montgomery-

Massingberd decided that the amount spent on forage for horses should be increased from £44,000 to £400,000, and this in contrast to the sum for motor fuel, which he considered should be raised from £12,000 to £121,000.

Presumably to justify the forage, while at the same time making amends for having mentioned petrol, this same field-marshal laid down that in future all cavalry officers should be provided with *two* horses, and that horses should also be provided for officers of the Royal Tank Corps, presumably in a prophylactic role.

It would be unfair to suggest that the C.I.G.S. was alone in this romantic behaviour. Others shared his prejudices. Just below the surface was another voice no less reactionary and hardly less influential, that of Sir Philip Chetwode, Commander-in-Chief India. Despite the proved success of tanks on the North-West Frontier, this old cavalryman made the surprising pronouncement that the Army in India would be unlikely to adopt tanks for a very long time, and then only to keep up the momentum of horsed cavalry! Even more remarkable was his response to the rumour that the Germans had invented an armour-piercing missile—the notorious Halgar-Ultra bullet.* Instead of greeting this 'news' with a modicum of concern, Chetwode reacted like a schoolboy who has just been told that there will be no more lessons because his school-books have been destroyed in a classroom fire. Here at last was agreeable evidence that it would be a waste of time to replace horses with tanks. The fact that a horse presented a large and easily penetrated target to *all* descriptions of bullet, let alone the 'Halgar-Ultra', conveniently escaped his notice.

No less forceful were the pontifications of General Edmonds, Chief of the Military Branch of the Historical Section of the Committee of Imperial Defence. Having considered all the evidence, he wrote: 'Any tank which shows its nose will in my opinion be knocked out—the wars you [Liddell Hart] and Fuller imagine are past.'[4] This implied inversion of the real chronology of military technology is surprising to say the least.

Along with high hopes of armour-piercing bullets, the military establishment, according to Liddell Hart, set store by their professed love of horses, backed up by such vague concepts as the carrying over into war of noble deeds on the hunting field and the part which horses

* According to Liddell Hart, this rumour was a deliberate bluff on the part of Germany.

played in the training of young officers. Not everyone saw it quite like this. As one observer remarked: 'A love of the horse and of hunting seems to blunt all their reasoning faculties.' Yet others, taking cognizance of the fact that horses suffer terribly in war, have noted the curious paradox that those who professed the greatest love of horses should be the very ones with least regard for their welfare.

As for the alleged benefits of hunting, Duff Cooper, though fond of the sport, took issue with the proposition that it sharpened the mind. In his experience, most hunting people were *not* particularly quick-witted. In his opinion, driving a car down the Great West Road was a keener test of quickness of decision than anything encountered in the hunting field.

It would be wrong to suppose that this soldierly regard for horses was confined to Britain and the period before the Second World War. According to General Patton, the saddest day in his life was when he watched his old cavalry unit march by for the last time and stack their sabres, while as recently as 1960 General Hackett observed: 'It is unfortunate that the almost total disappearance of the requirement for equitation as a military skill should have been thought to justify its abandonment as an aid to education of the officer ... the growing technical complexity of war and the changed circumstances of the battlefield have driven out the horse but they have also developed an increasing requirement for a balancing element in an officer's education which equitation amply provides.'[5]

Few could take exception to these sentiments. No one would deny that horses are more lovable than tanks and require a greater sense of balance. What *is* extraordinary is that a love of horses should have apparently nullified any apprehension regarding events in Germany during the inter-war period, and the way this equanimity was sustained by the military establishment up to the eleventh hour. Thus even in 1938 one of the main preoccupations of the new C.I.G.S., Lord Gort, was how to get rid of Major-General Hobart, leading specialist in tank warfare. Removed from the War Office, where he was Director of Military Training, Hobart was eventually packed off to command a mobile division in Egypt. It was a case of out of the frying pan into the fire. The G.O.C. Egypt, General Gordon-Finlayson, was another who did not believe in tanks. He greeted Hobart with,' I don't know what you've come here for, and I don't want you anyway.'[6]

Related to these events was one other ingredient of military incompetence which came to the fore in the 1930s—deception, of self and of others. A particular instance concerned that other great legacy of the First World War—air power. Between 1914 and 1918 aeroplanes had been used with considerable success by both sides in reconnaissance, ground support and bombing sorties. At the time it seemed obvious that air power would be a decisive factor in future wars. Thanks to civilian enthusiasts and private industry the design and performance of aircraft had improved rapidly from year to year. With more powerful engines and stronger airframes came greater (and more accurate) firepower and bomb-carrying capacity. But for the Army and the Navy the notion of military aircraft aroused little enthusiasm, while that of the R.A.F. as an independent junior service was complete anathema.

For once the usual rivalry between the two older arms sank beneath their mutual dislike of the new upstart. If anything, the admirals waxed rather more negative about aeroplanes than did the generals, whose minds, as we have seen, were already discomfited by the issue of tanks. As mechanization threatened horses, so aircraft threatened battleships. But unlike horses in military minds, battleships were only the last of a succession of obstacles to progressive naval thinking. Before battleships it had been wood, and before that sail. Each relinquishment and transition had been bitterly resented, heavily opposed, and productive of such irrational thinking as tends to occur when dearly loved objects have to be renounced. When there was talk of iron replacing wood in the construction of ships one admiral was heard to remark that the idea was preposterous. Since iron was heavier than water the ships would be bound to sink! On this issue it has been calculated that

Of the twenty major technological developments which lie between the first marine engine and the Polaris submarine, the Admiralty machine has discouraged, delayed, obstructed or positively rejected seventeen.

The essential and necessary incorporation of these developments in the structure of modernization has been achieved by individual and sometimes undisciplined officers, by political and industrial pressures, or—and most frequently—by their successful adoption in rival navies.[7]*

* It is fair to point out that, by allowing other navies to experiment with new technologies, the British taxpayer was saved considerable expense.

As for battleships, whose future usefulness and, indeed, very existence was threatened by the advent of aircraft:

> ... to most admirals the respective value of battleships and aircraft was not basically a technological issue but more in the nature of a spiritual issue. They cherished the Battle-fleet with a religious fervour, as an article of belief defying all scientific examination. The blindness of hard-headed sailors to realities that were obvious to a dispassionate observer is only explicable through understanding the place that 'ships of the line' filled in their hearts. A battleship had long been to an admiral what a cathedral is to a bishop.[8]

It was such strong emotional attachments that led the admirals to deceive their political masters.* The practical issue was whether or not battleships could defend themselves against aircraft. Having formed the opinion that they could, the Admiralty decided to prove its point. In 1936 (while aircraft production by the Axis Powers was getting into top gear), the King was invited to a demonstration in which naval ships would attempt to shoot down a radio-controlled 'Queen Bee' target aircraft. Unfortunately the demonstration did not go well. Despite the fact that the plane was limited to 80 m.p.h. and flew provocatively up and down without jinking, while the ships were given a running start on a parallel course, thereby reducing the speed differential to something approaching 50 m.p.h., not a hit was scored. Dismayed but resourceful, the admirals played their last card, deliberately crashing the radio-controlled plane into the sea— thereby 'proving' at considerable cost to the British taxpayer that planes are no match for battleships when these are in the right hands.

While we are on the subject of naval problems, it seems that deception as practised by the Admiralty was of a rather different order from that manifested by the War Office. Although theoretically advantaged by having their feet quite literally on the ground, the military establishment displayed a quality of *self-deception* (as opposed to deliberate deception of others) far in excess of that practised by the Navy, who, beneath their more fantastic protestations, did on occasion show a

---

* The low standard of those R.A.F. aircrew who made up the Fleet Air Arm, the refusal of the R.A.F. to develop a dive-bomber, and fear of cuts in the Navy Estimates were other contributory factors. (Personal communication from Donald Macintyre.)[9]

surprising streak of realism, fostered perhaps by the age-old experience of being up against the hard facts of nature—and the dangerously low buoyancy of the human body.

This contrast between the total obscurantism of the Army and the underlying realism of the Navy is typified by the case of Malta. In 1935 worsening relations with Italy had put the British Fleet in Malta at serious risk from sudden air attack. Like the American Fleet at Pearl Harbor six years later, they were a tempting target for an enemy. Army reactions to this situation were typified by a letter to Liddell Hart from Sir Philip Chetwode, Commander-in-Chief India, whose vision on tanks we considered earlier.

He wrote: 'You have evidently been crammed up, as I fear that both the Government and the public at home have, by the Air propaganda ... There is only one way in which the Air can win a war and that is by bombing women and children; and that will never bring a great nation to its knees, but only inferior people. You know perfectly well that the Navy laughs at the Air now. They have got protected decks, and with their "blisters" and multiple machine-guns and multiple anti-aircraft guns, they do not fear them in the slightest.'[10]

However, notwithstanding the bold encouragement of far-off, land-locked Sir Philip, the Admiralty, for all its protestations as to the invincibility of battleships, promptly abandoned Malta 'for fear of what it hadn't feared'. As Liddell Hart said, it was remarkable how quickly the Sea Lords awoke to reality and swallowed their previously disdainful views about the effect of aeroplanes.

In later years, particularly after the disasters of 1940, it became fashionable amongst writers and cartoonists to heap ridicule upon the military leaders of this unhappy era. In fact they were the victims of three factors, at least two of which—the economy and public attitudes to war—were beyond their control.

The third factor, as discussed in later chapters, was something endemic to the military profession—the over-control of aggression. It is epitomized in Boyle's description of three service chiefs as suffering from 'meek irresoluteness', and in Vansittart's description of the same men as 'this amiable trio'.

Had they not become clichés of the military scene, one might wonder at the paradox which these epithets represent, for here were three professionals in violence, at the top of their calling, doing their level best to deny the forthcoming clash of arms and yet apparently bent upon sacrificing military preparedness for fear of offending their colleagues. This, and a nervous tendency to safeguard their own reputations by sitting tight and doing nothing (but wave the talisman of hallowed traditions to ward off the looming holocaust), seems strange, to say the least.

One explanation of this lemming-like behaviour has been given by Divine. He draws attention to the interesting fact that in the field of munition technology 'the tip of the sword' has always had to be blunted. A new gun is built bigger and more powerful than its predecessors but its barrel left unrifled; aircraft carriers are developed but equipped with hopelessly slow and obsolete aircraft; bombers are made with inadequate bomb-loads; bombs are turned out with insufficient explosive power; gigantic 'armour-piercing' naval shells break up on impact.* Always the tip is left blunted. It is tempting to see a parallel between these phenomena and the tendency to provide heroic armies with commanders like Elphinstone, Raglan, Simpson, Buller and French, men whose forte it was to blunt the sword of massed aggression.

Presently we shall discuss the fuller implications of this clue to military incompetence. But first there is another war to consider.

* Much to the relief of the German Navy at Jutland.

The Second World War

'Frankly we would welcome an attack … We are ready for anything they may start … The British Army is the finest equipped army in the world.'

GENERAL SIR EDMUND IRONSIDE, C.I.G.S., April 14th, 1940

'Their [the Germans'] success [in May 1940] could easily have been prevented but for the opportunities presented to them by the Allied blunders—blunders that were largely due to the prevalence of out-of-date ideas.' LIDDELL HART, *History of the Second World War*

After an appalling start in which the Allies were outfought, out-manœuvred and outstripped in the quality of their military thinking and equipment, the Second World War produced the biggest transition in military competence since the days of Wellington. It was born of necessity and may be said to have dated from Dunkirk. This jolt to one hundred years of military maundering and twenty years of blind complacency achieved three ends. Within a space of days it shattered many long-held and dearly loved illusions about the nature of modern war. It hastened the eclipse of the old, the reactionary and the un-talented. Finally, by rendering the Army temporarily impotent, Dunkirk put the most junior service in the centre of the stage.

For the very first time the continued existence of the Army and the Navy became totally dependent upon their protection by the R.A.F. While the bitterness of their pill may have been masked at the time by the common threat, the inescapable facts of the Battle of Britain meant that things would never be the same again.

Yet a fourth reason for the transition to greater competence was the heartfelt wish on the part of many senior military commanders to avoid the terrible wastage of manpower that characterized the First World War. This laudable urge towards conservation of men's lives found expression in far more thorough planning, based upon a much more careful gathering and sifting of intelligence than had occurred in previous wars.

As for the bad start, this was a legacy of factors touched on in the previous chapters: rigidity of thinking, over-confidence resulting from a pathetic belief in antiquated methods of warfare, and refusal to accept that enemy intentions may confound the armchair prophets. The following examples from Liddell Hart's *History of the Second World War* illustrate these shortcomings. They concern the capture of a German plan of attack, the attitude of the Allies to the use of cavalry, and the fall of Tobruk.

On January 10th, 1940, a German aircraft carrying the liaison officer of the 2nd Air Fleet lost its way and crash-landed in Belgium. By an extraordinary chance the officer was in possession of *the complete operational plan for Germany's attack on the West*. He tried to burn the plan but failed to complete this task before he was captured. In this way its contents became known to the Allies. Hitler's response was to devise a new plan, which involved attacking France through the Ardennes rather than through Belgium as originally intended.

This episode was damaging to the Allies for two reasons. Firstly, in the belief that the captured plan was a deliberate deception, they failed to modify their own plans. Secondly, contrary to advice received years earlier, they clung to the belief that the wooded area of the Ardennes was impassable to tanks. As a result, the strongest Allied forces remained poised for an attack through Belgium while the Germans suffered little resistance to their outflanking drive through the Ardennes.

On all counts the behaviour of the Allied military planners was irrational. The Germans would have been unlikely to practise a deliberate deception of this kind because, whether or not the Allies treated it as such, it could be used to strengthen their hand with the Belgians. Further, since the Germans would not necessarily know whether it had been taken as a deception, they would not know whether to institute a second plan or stick to the old one, i.e., they would not know whether the Allies had taken the captured plan as genuine, as a bluff or as a double bluff.

In short, had the Germans wished to delude the Allies into expecting an attack through Belgium, they would hardly have chosen a means so ambiguous in its possible results.

But if the captured plan was *not* a deception then the Allies would not know for certain whether the Germans realized what had happened to the papers carried by their liaison officer. Under the circumstances they could only assume that the Germans would put the worst construction on what had happened and adopt a new plan, which is exactly what they did.

Whichever way one looks at it, the policy of doing nothing was inept and, in its outcome, disastrous.

We have already seen how love of horses obstructed British tank development. Our allies were similarly afflicted. When Hitler invaded Poland in 1939, the Polish military authorities 'still pinned their trust in the value of a large mass of horsed cavalry, and cherished a pathetic belief in the possibility of carrying out cavalry charges. In that respect it might truly be said that their ideas were eighty years out of date, since the futility of cavalry charges had been shown as far back as the American Civil War—although horse-minded soldiers contrived to shut their eyes to the lesson.'[1] In the event Poland, for all her forty divisions and twelve large cavalry brigades, was overrun by Germany in less than one month.

Likewise the French, though possessing many tanks which were as good as, if not better than, those of the Germans, were steadfast in their belief that horsed cavalry could destroy German armour in the Ardennes. (For this reason they refused to accept the suggestion that felled trees might be used to delay the German advance.) Like the Poles, they were sadly disillusioned about the outcome of a conflict between horses and tanks.

While Dunkirk certainly marked a watershed in military endeavour, it did not, unfortunately, eradicate those fundamental causes of high-level military incompetence which are examined in later chapters of this book. Before the war was over these apparently enduring features of militarism made their presence felt in two further disasters of great psychological significance: Tobruk and Singapore.

The British retreat from the Gazala Line in 1942 which resulted in the loss of Tobruk, followed by a headlong flight back into Egypt, was

ON THE PSYCHOLOGY OF MILITARY INCOMPETENCE

the second worst disaster of the war, after Dunkirk. Tobruk cost Britain 35,000 casualties, and enormous losses in ground and material.

Why did it happen? A popular explanation was that Rommel had the advantage in equipment—better tanks and guns. This excuse lacks validity. The Eighth Army had a four-to-one advantage in tanks (including 400 in reserve) which were on average of superior quality to those of the Panzer Army, a three-to-two advantage in artillery, and six hundred as opposed to five hundred and thirty aircraft.

A truer answer is inadequate generalship. The Army commander, Major-General Neil Ritchie, a fine-looking man, has been described by his contemporaries in ways strikingly reminiscent of Elphinstone, Raglan and Buller:

> Ritchie was all haywire by then. All for counter-attacking in this direction one day and another the next. Optimistic and trying not to believe that we had taken a knock. When I reported the state of 1st Armoured Division to him at a time when I was planning to use it for counter-attack, he flew to see me and almost took the view that I was being subversive. (General Messervy)

> General Ritchie had a great air of decisiveness, yet was really rather indecisive. (General Godwin-Austen) (According to the same corps commander, '[he] had a tendency to ask your advice and having received it act in the opposite way.')

> Ritchie is not sufficiently quick-witted or imaginative. (Major-General Dorman-Smith in a report to General Auchinleck)

> A fine robust-looking man with charm and manner, but no aurora. (General Ramsden)

> Confident and decisive in his speech, but one did not always feel he was quite so confident and decisive in his mind. (General Messervy)[2]

For an example of this indecision, reminiscent of that shown one hundred years earlier by Elphinstone before the retreat from Kabul:

> I got via Corps an order: on no account was El Adem to be evacuated—they were to fight it out to the last. It was already

surrounded. I was told by Norrie that these were the Army Commander's personal orders. Then I had a message: it might be evacuated if I thought it couldn't be held. I said I was quite sure it could not be held for long; then I was told to pass this message on to 29th Brigade. Then I got another order—the Army Commander says it must be held. Then yet another: that it was to be evacuated if the brigade could get out. I passed this on to Denis Reid [the Brigadier], and they got out. This was an example of what was happening all the time. (General Messervy)[3]

Under the ineffectual leadership of this big, kindly, courteous, unimaginative, apparently complacent yet occasionally touchy general, the Army suffered a decline in organization, discipline and drive. It became 'flabby instead of taut, sluggish instead of agile'.

Once again that fatal amalgam of over-confidence and under-estimation of the enemy produced a dulling of military endeavour. 'In the British Military Headquarters there was a comfortable assurance that he [Rommel] could be dealt with at leisure, and was bound to surrender.'[4] Partly through inadequate intelligence (in the military sense of this word) and partly through an inherent distaste for subterfuge, the Army command allowed themselves to be duped by the foxy antics of the other side. Unhampered by similar inhibitions, coldly professional and inventive from necessity, Rommel, in a succession of feints, outflankings, pincer movements and encirclements, ran rings around his much more powerful, honest, courageous, but stolid and slow-moving adversary.

Rommel himself, in his diary, ascribed his success to the British predilection for frontal assaults—brave but costly charges by small groups in which the attackers banged their heads time and again against the hull-down German Panzers.

This bull-headed approach, which whittled down the reserves of British tanks from 400 to a mere 170, was made worse by the policy of sending in the armour piecemeal, in 'penny packets'—a further example of that curious pulling of punches so evident in previous campaigns.

Another, more specific, reason for the disaster, which, if Auchinleck had not held the Germans at Alamein, might have lost Britain the Middle East oil supplies and therefore the war, has been suggested —the superiority of German anti-tank guns. The British solid-shot 2-pounders, which were incapable of penetrating the armour of the latest German tanks, stood in sad contrast to the 50 mm. and 76 mm. (captured Russian guns) which the Germans fielded.

According to Liddell Hart this discrepancy between German and British anti-tank potential need not have been had we taken a hint from Rommel's use of 88 mm. anti-aircraft guns as anti-tank weapons. This argument may be questioned on two counts: first, our excellent 3·7-inch anti-aircraft gun did not, for technical reasons, lend itself like the 88 to an anti-tank role; second, those 3·7s then available were more urgently needed in the role for which they had been designed.

From the standpoint of human behaviour, human feelings, leadership and decision-making, the events of 1942 in North Africa exemplified in microcosm the major causes of military incompetence. Underneath his robust exterior, Ritchie, like Buller before him and Elphinstone before *him*, lacked self-confidence and seemed more concerned with proving himself to himself than with prosecuting the war. The presiding over interminable 'committee meetings' through which the Army was run, the seeking of advice and then not taking it, and the disingenuous way in which he managed to convince the commander-in-chief that he was protecting Tobruk while in reality leaving it to the mercy of the Germans, are the actions of a man beset by inner doubts. These doubts were skilfully but not perfectly concealed by his often inappropriate façade of monumental complacency. In his relationship to Auchinleck he stood as Randolph Churchill stood to Winston—the weak son to the powerful father—one minute obstinately refusing to take sensible advice* (because to do so would be to admit his dependency), the next anxiously seeking guidance and reassurance.

No wonder then that Ritchie remained fettered by the stolid and archaic attitudes of the military organization in which he rose to generalship, an organization in which, to quote one contemporary historian, 'cleverness, push, ruthlessness, self-interest and ambition were considerably less prized than modesty, good manners, courage [and] a sense of duty';[5] of an army described by the commander-in-chief, General Auchinleck, as 'too rigid and lacking in flexibility to be really adaptable to the conditions of modern quick-moving

---

* e.g., that the Army should fight as it had been trained, in divisions and not in bits and pieces.

warfare in the Desert, or even elsewhere'.[6] On June 25th, 1942, Ritchie was relieved of his command.*

Tobruk was a disaster, but, in terms of human misery, not half so great as one which had unfolded a few months earlier, in another theatre of the war.

---

* In considering what has been said by others about Ritchie, including those comments reproduced here, it must be emphasized that this otherwise and subsequently highly competent officer was, by being appointed to command an army in the middle of a battle, more sinned against than sinning.

I I

Singapore

*'One can sum up by saying that those responsible for the conduct of
the land campaign in Malaya committed every conceivable blunder.'*
MAJ.-GEN. WOODBURN KIRBY, *Singapore: the Chain of Disaster*

In the nine weeks between early December 1941 and mid February
1942, the 'impregnable' fortress of Singapore, Europe's gateway to
the East, with its thriving city, huge naval dockyard and strategically
vital airfields, fell lock, stock and barrel into the hands of the Japanese.
The invasion of this island stronghold, the complete defeat of the
combined British and Australian garrison, with its Army, Navy and
Air Force units, and the ultimate unconditional surrender of the whole
area were so rapid that even the Japanese were staggered, indeed one
might almost say nonplussed, by the ease, speed and enormity of their
success.

In the long run, the results of this disaster may be deemed incalcu-
lable. The myth of European supremacy over Asiatic peoples was
exploded for ever, and the prestige and competence of British military
endeavour in the eyes of the world in general, and America in particular,
were damaged beyond repair. In the short run, Britain lost her last
and strongest foothold in the Far East—an appalling set-back for the
global war effort. We lost thousands of lives, both military and
civilian, but worse perhaps than the loss of life, the military debacle
condemned thousands more to three and a half years' misery in
Japanese internment camps. Finally the economic loss ran into hundreds
of millions of pounds. We forfeited elaborate and expensive dock
installations, naval and other engineering facilities, military stores,
fuel, the major port for exporting urgently needed rubber, and two
new and first-class battleships. Most of these material assets fell virtually

intact into the hands of the enemy, thus in effect doubling the value of their loss to the Allies.

Clearly there is much here to answer for. Though dwarfed at the time by other world events, the fall of Singapore constitutes a more remarkable and disturbing phenomenon than the siege of Kut, the mishandling of the Crimean War or even the more recent Suez fiasco.

Like the other cases we have discussed, that of Singapore is essentially a human problem—a product of human behaviour, human intellect, human character and human error. No explanation in terms of geography, climate, broad political or military considerations can possibly do justice to the facts. At bottom (and at the top) we are confronted with issues that are primarily psychological and which only a reduction to psychological principles can possibly explain.

Let us state the problem in terms of a number of questions:

1. Why was the 'impregnable' fortress planned and serviced in such a way that while presenting apparently formidable defences on its southern side, its back, the northern shore, was no more of a resistance to a would-be invader than the back of Bournemouth?

2. Why was there an almost total lack of co-ordination and co-operation between those who had been entrusted with the job of defending the island?

3. Why, when it was clear that the Japanese could and would assault the island from its northern side, was nothing done to erect defences in their path?

4. Why did the General Officer Commanding Singapore, Lieutenant-General Percival, ignore the urgent advice of his subordinate, Brigadier Simson, and of his superior, General Wavell, to implement these defences?

5. Why, on the one hand, was so little done to protect the civilian population against air raids and, on the other, so much done to prevent their knowing the true facts of the situation as these unfolded?

6. Why did General Percival persist in believing the Japanese would attack from the north-east when confronted with overwhelming evidence that their assault would come from the north-west?

7. Why did the officer commanding the Australian forces on the island forbid his troops to escape, while secretly plotting his own getaway from the island?

8. Finally, and perhaps of greatest interest, how did the men who

could perpetrate such colossal errors of judgment ever reach a position where this was possible?

Let us try to answer these questions.

As intimated earlier, the loss of Singapore had its origins in much earlier events. In 1925 there was a protracted and acrimonious argument between Army, Navy and Air Force chiefs as to how Singapore should be defended. While the older services pressed for fortifications and heavy fixed guns to repel an attack from seaward, Trenchard for the R.A.F. advocated a large force of aircraft to repel any attack before it could come within range of the island. Needless to say, the Army and the Navy won their case at the expense of the more junior service. Heavy fixed armaments became the order of the day.

This debate, in which the R.A.F. had to concede defeat, had three unfortunate consequences. Firstly, the island was left exposed and undefended on its northern side. Secondly, senior Army commanders from that time on stubbornly clung to the dogma that no Japanese would ever advance on Singapore down the Malay Peninsula. Finally, the bitter inter-service quarrel which ensued resulted in an almost total lack of co-ordination between the three services.

A continuation of this state of affairs was ensured by siting the Army, Navy and Air Force headquarters in Singapore as far apart as possible. Just one, albeit fairly disastrous, consequence of this carefully planned lack of interaction was that the R.A.F. began constructing airfields without consultation with the Army who would have to defend them.

The guardians of Singapore defended their wrong decisions in a number of ways. One of these was to import official lecturers from England. Apparently oblivious of the scepticism of their civilian audiences, these 'experts' tried to turn black into white by reiterating that no army, let alone a Japanese army, could advance through the impenetrable jungle of the Malay Peninsula, that this same jungle was quite impassable to tanks, and that the Japanese military machine was a primitive affair not to be taken seriously.

Local people, rubber planters and the like, who had the advantage over the lecturers from London of knowing something about the Malay Peninsula and even perhaps something about the Japanese, questioned these assumptions but, presumably because they were only civilians, their objections went unheeded.

The authorities, however, did become increasingly concerned to prevent the civilian population from discovering anything new that might conflict with the official set of delusions which they themselves espoused. Thus when the *Malay Tribune* published the news that Japanese transports had been sighted off the southern tip of Indo-China, the editor was immediately castigated by the commander-in-chief of the Far East, Air Chief Marshal Sir Robert Brooke-Popham, who said: 'I consider it most improper to print such alarmist views at a time like the present ... the position isn't half so serious as the *Tribune* makes out.'

The form of his complaint is not without interest. Firstly, he did not deny the truth of the press release. He hardly could, since it had originated in a report by Reuter's which had been passed by the censor and which undoubtedly *was* true. Secondly, he managed to imply all in one breath that the situation was both not serious and yet likely to cause alarm. This is curious, for if the close proximity of Japanese forces was not serious then why should a truthful report to this effect be alarming? Moreover, if it was alarming, because true, then it must have been serious, in which case the sooner the civilian population knew about it, and could learn to adjust to the imminent danger which threatened them, the better.

Finally, his words exemplified a tendency, seen all too often, to talk down to a civilian population as a group who, through some weakness of intellect or lack of moral fibre, could not be trusted with information held by their elders and betters.

The causes of this arrogance are not hard to see. By divulging information a professional in-group may feel that it is losing some of its mystique, thereby weakening its image in the eyes of its public, and this loss will be greatest precisely when the in-group is most at a loss as to what to do next.

The behaviour may be likened to that of doctors or nurses who, having taken a patient's temperature, insist on keeping this interesting information to themselves. Bereft of solid facts, the patient has to be satisfied with a condescending smile and a patronizing 'Don't worry, we'll soon have you up and about again.' Needless to say, this preservation of a mystique will be strongest in the more immature and less self-confident members of an in-group.

The guardians of Singapore were prime exemplars of this motivation. After a long history of wrong thinking they could not afford to be found mistaken. The more events proved them to be wrong, the stronger their defences became against admitting this to be the case. Like insecure doctors, they covered their refusal to disclose the true facts in two ways. These were in the nature of panaceas, one for themselves, the other for the patients—in this case the civilian population of Singapore. For themselves they had the rationalization that disclosure of the true facts would be bad for civilian morale; and for their 'patients', they supplied false information.

Thus the commander-in-chief went on record as saying, of his hopelessly inadequate collection of obsolete aircraft: 'We can get on all right with the Buffaloes [known to members of the Singapore Flying Club as "peanut specials"] here. They are quite good enough for Malaya.'

This particular inaccuracy seems to have been a product of stupidity, arrogance and dishonesty. It was arrogant in its underestimation of an Asiatic fighting force, stupid in the wrong prognosis of its effects on the Singapore civilians, and dishonest in that even a man like Brooke-Popham could hardly have reached the rank of air chief marshal without knowing something about the aircraft under his command.

As Hitler's administration demonstrated, in its starkest form, suppression of the truth involves two procedures: on the one hand censorship, and on the other official communiqués. The high command in Singapore employed both measures. Take the Order of the Day released to the *Malay Tribune* a bare two months before Singapore capitulated. It reads:

We are ready. We have had plenty of warning and our preparations are made and tested ... we are confident. Our defences are strong and our weapons efficient. Whatever our race ... we have one aim and one only, it is to defend these shores, to destroy such of our enemies as may set foot on our soil ... What of our enemy? We see before us a Japan drained for years by the exhausting claims of her wanton onslaught on China ... Let us all remember that we here in the Far East form part of the great campaign in the world of truth and justice and freedom.

As the editor of the *Tribune* said, it was hard to believe that anybody could deliberately tell so many lies.

It is at this point that one encounters a curious paradox. In our culture it is thought more serious to accuse a man of dishonesty than of stupidity. The former is the individual's 'fault', the latter no more his fault than is the length of his feet. To accuse a man of dishonesty is considered libellous, to accuse him of stupidity only unkind or at the worst abusive. And yet these men at the very apex of their military careers opted for the most transparent of deceptions.

It may be thought that all this is being unduly uncharitable; that they were hopelessly misled by the serious shortcomings of their intelligence services; that when General Percival issued his communiqués he genuinely if mistakenly believed their contents. Unfortunately this hypothesis does not stand up.

On Monday December 8th, 1941, G.H.Q. issued its first war communiqué. This stated that the Japanese had failed in their attempt to land at Kota Bahru. This was followed shortly after by a second communiqué which stated: 'All surface craft are retiring at high speed, and the few troops left on the beach are being heavily machine-gunned.'

Naturally this was heartening news for Singapore's less well-informed civilians. As Noel Barber points out: 'It was not difficult for them to imagine (*because this of course was what they wanted to imagine and this of course was what their military commanders wanted them to imagine*) a moonlit beach, with a few khaki-clad Japanese left bewildered to their fate by cowardly comrades who were bolting "at high speed" in their boats.'[1] (My italics.)

In fact the communiqué was essentially untrue and deliberately misleading. Within the space of a few hours from the time of the Japanese landing, Kota Bahru was firmly in enemy hands. Having deposited their assault troops, the Japanese transports did quite naturally return to base as fast as possible—here was the grain of truth which G.H.Q. distorted for their own ends.

That the military were covering up for their own dereliction of duty receives support from General Wavell, who in 1948 admitted to the erstwhile governor of Singapore that the 'original sin' for the lack of preparation and all that this led to must be placed on the heads of the military.

These same heads had many sins to answer for, not the least being the lack of information and training given to their own troops. Shortly before the Japanese invasion, and even as enemy tanks were preparing to roll down the Malay Peninsula, thousands of leaflets, neatly tied in bundles, were found in a cupboard at military headquarters. They were official War Office pamphlets giving non-technical advice on how to deal with enemy tanks, a matter on which the local troops seemed woefully ignorant.

It does not need many guesses as to why they had never been distributed. Some other military leaders in Malaya *did not think that the Japanese would use tanks*. Even had this turned out to be the case it is hard to see what harm would have been done in passing on the information. After all, few believed the enemy would use gas but this did not prevent the issue of gas masks. One can only assume that the anti-tank leaflets were left to moulder in a cupboard because they were tactless enough to proclaim a heresy. It is a feature of strongly held dogmas that they steadfastly resist not only unpalatable truths but even the faintest suggestion of the barest possibility of the most tangential reference to an unacceptable fact. Better that men should die and cities be overrun than that the sacred teachings should be found wanting.

Up to this point the impression may have been given that stupidity, obstinacy and wrong decisions were the prerogative of Army commanders. That this was not so was amply demonstrated by the Navy. For a start, naval chiefs had been in the forefront of those responsible for the fact that Singapore's sole defences were some fixed, seaward-facing 15-inch guns, backed up by a number of 9-inch guns for each of which there were only 30 rounds of ammunition—enough for one round per gun per day for a month.

But there was worse to follow. As a desperate measure, two battleships, the *Prince of Wales* and the *Repulse*, were sent to Singapore to create an eleventh-hour presence. They were under the command of Admiral Sir Tom Phillips, in the words of one who met him 'a real old sea-dog bluff and tough'. Unfortunately he too lacked sufficient perspicacity. Despite strong warnings that he could not expect adequate air cover, he was soon off with his two ships 'in search of trouble'. At first all went well as the ships steamed reassuringly up the east coast.

In pursuing this course of action Admiral Phillips turned the proverbial blind eye to a second warning which reached him. This stated quite categorically: 'Fighter protection ... will not, repeat not, be possible.' He felt safe because the weather was bad and the skies overcast. But then, quite suddenly, the skies cleared. With a flash of prudence the admiral turned his ships and signalled Singapore that he was returning to base. This was the last that was ever heard from him.

It seems that on his return journey he received a report (subsequently proved false) that the Japanese had launched an attack on the town of Kuantan. Without informing Singapore he decided to 'go in and help'. It was a fatal decision. His ships were spotted by the Japanese Air Force, torpedoed and sunk with a total loss of 840 officers and men. By all accounts Phillips was a brave and conscientious officer, but his braveness bordered on foolhardiness and his errors of judgment not only had a devastating effect on that much cherished commodity, the morale of Singapore civilians, but also sealed the fate of their city. Now there was nothing left with which to protect it. Naturally one is forced to ask how could a man so excellent in some respects but so limited in others ever have reached a position to inflict such grievous loss upon his fellow countrymen?

So much for the Navy. In their chosen field the senior command of the R.A.F. acquitted themselves little better. It has already been seen how Air Chief Marshal Brooke-Popham underestimated Japan's air strength in comparison with his own ill-assorted group of obsolete aircraft. This same sixty-three-year-old officer, whose most notable characteristic was a tendency to fall asleep on the slightest pretext, showed such disastrous hesitancy and indecision in his capacity as C.-in-C. that as the official history was moved to state: 'It is possible that he did not fully realize the importance of speed ... The need for a quick decision was not apparently realized at Headquarters Malaya Command.'

But it was in the matter of Japanese air raids that the R.A.F. command gave the first clear demonstration of its limitations. Although the inevitability of air raids must have been obvious for some time, the first night raid was marked by a complete absence of blackout or night fighters and this despite the fact that, as Air Vice Marshal Maltby was later to admit, the R.A.F. had a clear 30 minutes' warning of the approaching aircraft. It seems the Japanese had committed the unforgivable faux pas of attacking at night: unforgivable because it conflicted with the official dogma that the Japanese were unable to fly their planes during the hours of darkness. This particular *idée fixe* cost Singapore 61 dead and 133 injured.

But the recklessness of the admirals and the dithering of the air marshals were as nothing to the intransigence of the generals. It seemed that nothing could move them, not even the pleading of their fellow officers. As Noel Barber says in his book, *Sinister Twilight*:

When Brigadier Simson, the Chief Engineer, went to see Major-General Gordon Bennett (commanding the 8th Australian Division) he found it impossible to make him realize that there was an urgent need for anti-tank defences. 'At first he did not wish to discuss the matter at all,' Simson noted after the meeting. Simson was horrified. Could not the Australian general understand that there was nothing on the long road to prevent the enemy reaching Johore? Apparently Gordon Bennett could not, for in his diary that night he wrote, 'Malaya Command sent Brigadier Simson to discuss with me the creation of anti-tank obstacles for use on the road ... Personally I have little time for these obstacles ... preferring to stop and destroy tanks with anti-tank weapons.'

No wonder that the Japanese never slowed down, no wonder that time after time ... troops were annihilated by skilful Japanese enveloping tactics. On the British side wrong decisions were made. Communications broke down ... Whole pockets of troops were cut off. The first Japanese tanks appeared and 'came as a great surprise' to the British who had not one single tank in Malaya. In a jungle country where the British had insisted that tanks could never operate, the Japanese tanks moved easily between the spacious rows of rubber trees.[2]

Major-General Gordon Bennett was not, to use the appropriate vernacular, an isolated pocket of resistance nor did he hold the record of obstinacy. In Barber's words: 'Attempts [by Brigadier Simson] to improve and add to the defences had been baulked at every turn, largely by General Percival who seemed to have a fixation against such measures ... Nothing had been done, nothing was being done despite many previous pleas.'

A hazard of belonging to any rigidly authoritarian hierarchical organization is that, from time to time, the individual, out of dire necessity or from strong personal conviction, feels compelled to apply pressures to those *above* him. It is a hazard, because the ethos of the organization, whether it be a Victorian family, an English boarding school or the British Army, demands that pressure always moves in one way only, downwards rather than upwards. To buck the system, by prodding those above, can have unpleasant consequences.

Hence it is a measure of the seriousness with which Brigadier

Simson regarded the situation that he made one last attempt to move his general.

He drew a deep breath and announced that he would like to take this opportunity of a heart-to-heart talk on the subject of defences. Percival looked a trifle startled but sat down with a tired expression and listened. The general was a difficult man to 'warm up'. Tall, thin, with two protruding teeth, he was a completely negative personality and his first instinct when faced with a problem was that it couldn't be done—in direct contrast to Simson whose first thought was always 'Well—let's try.' This was why Simson had elected to stay and risk all at this strange meeting in the dead of night, and now he spoke with the passionate eloquence of the professional. Defences were his main job. He believed implicitly in their value which history had repeatedly proved in modern war. And he had all the materials to hand ... He had the staff and materials, he said to Percival, to throw up fixed and semi-permanent defences, anti-tank defences, underwater obstacles, fire traps, mines, anchored but floating barbed wire, methods of illuminating the water at night ... To the Brigadier's dismay, Percival refused his pleas.[3]

It seems that Simson was past taking no for an answer, for he said to the general: 'Sir—I must emphasize the urgency of doing everything to help our troops. They're often only partially trained, they're tired and dispirited. They've been retreating for hundreds of miles. And please remember, sir, the Japanese are better trained, better equipped, and they're inspired by an unbroken run of victories ... and it has to be done now, sir ... once the area comes under fire, civilian labour will vanish.'

The plea was forceful, respectful and logical but, amazingly, the general remained unmoved. Simson, his anger rising, said: 'Look here, General—I've raised this question time after time. You've always refused. What's more, you've always refused to give me any *reasons*. At least tell me one thing—why on earth *are* you taking this stand?'

At long last the General Officer Commanding Malaya gave his answer. 'I believe that defences of the sort you want to throw up are bad for the morale of troops and civilians.'

As Barber comments: 'Simson was "frankly horrified" and remembers standing there in the room suddenly feeling quite cold, and realizing that, except for a miracle, Singapore was as good as lost. As he put on his Sam Browne, Simson could not forbear to make one last remark – "Sir, it's going to be much worse for morale if the Japanese start running all over the island." '

To most people, Percival's excuse for refusing the request of his Chief Engineer must seem illogical, to say the least. Does it lower an airman's morale to give him a parachute or a householder to give him a burglar-proof lock for his front door?

Barber has an interesting footnote on this issue: 'Why was General Percival so biased about defence works? Simson believes that, like some other commanders in Malaya who were indifferent, "Somewhere in their military education such a dictum on morale had been impressed upon them or they possibly misunderstood the true value of defences in the circumstance such as now existed." '4

This is certainly not beyond the bounds of possibility. But if so, why should a proper emphasis on the importance of morale during the teaching of officers become an irreversible, unmodifiable, conditioned reflex for every subsequent situation however inappropriate? And why else should a man resist the notion of defences? After all, it does not really require any unusual feat of intellect to appreciate that it is more difficult to walk through barbed wire than across open country, that an enemy held up by wire or anti-tank ditches becomes an easier target than one without these impediments to his progress, or that it is easier to repel a waterborne invasion made under cover of darkness if the attacking forces can be blinded and illuminated by searchlights?

A further observation by Barber points up another curious feature of these phenomena and shows that they were by no means confined to the Singapore 'campaign'. 'When Hore-Belisha was Secretary for War he visited the B.E.F. in 1939 and was aghast at the lack of defence works, and plainly showed his annoyance, with the result that according to his diary of December 2nd, 1939, "Ironside [C.I.G.S.] after his visit to B.E.F. came to see me and with great emphasis told me that the officers were most upset at the criticisms made about lack of defences ... He said Gort was threatening to resign." '5

It seems that an inability to understand the value of defence was certainly not because of any indifference to criticism. There is also the sobering implication that even the Chief of the Imperial General Staff evidently regarded the lack of defences as rather less important than the fact that military feelings had been hurt. Can it be that some-where in the minds of some professional military commanders there lurks a natural distaste for defensive responses?

Defensive, as opposed to offensive, responses rank low in military esteem. Defensive activity is protective, womanly, one might almost say maternal. In sex, to use a particularly trite but apposite metaphor, it is the male who penetrates the fortress of the female; he is the attacker, she the defender. For the male to carry out elaborate preparations for his own safety is to some extent effeminate, an admission of weakness. For the male who has doubts about his own virility, whose life and choice of career are governed by unconscious doubts about his own masculinity and sexual adequacy, such effeminate activities may be anathema. By the same token it is possible that there is an affinity between the behaviour of the generals in Singapore and the refusal on the part of male industrial workers to wear protective clothing, ear-plugs and the like.* In both cases there is perhaps the feeling that it is unmasculine to defend oneself, that by so doing one appears to be a cissy.

In the case of Percival and Gordon Bennett, to erect defences would have been to admit to themselves the danger in which they stood. In other words, their professed anxiety about civilian morale was really displaced from anxiety about their own morale. Looking further into the story of Singapore one is struck by the compulsive element in this refusal of the military to defend itself. Such compulsive behaviour is typical of many who present an authoritarian personality and are 'reared' in an organization which traditionally deals with fear and danger by ritualistic means—'bull', drill, parades, etc. Comparable devices are to be found in the aggression-inhibiting ritual and display of many animals, their purpose being to keep intra-species aggression within bounds. Contemplation of instinctual mechanisms in lower animals suggests yet another contributory reason for the behaviour of these curiously inefficient generals. It is that of helpless resignation in the face of overwhelming same-species aggression. Whereas the brown rat, that most ferocious of fighters, will turn and attack any large predator, it makes no such attempt to defend itself against a concerted attack by fellow rats. Under these circumstances

* According to Donald Macintyre, the deafness of so many gunnery officers in the Royal Navy is due to the fact that they think it unmanly to wear ear-plugs.[6]

the alien rat which has unwittingly infringed the territorial rights of another colony lets itself be torn to shreds rather than fight back.

At a human level such behaviour could well be part and parcel of appeasement tendencies. According to Konrad Lorenz those ritualized behaviour-patterns which constitute appeasement gestures have evolved in a large variety of animals as the ultimate defence against the otherwise lethal effects of intra-species aggression. In trying to appease another member of the same species, the animal does everything to avoid stimulating its aggression. A cichlid, for instance, elicits aggression in another by displaying its colours, unfolding its fins or spreading its gill covers to exhibit its body contours as fully as possible. If the same fish wishes to appease a superior opponent it does exactly the opposite: it grows pale, draws in its fins, displays the narrow side of its body and moves slowly, stealthily, literally stealing away all aggression-eliciting stimuli.

The tendency to appease, to turn the other cheek, looms so large in human affairs, whether in the 'love your enemies' sense of Christian teaching or in the bowing of subordinates to their superiors, that there is no reason to suppose that it does not also occur in certain sorts of military commanders when confronted by such seemingly hopeless odds and fighting far from home.

But to continue with the story: shortly after his conversation with Simson, General Percival was visited by the Supreme Commander of Allied Forces in the Far East, General Sir Archibald Wavell. Of this incident Wavell wrote that he had been 'very much shaken that nothing had been done' and, 'speaking with some asperity', demanded to know why. Percival gave the same answer that he had given to Simson: to preserve civilian morale. Wavell retorted that it would be very much worse for morale if the troops on the peninsula were driven back on to the island.

The upshot of this encounter was a directive from Churchill giving detailed instructions on how to defend the north shore. The measures listed were precisely those which had been advocated by Simson. But despite these pressures, transmitted to him via Wavell, Percival still did nothing. When he eventually issued a plan it was already too late, for the necessary civilian labour was no longer available.

On the disposition of his forces Percival's thinking seemed no less deranged. Rather than hold a force in reserve that could be rushed quickly to wherever the Japanese eventually chose to land, he decided to spread his troops thinly over a long front. In other words, he decided now 'because it would be good for morale' just exactly what he had refused to do earlier because it would be *bad* for morale.

In his 'Battle of Singapore' Order of the Day, Percival made great play of phrases like 'the enemy within our gates', 'loose talk' and 'rumour-mongering', all calculated to alarm civilians (for in fact there was virtually no fifth column). And this from the man who had laid such stress on the importance of civilian morale.

Of this period Barber writes: 'In all the catalogue of ineffectual leadership ... nothing is quite so puzzling as the virtual absence of any deterrent action during the last precious hours of daylight before the Japanese attacked ... it is hard to believe that a modern general could so easily ignore what was happening around him.'[7]

By this time everyone, including Wavell, predicted quite correctly that the Japanese attack would come from the north-west. There were good reasons for this prediction. However, Percival immediately ordered that vast quantities of defence stores should be shifted *from* the north-west to the north-east corner of Singapore, and this despite the fact that a reconnaissance had shown that there were insuperable obstacles to an attack from the north-east.

However, having moved his stores to the north-east Percival then learnt that the enemy were massing in the north-west. He promptly ordered all the stores moved back again. But by now it was once again too late.

For the Allies it was a week of chaos and confusion unrelieved by any vestiges of competent generalship. Thanks to the absence of defences, including a failure to use the searchlights which had been assembled to blind and make targets of the attackers as they paddled their way across the Johore Straits, the Japanese landed almost unmolested. Despite a devastating barrage from Japanese artillery, British guns, instead of pounding the enemy's point of embarkation, remained mute, awaiting orders that never came. Despite the weeks of warning, Allied ground forces were speedily outflanked, encircled, cut off or routed.

In all this, Percival's want of generalship has been summed up thus: 'It seems evident that, while Percival paid lip-service to the need to gain time, he failed to take the only step which might have enabled him to do so. Had defences been constructed in Johore in December and early January, there might well have been a chance for Singapore to survive for sufficient time for the Australian reinforcements from the Middle East to arrive.'[8]

In the event 138,708 British, Indian and Australian soldiers either died or went into captivity.

Of all the instances of military incompetence considered in this book, it is the fall of Singapore which most clearly gives the lie to the so-called 'bloody fool' theory of military ineptitude. Percival was in fact highly intelligent and had shown himself in previous years to be a brilliant staff officer. What he shared with other, earlier, military incompetents were passivity and courtesy, rigidity and obstinacy, procrastination, gentleness and dogmatism. Can these really be the common denominators of military incompetence? Or is there yet another, deeper cause underlying all the rest? Perhaps our next and final example will provide at least a clue as to what it might be.

12

Arnhem

'All the accumulated evidence confirms that, like Gallipoli, this was a British disaster where naked courage lacked the bodyguard of competent planning, competent intelligence, competent technology. Yet war's object is victory, not the Victoria Cross, and it was shameful that by the autumn of 1944 we could still be so amateur.'

RONALD LEWIN

'It began to seem to me that the generals had got us into something they had no business doing.' A PRIVATE SOLDIER AT ARNHEM

If it achieved nothing else, 'Operation Market-Garden', Montgomery's plan to capture and hold a bridgehead across the Rhine in northern Holland, at least demolishes the myth that military incompetence stems from stupidity. For sheer initiative, quickness of mind, fortitude and selfless heroism, the conduct of those who actually fought the battle has never been surpassed. By the same token, the men who planned and administered the operation were probably as intellectually gifted, well trained, professionally competent, dedicated and conscientious as any military planners have ever been. And yet the unfolding of 'Market-Garden' revealed all the symptoms of high-level military incompetence.

In its conception the plan was a high-risk venture which, if it had paid off, might have shortened the war by several months. A secondary feature of the plan was that it promised to gratify Montgomery's wish that his armies would win the race for Berlin. In the event this secondary incentive took precedence over the first, with calamitous results.

'Market-Garden' was a two-stage operation. In the first stage a massive airborne drop on northern Holland was timed to coincide with the invasion of southern Holland by land forces of British 2nd Army. In the second stage the paratroops and glider-borne forces of 1st Airborne Division were to capture and hold the great road bridge at Arnhem while the tanks of 2nd Army's XXX Corps raced across Holland to consolidate their gains. Success depended upon an absence of serious enemy resistance in the Arnhem area; the capture of the bridge before the Germans had time to blow it up or bring up reinforcements; successive waves of airborne reinforcements from England, to back up the initial drop; and, finally, the arrival at Arnhem of XXX Corps *within forty-eight hours* of the drop.

The failure of the operation resulted from a concatenation of the following factors:

1. As a result of his neglect to open up the port of Antwerp by clearing the Schelde estuary, Montgomery allowed the German 15th Army to escape into north Holland, where it was available to defend the approaches to Arnhem.

2. The arrival at Arnhem of XXX Corps depended upon them advancing across 64 miles of enemy-held territory on a one-tank front along elevated, unprotected highways, flanked by a soft and sodden tank-proof landscape, interspersed with waterways. Any delay—a blown bridge, an enemy ambush, a blocked road—and the entire column would be stopped. Any delay and the Germans would have more time to bring up reinforcements. In the event it is hardly surprising that XXX Corps never did reach Arnhem—that they could not achieve even in 9 days what had been scheduled to take 48 hours.

3. As might have been expected from what is known of English autumns, the mists, if not the mellow fruitfulness, of an English late September delayed the departure of subsequent gliders and paratroops for the reinforcement of 1st Airborne Division.

4. 'Market-Garden', perhaps more than most military operations, necessitated good communications between the various units and commanders of the attacking force. But here technology failed them. Though it was now fifty years since Marconi had succeeded in sending messages by wireless, the radio sets carried by the invasion force proved useless. Unless within earshot of each other, no one knew what anyone else was doing.

5. Since the airborne assault was to take place in daylight, and because it was vital that XXX Corps should complete their journey

within 48 hours, the whole enterprise depended upon an absence of strong German forces both in the Arnhem area and on the approach route from the south. Hence it came as something of a jolt when SHAEF received reports from the Dutch underground that two S.S. Panzer divisions which had mysteriously 'disappeared' some time previously had now reappeared almost alongside the dropping zone. This information, passed on to Montgomery, received support from British aerial photography of German tanks in the Arnhem area. Meanwhile forward troops of British 2nd Army reported a build-up of German forces along *their* intended line of advance.

This was the moment to reassess the risks involved. But since these ugly facts did not accord with what had been planned they fell upon a succession of deaf ears. Taking a lead from Montgomery, who had described the SHAEF report as ridiculous, British 2nd Army Head-quarters were quick to discount it also. When one of his intelligence officers showed him the aerial photographs of German armour, General Browning, at First British Airborne H.Q., retorted: 'I wouldn't trouble myself about these if I were you … they're probably not serviceable at any rate.' The intelligence officer was then visited by the Corps medical officer, who suggested he should take some leave because he was so obviously exhausted. And at First Allied Army H.Q. the Chief Intelligence Officer, a British lieutenant-colonel, decided there was no direct evidence that the Arnhem area contained 'much more than the considerable flak defences already known to exist'. As Ryan puts it: 'All down the Allied line of command the evaluation of intelligence on the Panzers in the Arnhem area was magnificently bungled.'

Finally, just in case there were any residual doubts, the intelligence staff of 2nd Army came up with the reassuring opinion that any German forces in the Arnhem area were 'weak, demoralized, and likely to collapse entirely if confronted with a large airborne attack'.

'Market-Garden' went ahead—but not quite as planned. Instead of encountering a few old men who collapsed or ran away, 1st Airborne Division fell upon a hornets' nest of German armour. Far from being demoralized, the enemy fought like tigers to defend the gateway to their homeland. And far from sweeping across Holland to aid the hard-pressed paratroops, the tanks of 2nd Army's XXX Corps were reduced to a crawl by the combination of unsuitable terrain and a determined opposition.

Defeat was absolute and terrible. Short on everything but courage, the men of 1st Airborne Division held on until their numbers had been reduced from 10,005 to less than a quarter of that figure. Total Allied losses—in killed, wounded and missing—exceeded 17,000, some 5,000 more than those who became casualties on D-Day. Dutch civilian casualties from 'Market-Garden' have been estimated at between 500 and 10,000.

Apart from what one American historian has described as 'a fifty-mile salient—leading nowhere', nothing had been gained beyond a lesson for posterity; though even this had its impact weakened by Field-Marshal Montgomery's subsequent description of 'Market-Garden' as a ninety per cent success, a sentiment which drew from Prince Bernhard of the Netherlands the comment: 'My country can never again afford the luxury of a Montgomery success.'[1]

For the student of military disasters, the attack on Arnhem ranks with Kut and the Bay of Pigs fiasco (see page 397). Through inappropriate risk-taking, underestimation of the enemy, the neglect of unpalatable information and a failure of technology, military decisions by able brains, at high levels of command, brought down misery and chaos.

PART TWO

'It is better to struggle with a stallion when the problem is how to hold it back, than to urge on a bull which refuses to budge.'
GENERAL MOSHE DAYAN

AUTHOR'S NOTE

Because of their juxtaposition, the preceding accounts of military disasters may have given the impression that they are typical of military endeavour. This is not so. The very fact that they have provided material for so many books and plays attests to their comparative rarity. By the same token, incompetent senior commanders are outnumbered by their competent brethren.

However, as we have seen, military incompetence, when it does occur, can be immensely costly; which is why the next part of this book was written with no holds barred.

Just as a book on cancer could not afford to be mealy-mouthed about malignancy, or gloss over the hazards of carcinogens and dangerous patterns of behaviour, so a work on military incompetence would not be worth the paper it was written on if it did not delve deep and expose what is unpleasant. Military incompetence, like cancer, is part of the price paid for complexity. Whatever else it does, the analysis of the potential for military incompetence, like that for cancer, serves to emphasize that most armies and navies, like most bodies, perform their numerous and difficult functions with some efficiency most of the time.

13

Is There a Case to Answer?

Now that we have completed a survey covering over a hundred years of military mishaps, what conclusions can be drawn regarding the incidence of military incompetence?

There are a number of possible answers. Firstly, it could be argued that so-called incompetence at high levels of command is really a figment of the imagination of vindictive, inaccurate or untruthful historians. In parenthesis, it might also be argued that if, on rare occasions, some commanders have shown minor lapses in their generalship these are not matters to be spoken of, let alone made the subject of extensive analysis.

A second conclusion might be that what seems to have been military incompetence was really due to other, non-military factors, such as governmental stinginess, vagaries of the weather and sheer bad luck.

A third conclusion might be that since every military action is an uncontrolled experiment, in the sense that it can never be known what *would* have been the outcome had decisions been different, there remains the almost unimaginable possibility that things might have been worse, that what *was* done did represent the least disastrous of possible courses open. Many of the costly actions of the First World War fall into this category.

No one would deny that there is more than a grain of truth in all these propositions. Facts do get distorted in the telling. Disasters are indeed more newsworthy than successes. Writers undoubtedly do enjoy painting the worst possible picture of their particular bêtes noires, many generals have had to contend with the ineptitude, uninformed interference and stinginess of their political masters, and of course things might have been worse.

There *are* counter-arguments, however. Because they are surrogate father-figures, people are only too ready and anxious to love their admirals and generals, particularly in time of war. One has only to read Lowis's *Fabulous Admirals* to realize that even the most outrageous eccentricities (eccentricities which would not be tolerated in any other walk of life) are considered amusing if not actually endearing when part and parcel of some famous warrior. After all, even Sir John French still has his circle of devoted admirers. As Alfred Vagts puts it: 'A very large part of military history is written, if not for express purposes of supporting an army's authority and prestige, at least with the intention of not hurting it, not revealing its secrets, avoiding the betrayal of weakness, vacillation or distemper.' 'The historical record of warfare is thus dependent on the writer's desire to *preserve* reputations.'[1]

All in all then, the case for maintaining that there is a bias towards exaggerating the incidence and extent of military incompetence is perhaps compensated for by some quite contrary tendencies.

Under the circumstances, this book takes the view that certain sorts of incompetence have been an enduring feature of the military scene and that amongst the millions of officers and men who have fought heroically and efficiently, often under the most trying conditions, there have marched a small but influential number whose ability has fallen far short of that required by the positions which they held. Two questions then occur. Is there any common pattern to this incompetence and, if there is, whence does it arise?

As a first step towards answering these questions let us try and summarize the data contained in the foregoing chapters. In brief, then, military incompetence involves:

1. *A serious wastage of human resources and failure to observe one of the first principles of war — economy of force.* This failure derives in part from an inability to make war swiftly. It also derives from certain attitudes of mind which we shall consider presently.

2. *A fundamental conservatism and clinging to outworn tradition*, an inability to profit from past experience (owing in part to a refusal to admit past mistakes). It also involves a failure to use or tendency to misuse available technology.

3. *A tendency to reject or ignore information* which is unpalatable or which conflicts with preconceptions.

4. *A tendency to underestimate the enemy* and overestimate the capabilities of one's own side.

5. *Indecisiveness* and a tendency to abdicate from the role of decision-maker.

6. *An obstinate persistence in a given task* despite strong contrary evidence.

7. *A failure to exploit a situation* gained and a tendency to 'pull punches' rather than push home an attack.

8. *A failure to make adequate reconnaissance.*

9. *A predilection for frontal assaults,* often against the enemy's strongest point.

10. *A belief in brute force* rather than the clever ruse.

11. *A failure to make use of surprise* or deception.

12. *An undue readiness to find scapegoats* for military set-backs.

13. *A suppression or distortion of news* from the front, usually rationalized as necessary for morale or security.

14. *A belief in mystical forces*—fate, bad luck, etc.

Some or all of these several aspects of incompetence have played a significant part in the military mishaps considered in earlier chapters.

It remains now to show that they have a common aetiology and can be understood in terms of a complex interaction between the nature of military organizations and certain features of human personality. By way of a starting-point let us first consider the question of wastage of human resources. As we shall see, this is perhaps the best single key to all that we need to understand.

On logical if not humanitarian grounds the maintenance of an efficient force should be the first consideration of a military commander. Other qualities of generalship will avail him nothing if he has no one left to do the fighting. Excessive loss of life and high casualty figures would therefore seem a likely indicator of military incompetence. In reality the situation is far more complex. At least three situations have to be distinguished. Firstly, there are the well-known cases of what seem to be purely administrative incompetence, as for example in what John Laffin has described as the imbecile Walcheren Expedition of 1809. Though the purpose of this expedition was to attack Antwerp, the troops were in fact kept waiting for eight weeks on unhealthy Walcheren Island in Zeeland. In the event, and owing to the procrastination of the military commander, Lord Chatham, and the naval commander,

Sir Richard Strachan ('a dull hesitant incompetent pair'), 7,000 men died, 14,000 had their health permanently ruined and thousands more became ill, mostly from malaria. Only 217 were killed in action. While dying the men were given no attention and little to eat. As Laffin remarks: 'Sick men were expendable.'[2]

Of the same genre was the appalling wastage of human life in the Crimean War. During this campaign the Army suffered a thirty per cent decline in strength through disease, malnutrition and exposure. The four main factors appear to have been ignorance, lack of initiative and inventiveness, a chilling disregard for the welfare of junior ranks, and a fear of offending higher authority. (The nearest approach to this in modern times was the bumbling incompetence and callous indifference of London Borough Councils to the problems posed by air raids in the blitz of 1940 and 1941 and ascribed to laggard bureaucrats obsessed only with their own prestige.[3])

The second class of manpower wastage is that involving casualties from enemy action as a result of the incompetent planning of senior military commanders. The men who perished in the attack on Fort Rooyah in the Indian Mutiny, the thousands of casualties from the Germans' use of gas in 1915, the 13,000 who went into captivity following the siege of Kut, the 138,000 casualties of Singapore, the 8,500 Americans who died in the Ardennes offensive of 1944 and the 17,000 British, American and Polish who were killed, wounded or reported missing at Arnhem fall into this category.

There are of course several cases of gross wastage which fall into both the aforementioned categories. Thus in the ill-fated retreat from Kabul (page 75) the loss of an entire army of 4,500 men and 12,000 camp-followers is partly attributable to climatic factors (several thousand died of cold) and partly to depredations by hostile tribesmen. In both cases their demise resulted from the monumental feebleness and indecision of their commander-in-chief, General Elphinstone.

The third and most costly type of manpower wastage is that resulting from a deliberate policy of attrition adopted by commanders who regarded soldiers as wholly expendable; generals for whom the conservation of human life ranked lower in importance than various other criteria which were governing their actions. A good early example of this phenomenon is to be found in Napoleon's campaigns, as suggested by his remark: 'A man such as I am is not much concerned over the lives of a million men.' The million or so who fell on the Somme, at

Verdun and at Passchendaele were victims of this same attitude of mind.

This formidable but by no means exhaustive list of human casualties suggests that over the years a handful of senior military commanders have been careless—to say the least—with the fighting forces with which they were entrusted. Upon closer analysis, however, it seems that in their failure to fulfil the primary function of a commander, the culprits fall into two categories. First there are those like Elphinstone, Raglan, Buller and Percival: mild, courteous and peaceful men who, though no doubt caring deeply about the fearful losses which their armies suffered, seemed quite incapable of ameliorating the situation.

Thus Elphinstone *could* have occupied the fortress of Balla Hissar. He *could* have allowed his troops to prevent frostbite by wearing make-shift puttees. He *could* have started the retreat earlier and completed it more quickly. Likewise Raglan *could* have exerted pressure on the Government of the day to supply the crying needs of his troops. He *could* have taken direct action to procure firewood for the thousands who were dying of exposure. He *could* have galvanized his staff into some sort of remedial activity. He did none of these things, preferring to withdraw into the relative comfort of his farmhouse headquarters. Percival *could* have instituted defences for Johore and the island of Singapore. He *could* have prepared adequate air-raid defences and shelters for the citizens of Singapore. He chose not to.

Buller, perhaps the most interesting case of all, *could* have saved life by a more generous deployment of his forces. So deeply did he feel for the suffering of his men that he carried the principle of economy of force to such a ludicrous extent that thousands died for want of help from the tens of thousands kept idly standing by.

While the incompetence of these men seemed to spring from a crippling passivity and lack of what General Gordon Bennett has called 'aggressive spirit', they stood in sharp contrast to those of a second group whose besetting sin was overweening ambition coupled with a terrifying insensitivity to the suffering of others. These, men like Haig, Townshend, Walpole, Nixon and Joffre, seemed dedicated to one goal—self-advancement. Vain, devious, scheming and dishonest, they were certainly not inactive in the courses they pursued, nor of course were they necessarily without military talents.

In all this we are anticipating a theory of military incompetence rather different from that held by proponents of the so-called 'bloody fool' theory. Perhaps we are being too complicated – perhaps intellectual deficit could explain the data? Let us then, before considering the other factors contributing to military incompetence, first examine this older and more favoured hypothesis.

14

The Intellectual Ability of Senior Military Commanders

'I feel a fundamental crippling incuriousness about our officers. Too much body and too little head.'   T. E. LAWRENCE

'Most British defeats have been caused by stupidity.'
     CORRELLI BARNETT, *The Desert Generals*

What grounds are there, then, for the most popular explanation of military incompetence—stupidity?

There is the suggestion that the armed forces do not attract the best brains. A recent call-up survey in the United States put the status of Army officers below that of professors, physicians, clergymen and school-teachers. As Morris Janowitz remarks: 'A liberal ideology ... holds that since war is essentially destructive, the best minds are attracted to more positive endeavours.'[1] According to Field-Marshal Montgomery this has been true for some time. Writing of 1787, he noted that the Army was the 'normal career of the less bright younger sons'.[2] And of 1907 he wrote: ' ... in those days the Army did not attract the best brains in the country.'

Then again there is the *supposedly* low quality of officer cadets entering the services. According to Janowitz: 'The impression exists among educators that the intellectual level of those entering the military profession via the service academies reflected the adequate effective and adequate minimum standards rather than any extensive concentration of students at the upper end of the intelligence continuum.'[3]

The same writer notes that in Britain sixty per cent of all entrants to the Royal Military Academy are graded *before arrival* as likely to make a below-average officer.

In training for generalship, it seems that intellectual ability has not always counted for very much. Even Haig, 'the educated soldier', became Commander-in-Chief of the British Army in the First World War despite a poor academic record. This dour lowland Scot, described by Duff Cooper as the dunce of the family, by Lloyd George as 'utterly stupid', and by Briand as '*tête du bois*', had the very greatest difficulty in passing the Sandhurst entrance exam, and then only with the help of a crammer whose knowledge of the academy's methods virtually guaranteed a scrape-through for even the dullest candidate. Sir Henry Wilson, who later became C.I.G.S., had the distinction of failing three times in his attempts to enter the R.M.C.

It might well be asked why they bothered with a selection procedure which purported to test for intellectual ability when its end-purpose could be so easily circumvented. Major-General Fuller, according to Sir John Smyth, went to a crammer who had him learn by rote the answers to twelve likely questions. Since fifty per cent of the predicted questions came up in the examination, Fuller managed the record mark of 497 out of 500![4] So unaccustomed were the R.M.C. examiners to any sort of intellectual effort that when one cadet memorized the set book on the Peninsular War he was accused of cheating!

That intellectual performance counted for little when it came to subsequent promotion in the Army might seem suggested by the scholastic progress of Field-Marshals Montgomery and Auchinleck, both of whom scraped into and passed out of Sandhurst with low marks.

Commenting on these phenomena, Smyth makes the point that 'it says quite a lot for the stiffness of the Sandhurst extrance exam, and for the high standard of competition whilst at the College, that both of these very capable officers should have passed both in and out of College with such a comparatively modest placing.'[5] He may be right, though in the light of other evidence a more logical explanation would be that the academic requirements of the R.M.C. are not wholly relevant to those actually required for competent generalship. Certainly, a brilliant performance in military schools is no guarantee of subsequent ability. General Colley, whose succession of defeats culminated in 1881 in his own demise at Majuba Hill, had the distinction of passing out of Staff College with the highest marks on record. The irrelevance of early scholarship to subsequent generalship also finds support in Napoleon and Wellington both of whom achieved very low grades at school.[6] And in more recent times the early academic brilliance of Lieutenant-General Percival evidently availed him little at Singapore.

However, while military history is replete with examples of what has been dubbed stupidity, there are grounds for believing that such explanations merely reflect a preference for simple theories of what are really very complex phenomena. If the complex phenomenon is unpleasant and the simple explanation abusive, then so much the better.

The view taken here is that those intellectual shortcomings which *appear* to underlie military incompetence may have nothing whatever to do with intelligence, but usually result from the effect upon native ability of two ancient and related traditions. The first of these, originally founded in fact, is that fighting depends more upon muscle than brain, the second that any show of education is not only bad form but likely to be positively incapacitating. The prevalence of these traditions was mentioned by H.M. Commissioners in their report upon military bungling in the South African War. Not only did they take a very pessimistic view of the educational standard and intellectual ability of officers from the most junior to the most senior; they also noted to their dismay that 'keenness is out of fashion and not the correct form'.[7]

One has only to read subsequent descriptions of military and naval training establishments for officers to realize that their words seemed to have fallen on deaf ears. In such places size, muscle and prowess at games still constituted the main criteria by which a man was judged. When writing of personalities who passed through Sandhurst, Brigadier Smyth, V.C., dwells chiefly upon their physical attributes. He says of Field-Marshal Alexander that he was a good long-distance runner; a cadet who won the Sword of Honour was 'a fine athlete and captain of the rugger team', and an officer who was killed in 1915 'possessed a fine physique and was conspicuous in every sport he took up'.[8] In the light of these and similar comments the uninitiated might be forgiven for thinking that the main purpose of the Royal Military College was to turn out athletes or male models rather than brains capable of mastering the intricacies of war.

Fortunately there are some who have seen the threat to originality and intelligent thinking in this approach. As recently as 1955, at the Sovereign's Parade, the Duke of Edinburgh felt it necessary to say:

'Finally, as you grow older, try not to be afraid of new ideas. New or original ideas can be bad as well as good, but whereas an intelligent man with an open mind can demolish a bad idea by reasoned argument, those who allow their brains to atrophy resort to meaningless catch-phrases, to derision and finally to anger in the face of anything new.'[9]

If the intellectual level of future military leaders was below the average for other comparable professions, it seems that subsequent training did little to redress the balance. There have been occasions for believing in a conspiracy to keep it that way. In 1901 the Akers-Douglas Committee, inquiring into R.M.A. affairs, advocated doing away entirely with civilian instructors and using only Army officers. This inbreeding of the uneducated was, however, resisted by the later Massey Committee, who, depressed by what they found, considered that:

1. The general education of cadets should be continued (an impossible state of affairs if the instructors were themselves uneducated).

2. Few young officers showed any capacity for command.

3. There was too much drill, too much rigid discipline and too much cramming for marks.

4. The instructors were mediocre and selected for prowess at games and smartness rather than for their knowledge of the subject they had to teach or their qualifications as teachers.

Much the same picture has been painted of Britannia, forerunner of the Royal Naval College at Dartmouth. Again, the emphasis was on blind obedience, sport and ceremonial, with scant regard to intellectual pursuits and little pride in knowing one's job.

The academies which produced the men whose incompetence has sometimes cost society dear cannot be entirely held to blame for the ethos and values which they maintain. Equally culpable are those who encouraged them. Even as recently as 1949, after a war which was nearly lost through the effect of the cerebral millstone of archaic tradition, Admiral of the Fleet the Earl of Cork and Orrery, having taken the salute at Dartmouth, urged the cadets to 'absorb tradition' — rather as one might adjure a sufferer from rheumatism to absorb some ancient well-tried liniment. And for the Army there is the account of how one famous general, after inspecting the Royal Military Academy, confined himself to praising the drill rather than dilating on the importance of knowing the job of modern soldiering.

Perhaps the clearest indication of the intellectual abyss created in officer training comes from rueful comments by the men themselves. As E. S. Brand remarked of Dartmouth College: 'Unhappily, in spite of all their efforts we* were not well educated in contrast to the public-school cadets whose outlook and knowledge were so much broader than my own.'[10] In this context, Janowitz makes the point that only the most exceptional senior commanders can allow themselves to question their choice of a military career. Thus General Robert E. Lee admitted that 'the greatest mistake of my life was taking a military education', and General Stilwell said: 'It is common knowledge that an Army officer has a one-track mind, that he is personally interested in stirring up wars so that he can get promotion and be decorated and that he has an extraordinarily limited education with no appreciation of the finer things of life.'[11]

Even as recently as 1972 the Sandhurst course for officer cadets was reduced from two years to one, an event which drew from Geoffrey Sale, former Director of Studies at the R.M.A., the comment: 'That a professionally trained officer corps can be produced able to cope with all situations and to understand the why and wherefore of his profession in 12 or 13 months is plainly ridiculous. It is to condemn the long-term officer who is not a graduate to being semi-educated, a worthy object of the B.B.C.'s caricatures.'[12]

Whether or not intellectual shortcomings lie at the heart of much military incompetence, the fact remains that a deliberate cult of anti-intellectualism has characterized the armed services. While its origins relate, as we shall see, to much deeper reasons for military mishaps than mere ignorance or slowness of mind, the fact remains that its effects have not been helpful. That generals and admirals between the wars denigrated progressive thinkers and poured scorn on men who wrote books which challenged existing practices must surely have tended to stifle any exercise of the intellect by those who wanted to get on, and deterred the gifted from ever seeking a military career. As Robert McNamara once remarked: 'Brains are like hearts, they go where they're appreciated.'

Even as late as the Second World War the baneful effects of anti-intellectualism were taking a toll of much-needed brain-power. A classic example is that of Major-General Dorman-Smith. This out-spoken but exceptionally gifted officer whose talents were appreciated

* Cadets entering direct from preparatory school.

(and used) by such unequivocally great generals as Wavell, Auchinleck and O'Connor created such resentment of his intellectual abilities in the military establishment that when he was relieved of his post as Deputy C.I.G.S. Cairo after the first battle of Alamein, his fall was final. 'There was no recovery. All those in the Army who bore him ill-will, who had never forgiven him his brilliance and unorthodoxy, saw to that ... the word quietly filtered through the military "Establishment" that Dorman-Smith was not to be given a chance to rebuild his career.'[13]

The saddest feature of anti-intellectualism is that it often reflects an actual suppression of intellectual activity rather than any lack of ability. This is suggested by the rapidity with which so many military men rush into print as soon as they have retired. Evidently there was something waiting to get out. Unfortunately, as Liddell Hart points out, a lifetime of having to curb the expression of original thought culminates so often in there being nothing left to express. Recent research on the relationship between mental activity and cerebral blood-flow[14] adds point to the old belief that the brain, like muscle, atrophies from prolonged disuse. But perhaps *this* touches upon the real cause of military incompetence—age? Since traditionally promotion has depended upon seniority, commanders, generals and above have tended to be old, and since thinking, memory, intelligence and the special senses all deteriorate with age, then maybe bad generals are just old generals? Certainly it can be said that age will intensify most defects of the mind,[15] and over the years the quality of generals has seemed not unrelated to retiring age. At no time was this more in evidence than during the Crimean War (see page 37) and then again in the 1930s. Another contribution to incompetence tied up with age was the unhelpful tendency to sack, forcibly retire or otherwise curtail the promotion of those *young* officers who unwisely failed to conceal their lights beneath bushels of conformity.

Such was the case of Major-General J.-F. C. Fuller. On December 13th, 1933, Fuller, one of the most intellectually gifted men ever to serve in the British Army, was placed on the retired list. This waste of talent resulted from the prejudice aroused by his fully-borne-out prophecies, and the fact that he had dared to criticize those less gifted than himself.

Now at the time Fuller had a champion in the then relatively junior General Ironside, who regarded the retirement as a scandal and opined that Fuller had 'the best brain in the Army'. When, however, Ironside was in a position to reinstate Fuller, he immediately cried off, saying: 'Oh, I couldn't do that—it would upset the turn of promotion.' As Liddell Hart remarked: 'This was a sadly revealing example of how even a progressively minded soldier tended to be subservient to the law of "Buggins' turn".' In fairness to Ironside it should be added that in 1939 he did try to reinstate Fuller, but this time, just one year before the Army was nearly annihilated by the sort of armoured forces which Fuller had been advocating, the reinstatement was scotched by the War Office.

Yet another way in which age determines incompetence is through the voluntary resignation of intelligent young officers. According to Janowitz, a recent study of U.S. Army lieutenants suggests that the brighter ones resign as soon as they have completed their obligatory service, while those less well equipped remain. This natural selection would militate against finding the brighter people in the upper echelons of the military establishment.

But notwithstanding these considerations, age is far from being a complete explanation of military incompetence, for there have been plenty of able old generals and some remarkably inept young ones. As Vagts has noted: 'Generals of eighty, generals who were sick of body and even in mind, have won important victories.' Moreover, the very complex nature of military incompetence defies any explanation as simple as that of senility. Indeed, there are grounds for regarding the age factor as a symptom rather than a cause. But let us look at another aspect of what appears to be intellectual incompetence—the urge to pontificate.

The relation between the roles of ignorance and pontification in military incompetence is not entirely simple. Firstly, in accordance with the principle that nature abhors a vacuum, ignorance tends to evoke pontification in those who wish to conceal their lack of knowledge, or for whom ignorance of the facts means that they feel free to express strongly held beliefs of a contrary nature. For harmless enough examples of the first of these two classes of pontification there is the anecdote of the Dartmouth cadet who, when he asked why π was 22/7, was told by his mentor: 'It is not for us to question the wisdom of the Admiralty.'

Simon Raven has given an amusing illustration of the second class of pontification. 'I have never forgotten the trouble I got into for contradicting a general who announced that sodomy had rotted the Roman Empire; the fact that this officer scarcely knew a word of Latin and by his own confession had never read a line of Gibbon was held to be irrelevant.'[16]

Rather more serious are those pontifications which aim to make nasty facts go away by the magical process of emitting loud noises in the opposite direction. Here are some utterances of this kind.

Field-Marshal Montgomery-Massingberd, Chief of the Imperial General Staff from 1926 to 1933: 'There are certain critics in the press who say we should organize the Army again for a war in Europe ... the Army is not likely to be used for a big war in Europe for many years to come.'

Sir Ronald Charles, Master General of Ordnance: 'There is no likelihood of war in our lifetime.' This was said at the time of Hitler's accession to power. And, also before the last war, Sir Hugh Elles, Director of Military Training: 'The Japanese are no danger to us and eager for our friendship.'

In a calling where the accuracy of a communication may be a matter of life or death, the predisposition to pontificate is a dangerous liability. Unfortunately such a predisposition will be strongest in those like headmasters, judges, prison governors and senior military commanders who for too long have been in a position to lord it over their fellow men. Unfortunately such a predisposition will also be strongest in authoritarian organizations where the preservation of apparent omniscience by those above may be deemed of more importance than the truth.*

But the important thing about pontification is that though an intellectual exercise its origins are emotional.

Closely allied to pontification and no less hazardous is 'cognitive dissonance'. This uncomfortable mental state arises when a person possesses knowledge or beliefs which conflict with a decision he has made. The following hypothetical situation should make the matter plain. A heavy smoker experiences dissonance because the knowledge that he smokes is inconsistent with the knowledge that smoking causes cancer. Since he finds it impossible to give up cigarettes, he tries to reduce dissonance (i.e. tip the balance towards peace of mind) by con-

* According to research by Chaubey, fear of failure increases after middle age.[17]

centrating on justifications for smoking and ignoring evidence for its risks. He may tell himself that the revenue from tobacco helps the Government (i.e. he is therefore being patriotic), that it helps keep his weight down and that it is a manly, sociable habit. At the same time he may well refrain from reading the latest report on the relationship between smoking and lung cancer. If on the other hand he cannot avoid being confronted by tiresome statistics, he may well strive to reduce dissonance by telling himself (and others) that the correlation between smoking and cancer could just as well be taken to signify that people who are going to get cancer anyway tend to smoke in order to ward off the disease.

Since it was first propounded by Festinger in 1957, Dissonance Theory has given rise to a large number of empirical studies. Though the precise nature of the underlying psychological processes is far from clear, there are certain conclusions which could have serious implications for military decision-making. They may be summarized by saying that: 'Once the decision has been made and the person is committed to a given course of action, the psychological situation changes decisively. There is less emphasis on objectivity and there is more partiality and bias in the way in which the person views and evaluates the alternatives.'[18]

In other words, decision-making may well be followed by a period of mental activity that could be described as at the very least somewhat one-sided.

Since the extent of dissonance experienced is a function of the importance of the decision made, it is likely that many military decisions eventuate in fairly severe forms of mental disquiet. But a military commander cannot afford to reduce dissonance when this involves closing his mind to or 'reinterpreting' unpalatable information. The dire consequences that might follow such an attempt were only too evident after the Cambrai offensive and again during Townshend's advance on Ctesiphon. In both instances the ostrich-like behaviour of senior officers cost the Army dear.* The same may be said of the Ardennes counter-attack in 1944 and of Montgomery's failure, in the light of subsequent intelligence reports, to think twice about his decision to capture the bridge at Arnhem.

While the costs of dissonance resolution by some military men may be inordinately high, the probability of these costs occurring is also

* Byng in the first instance and Nixon in the second.

very high. There are three reasons. In the first place, military decisions are very often irrevocable. Secondly, they involve large pay-offs — much hangs on their outcome, *including the reputation of the decision-maker*. Finally, those commanders with weak egos, with over-strong needs for approval and the most closed minds will be the very ones least able to tolerate the nagging doubts of cognitive dissonance. In other words it will be the *least* rational who are the *most* likely to reduce dissonance by ignoring unpalatable intelligence. Research on individual differences in cognitive dissonance suggests that its effects are likely to be strongest in those afflicted with chronic low self-esteem and general passivity.[19]

More recent research on cognitive dissonance has emphasized another variable of some consequence for military behaviour: the degree of justification for the initial decision. Experiments by Zimbardo and others have shown that the less justified a decision, the greater will be the dissonance and therefore the more vigorous its resolution. No better example is afforded than that of Townshend's occupation of Kut. Since his advance up the Tigris was totally unjustified by facts of which he was fully aware, his dissonance, when disaster struck, must have been extreme and, to a man of his egotistical nature, demanding of instant resolution. So, again in the face of much contrary evidence, he withdrew into Kut. The wiser and possible course of retreating to Basra would have been a greater admission of the lack of justification for his previous decision. By the same token, once inside Kut nothing would budge him, because to break out, even to assist those who had been sent to release him, would have emphasized his lack of justification for being there in the first place. In short, an inability to admit one has been in the wrong will be greater the more wrong one has been, and the more wrong one has been the more bizarre will be subsequent attempts to justify the unjustifiable.

We can see now the relationship between pontification and cognitive dissonance. Pontification is one of the ways in which people try to resolve their dissonance. By loudly asserting what is consistent with some decision they have made and ignoring what is contrary they can reduce their dissonance. Clearly this particular concatenation of intellectual processes may prove very hazardous in a military context.

But there is another aspect of decision-making no less hazardous — its 'riskiness'. Recent research has shown that people vary in the degree to which they adjust the riskiness of their decisions to the realities of the external situation.[20] Individuals who become anxious under conditions of stress, or who are prone to be defensive and deny anything that threatens their self-esteem, tend to be bad at judging whether the risks they take, or the caution they display, are justified by the possible outcomes of their decisions. For example, they might well adopt the same degree of caution whether placing a small bet, getting married or starting a nuclear war. There is a sad irony about this state of affairs, for it means that those people who are most sensitive to the success or failure of a decision will be the very ones who make the biggest mistakes. Conversely, less anxious individuals will act more rationally because able to devote greater attention to the realities with which they are confronted.

Obviously these findings have considerable and alarming implications for the military scene. For as one psychologist has said: 'Under stress men are more likely to act irrationally, to strike out blindly, or even to freeze into stupid immobility.'[21] Others have remarked: 'The presence of relatively high levels of rationality in decision-making may characterize but a minority of men ... we are burdened by a nagging curiosity about how those persons controlling our destiny would distribute themselves within the personality-groups outlined.'[22] But why should anxious and defensive individuals, those who have the most to lose, act more irrationally than those less afflicted by neurosis? Two reasons have been advanced. The first has been well stated by Deutsch: 'Nervousness, the need to respond quickly because of the fear that one will lose either the desire or ability to respond, enhances the likelihood that a response will be triggered off by an insufficient stimulus, and thus makes for instability.'[23]

The second reason why a proportion of people will make irrational decisions whose riskiness is unrelated to reality is because, being neurotic, they will strive to maintain an image of themselves as either 'bold and daring' or as 'careful and judicious decision-makers', and the urge to sustain their particular conceit will take precedence over the need to behave realistically. Townshend's risky bid to capture Baghdad is consistent with this principle.

This chapter started with the intention of examining the oldest theory of military incompetence: namely that inept decisions occur through intellectual disabilities. The simplest form of this theory is that some military commanders (like some psychologists) are just plain stupid and that their faulty decisions spring from lowly intelligence.

Since decision-making is, by definition, a cognitive process then obviously the oldest theory is in one sense a truism, but it by no means follows that the simple hypothesis of low intelligence fits the bill. On the contrary, by looking further into the nature of decision-processes we are compelled to entertain another rather different possibility: namely, that the apparent intellectual failings of some military commanders are due not to lack of intelligence but to their feelings. Cognitive dissonance, pontification, denial, risk-taking and anti-intellectualism are all, in reality, more concerned with emotion than with intelligence. The susceptibility to cognitive dissonance, the tendency to pontificate and the inability to adjust the riskiness of decisions to the real situation are a product of such neurotic disabilities as extreme anxiety under stress, low self-esteem, nervousness, the need for approval and general defensiveness. These, it seems, over and above his level of intelligence, are the factors which interfere with what a man decides to do in a given situation.

15

Military Organizations

'Everything is very simple in war, but the simplest thing is very difficult. These difficulties accumulate and produce a friction which no man can imagine exactly who has not seen war.'

C. VON CLAUSEWITZ, *On War*

Military organizations make for military incompetence in two ways — directly, by forcing their members to act in a fashion that is not always conducive to military success, and indirectly, by attracting, selecting and promoting a minority of people with particular defects of intellect and personality.

The root-cause of all this is that since men are not by nature all that well equipped for aggression on a grand scale, they have had to develop a complex of rules, conventions and ways of thinking which, in the course of time, ossify into outmoded tradition, curious ritual, inappropriate dogma and that bane of some military organizations, irrelevant 'bullshit'. We are talking of 'militarism', a sub-culture which, in the end, may well hamper rather than facilitate warring behaviour. Three factors contribute to its growth. The first is that the origins of fighting are instinctive—so-called intra-species aggression. The second is that fighting was originally more a trial of strength than of wits. And the third that it is something which, in its original form, many lower species can do rather better than we can. Let us consider these points in a little more detail.

Broadly speaking, human activities may be regarded as falling into one or the other of two main groups: those which are directly instinctual and those which are not. Into the first, which involves what have been succinctly described as the 'three Fs'—feeding, fighting and 'reproduction'—fall such robust pastimes as pugilism, professional pie-eating, prostitution and soldiering. Into the second group fall all those other vocations which, though sometimes subserving the basic drives, do not have as their end-product the original consummatory response.

Besides this most important difference, the instinctual vocations have three other characteristics which differentiate them from those in the second category. They may involve unlearned patterns of behaviour, are motivated by crude if powerful emotions—fear, lust, rage—and are designed to culminate in an unlearned response of a distinctly physical kind.

Now attempts to professionalize instincts may be comparatively easy, as in the case of pie-eating or prostitution, or fraught with difficulty, as in the case of fighting. Prostitution is easy because the transformation of an unlearned drive into a money-making career is more a matter of realizing a potential than seriously modifying nature. As one of Wayland Young's interviewees remarked: 'I'd been working in that factory five years before I realized I was sitting on a fortune all the time.'[1] Thenceforth the entrepreneurial woman of easy virtue has merely to apply some scent, do clever things with her mascara, and the rest follows. Over the years the profession, like the art of love, has changed little. But for the professional soldier, progress has hardly been so smooth, largely because, unlike prostitution, the consummatory response has changed. The original purpose of intra-species aggression is not destruction but distribution. In lower forms of life the instinct of aggression is controlled by a language of signs and countersigns, so that everyone remains spread out with a minimum of bloodshed. Moreover, those animals best equipped to do each other an injury are also those with the most effective controls against so doing. A dog tactless enough to encroach upon a rival's territory may become involved in a noisy scuffle but has only to drop his tail, roll over and urinate to terminate the attack upon his person; the most he loses is face (and water). But for humans such natural safeguards (besides being embarrassing) have proved rather less effective.* In the first place, being omnivorous, they have not evolved such foolproof aggression-inhibitors as have the carnivores. In the second place, they have made up for their lack of natural weapons by acquiring some far

* It could be argued that in humans involuntary urination evoked by fear ceased to be effective as a gesture of appeasement with the invention of trousers!

more deadly artificial ones. These have extended the killing-distance far beyond the point at which any inborn signalling-system could be expected to work (natural signs of appeasement depend upon proximity), and have reduced to vanishing-point the natural instigators of aggression. It is much more difficult to feel spontaneously hostile towards an enemy you cannot see.

Yet other difficulties have been posed by the sheer size of human warring groups. With the transition from small parties of hostile tribesmen to large mercenary armies came problems of motivation and control. Since the history of warfare is largely that of the many who, through poverty or the press gang, were forced to take up arms for a cause which few could even comprehend, the evoking and direction of aggression called for special measures. These included devices to ensure group cohesion, to incite hostility, to enforce obedience and to suppress mutiny. They also included means whereby the intentions of leaders could be translated into a concerted action by followers. In short, it called for two other components of militarism — firstly, a system of rewards and punishment, of rank, medals, battle emblems and prize money, of confidential reports, courts martial and the lash; and secondly, a system of orders and over-learned drills whereby complex patterns of behaviour could be set in motion by the briefest of instructions.

No less important for a theory of military incompetence is the means whereby militarism is administered and its continuity ensured. This is the problem of 'who bosses whom' in the military hierarchy, and the sorts of criteria which determine a person's position in the pecking order.

Originally, since combat was largely a matter of brute force, we must suppose that the strongest came to the top. In fighting, as in prostitution, vital statistics gained the day — a sort of natural selection according to criteria that were essentially physical. But in the course of time the growing number of personnel involved, and improvements in technique, required some revision of earlier criteria. A distinction became necessary between organizers and the organized, between brains and brawn. To this end civil government might have been expected to construct armies in which such dichotomies obtained. One might have expected that officers would have been chosen for their brains, and the hierarchy of command based upon merit and professional expertise. In Britain, civil government did nothing of the kind. For very good reasons military power was vested in those who, because well satisfied with their position in society, would do their best to maintain the status quo—in short, the rich and the highly born. By the methods of purchase and nomination; the control of the Army was given over to men who, with nothing to gain from revolution, would remain the loyal, apolitical supporters of the existing regime. Professional ability, energy and dedication to the job counted for little. If you were rich and well connected you were in; if you were not, you were not. This state of affairs came to full fruition in the Victorian Army and was still in evidence at the outbreak of the First World War. It did little for military competence, but was eminently successful in other ways. Few countries can boast of such an absence of military coups as Britain.

However, following the disastrous events of the Crimean War, there occurred a gradual change in the make-up of the British officer corps. The abolition of the purchase system and the growth of public schools produced a decline in the numbers of rich aristocrats seeking commissions, and a rise in the numbers of officers drawn from the bourgeoisie. Moreover, standards of competence were improved by selection based upon examination. But while these reforms undoubtedly raised the quality and expertise of those occupying higher positions in the military hierarchy, they did no violence to the old policy of confining military power to those without political aspiration, nor did they do much to change the class-consciousness which has characterized not only the British Army since the time of Cromwell, but also the armed services of France and Germany.

The essential nature of militarism should now be clear. We see it as an ever-increasing web of rules, restrictions and constraints, presided over by an elite, one of whose motives was to preserve the status quo. We see it, in the case of the older European powers, as the natural product of a fundamentally jealous, class-conscious hierarchy whose nostalgia and basic conservatism ensure that the present must always bear the hallmark of the past. And we see it as remarkably similar in many respects to the ethos of the prototypical Victorian upper-class family group, where absolute obedience and submission to authority are traded for security and dependence.

Obviously, there is much here to make for incompetence in warfare. But this incompetence is augmented by another factor, namely the characteristics of some of those attracted to the military. Let us examine this hypothesis.

By modern standards, and viewed from the outside, the nature of militarism may not seem very attractive, including as it does a number of attributes which are positively repellent to those who value personal freedom, egalitarianism and creative as opposed to destructive ends. This distaste is common to both C. P. Snow's cultures. It is as strongly voiced in Einstein's comment— 'This subject brings me to that vilest offspring of the herd mind—the odious militia. The man who enjoys marching in line and file to the strains of music falls below my contempt; he received his great brain by mistake—the spinal cord would have been amply sufficient'[2]—as it is in Thackeray's—'A man one degree removed from idiocy with brains sufficient to direct his powers of mischief and endurance may make a distinguished soldier.'

Why then do people join the Army, and are there some characteristics of the military which have a positively magnetic attraction for those whose subsequent performance may be deemed incompetent?

An answer to the first question would include reasons which range in nobility from the need for selfless devotion to duty in a patriotic cause to the desire for upward social mobility. They would include a need for an exciting and varied career interspersed with plenty of leisure given over to gentlemanly pursuits, a penchant for violence and a propensity to follow in father's footsteps.* They would also include a distaste for other professions of comparable social status.

The second question is rather more difficult to answer but one might say that since at a deeper level of analysis militarism constitutes a number of defences against certain anxieties, people who share the same anxieties and have a predisposition towards similar sorts of defence will be drawn towards membership of the military, rather as an alcoholic might be drawn to join Alcoholics Anonymous. In other words, an individual with particular problems of a psychological kind may be expected to gravitate towards a group which he recognizes not only as containing fellow sufferers, but also as having developed effective ways of dealing with the special needs of its members. The therapeutic gain from such behaviour during the Second World War has been noted by Robert Holt. He wrote:

* According to C. B. Otley, the single biggest occupational group from which officer cadets come is the military profession.

It was a common clinical observation during the war that military service was an unusually good environment for men who lacked inner controls ... The combination of absolute security, a strong institutional parent-substitute on whom one could lean unobtrusively, and socially approved outlets for aggression provided a form of social control that allowed impulses to be expressed in acceptable ways.[3]

In following this line of thought, we start with the apparent paradox that whereas the military way is concerned with defence against the external enemy, a large part of militarism concerns defences against the anxieties and aggressive impulses of its member subscribers. Much that we have discussed under the heading of militarism can be legitimately viewed as devices so to control aggression that it is projected only upon legitimate targets while keeping other outlets blocked. In this respect there is a close parallel between aspects of militarism and the group behaviour of some subhuman species. Even a troop of baboons contrives a rigid dominance-hierarchy wherein each male knows his place. As K. R. L. Hall remarks of these animals: 'Controlled aggression is a valuable survival-characteristic in that it ensures protection of the group and group-cohesion.'[4]

At a human level, armies resemble the authoritarian family group. Just as the ethos of an upper-class Victorian family totally forbade any show of aggression by the child towards its parents, but encouraged organized aggression towards contemporaries in such school pursuits as boxing and sanctioned bullying, so in the Army the slightest hint of insubordination (i.e., aggression directed towards a superior) is severely punished, while aggression towards the enemy is encouraged and rewarded. Obviously such redirection of aggression is entirely consistent with the purpose of military organizations. By the same token, a tight rein on aggression is mandatory in a profession whose stock in trade and solution to most problems is physical violence. The My-Lai massacre and similar atrocities show only too clearly how quickly things can get out of hand. As I. L. Janis has remarked: 'The military group provides powerful incentives for releasing forbidden impulses, inducing the soldier to try out formerly inhibited acts which he originally regarded as morally repugnant.'[5] From a psychological point of view, therefore, militarism strives to maintain that paradoxical state of affairs where feeling angry may well be totally split off from aggression, one in which a soldier is required to suppress his aggression towards his superiors whom he may loathe, while venting it upon a hypothetical enemy towards whom he may well entertain no hostile feelings.

It is a situation fraught with the possibility of breakdown. On the one hand there is the anecdotal evidence of soldiers who, in the heat of the battle, shoot their own N.C.O.s and officers in the back, or who when firing sten guns on a range turn round to ask a question without remembering to ease their finger on the trigger. Such mishaps suggest that even the strongest defences against tabooed aggression may fracture under pressure.* On the other hand, there are those embarrassing occasions, such as occurred during the American Civil War, when soldiers so far forget themselves as to become friendly with the enemy. The classic example of this unorthodox behaviour occurred on Christmas Day, 1914, when British and German troops joined together for convivialities in no man's land. Needless to say, these reprehensible flickerings of humanity were quickly stamped out by the generals on both sides. Fortunately (for the generals) no lasting harm was done, but the episode did highlight the necessity for those aspects of militarism which ensure that aggression does not wilt through want of hate.

It is just because the business of a soldier is destruction and violence that the need to take general precautions against disorder becomes so pressing. In this respect aspects of militarism are analogous to those precautions against heat, vibration or matter in the wrong place which might be taken by any imaginative explosives expert. The aspects in question may be subsumed under the general, if faintly impolite, heading of 'bullshit'. So important is this curious phenomenon that it deserves a section to itself.

* In the first six months of 1971 more than a hundred American officers were 'fragged' by their men. According to one authority, the word 'frag' 'derives from the ordinary fragmentation grenade which troops use to booby-trap—and maim or kill—officers and non-coms who are too keen to engage in combat'.[6]

16

'Bullshit'

'I have been in the Army for nearly five years and I cannot see how polishing brass, floors, and anything else the N.C.O. thinks of, makes a man of you, nor jumping to attention all the time, and marching round like a load of chorus girls and asking permission to go to the toilet. The truth is that when you join the Army you give up your freedom, both physical and mental, you are just to obey orders.' A REGULAR SOLDIER

Why one of the more striking features of militarism should be associated with bovine excreta remains a matter for debate. Certainly it accords with the principle of disowning certain sorts of behaviour by associating them with some other species or nationality. In this respect 'bull' shares the fate of 'Dutch' courage, 'French' leave and 'Swedish' massage.

According to Eric Partridge, the word was coined by Australian soldiers in 1916. Coming from a country whose armed forces have always been relatively free from this element of militarism, they were evidently so struck by the excessive spit and polish of the British Army that they felt moved to give it a label. Going further back, it is possible that the expression has its origins in 'a bull', the false hairpiece worn by women between 1690 and 1770. This would be consistent with the fact that modern dictionaries define 'bull' as 'a ludicrous jest, a self-contradictory statement, to cheat, empty talk, absurd fussiness over dress'. Whatever its etymological significance, such definitions certainly capture the nature of military 'bull'—one of the most astonishing, apparently irrational and yet significant aspects of militarism, one which connotes an attitude of mind, a pattern of behaviour and an end-product.

As implied by the old jingle,

> If it moves, salute it.
> If it doesn't move, pick it up.
> If you can't pick it up, paint it!

the phenomenon involves ritualistic observance of the dominance–submission relationships of the military hierarchy, extreme orderliness and a preoccupation with outward appearances. In this latter respect it is the extension of a commonplace tendency in most human societies — that of taking outward show as the criterion according to which most judgments are made. This reliance on externals and constant urge to 'keep up appearances' may well have its origins in three features of very early childhood. The first is that perceptual and sensory processes mature sooner than those underlying the capacity to think. The second is that early unlearned responses, so-called instinctual patterns of behaviour, are set in motion by specific features of the sensory impression and do not require anything in the nature of an intervening thought. And the third is that children adopt the values and attitudes of their parents, who themselves set great store by appearances. Many people even choose their mates because of what they happen to look like. Huge industries are geared to the sustaining of this particular source of possible deception, with results which may well end in the very antithesis of connubial bliss.

In the military, manifestations of 'bull' range from such minor apparent absurdities as the polishing of the backs of cap badges, through the blancoing of trees for a forthcoming general's inspection, to such grandiose schemes as the décor of Speer's Reich Chancellery. For less pleasant manifestations there are those drills and uniforms which have plagued the life of soldiers and, from time to time, inflicted rather more suffering upon them than the enemy. Classic examples are the stock, a high leather collar which held the head like a vice, and the queue, a stiff pigtail for which the scalp had to be dragged back so far as to prevent a man from closing his eyes.

Besides its emphasis on appearances and its constraining aspects, 'bull' also involves a compulsive concern with cleanliness. In this respect alone it may achieve impressive levels of irrationality. To make it white, webbing equipment may be boiled almost to the point of destruction, while the blankets that the owner sleeps in stay unwashed for weeks.

There are, of course, good arguments for 'bull': that it ensures a level of orderliness, cleanliness, discipline, personal pride, obedience and morale which, so it seems, could not be reached by any other means, i.e., by reasoned as opposed to compulsive behaviour. By the same token it achieves a level of uniformity that makes for solidarity and group cohesiveness.

However, the case against it is also strong. It is time-wasting, excruciatingly boring for all those with more than the most mediocre intellect, and a poor substitute for thought. Since it aims to govern behaviour by a set of rules and defines a rigid programme for different occasions, it cannot meet the unanticipated event. This may have fatal consequences, such as the possibility of admirals standing stiffly to attention, hands at the salute, while their battleships sink slowly under them. But for 'bull' they might have done something rather more useful in their last remaining moments of buoyancy.

Like any compulsive symptom, 'bull' and its close cousins, ritual, dogma and superstition, have put themselves so far beyond reasoned thought that they create resistance to change and the acceptance of new ideas. Take military drill. This starts as a skill adapted to a reality-situation. It develops into a rigid pattern of behaviour which, by becoming automatic, takes the load off memory. Once learned, it is directed by processes of which we are scarcely conscious, and which leave the limited channel-capacity of conscious experience mercifully free to deal with other and more pressing events. It is drill, in such a sense, which ensures that most motorists let off the handbrake before engaging the clutch, and that most speakers construct their sentences according to the rules of language.

Military drills started in this way. They were devices which could eventually weld together a heterogeneous miscellany of uneducated peasants into a single corporate homogeneous machine that did as it was told.

This was all to the good but for one thing—ritualization, implying the tendency to transform means into ends. Thus the battle drill of one era becomes the ceremonial drill of another. What started as a functionally useful manœuvre becomes a highly stereotyped pattern of movements on the barrack square. In itself this may be no bad thing. Ceremonial can be pleasing to the eye, an anodyne for tax-payers and even, on occasions, a device for raising charitable funds. But unfortunately ceremonial drill, like other forms of 'bull', is addictive and, by being so, usurps the time and energy which should be devoted to other more adaptive pastimes. It then becomes a substitute for doing something else, as when the conservative element in the Brigade of Guards resisted the adopting of the new battle drill *because* it would interfere with their existing ceremonial procedures. When it is considered that it was this same new drill which was being studied by the German General Staff and the Russian high command in preparation for the last war, one can appreciate the price that may be paid when, to use a military expression, 'bullshit baffles brains'!

As a factor in fighting efficiency, 'bull' has also been unhelpful in the Navy. If we assume that one of the main purposes of a navy is to defeat the enemy, and that this, in the past anyway, was achieved by shell fire, it might be supposed that much time would have been spent on practising gunnery. But in the British Navy in the years before the First World War, ship commanders were actively discouraged from gunnery practice because the smoke might mark the paintwork and soil the gleaming decks. The price for this was paid at Jutland.[1]

In parenthesis, it is no accident that 'bull' is so closely linked to conservatism, for its very nature is to prevent change, to impose a pattern upon material and upon behaviour, and to preserve the status quo whether it is that of shining brass or social structure. It is no accident that 'bull' in civilian life, that of the bowler hat, rolled umbrella and striped trouser, or that of the garden party, should flourish in those sections of society renowned for their conservatism.*

Now that we have touched upon some of the more obvious manifestations of the phenomenon, let us examine its deeper causes, and relevance to the central thesis of this book.

For a start, it seems to be a natural product of authoritarian, hierarchical organizations. Secondly, though its outward and visible signs are manifold, they have three common denominators. The first is *constraint*; the second, *deception*; and the third, *substitution for thought*. In a sense each follows from the others. It is essentially by constraint that 'bull' seems to combat disorderliness, whether this be of appearance, conduct or thought, but in so doing it necessarily conceals what is really the case. It is worth noting that this aspect of behaviour marks

---

* Conservatism with a small 'c' is not confined to the political Right; hence 'bull' is just as evident in some communist armies as it is in those of the West.

yet another point of similarity between the oldest profession, militarism, and the second oldest, prostitution.

However, where there are similarities there are also differences. While rouge and false hair, like blanco and busbies, are both concerned with outward show, the reasons for the deception are obviously very different. Whereas those aids to beauty affected by a woman of easy virtue aim to attract custom and stimulate desire, those of the soldier reflect a more complex set of motives.

As pointed out earlier (page 170), fighting, like sexual behaviour, is an instinctual activity, and as such prone to control by what ethologists call sign stimuli—particular shapes, colours or patterns of response which are specific to all members (of one or both sexes) of any given species. Thus all robins have red breasts, and all herring gulls have red spots upon their beaks. Such distinctive labels serve a simple communicative function, in as much as their perception by another member of the same species automatically *releases* instinctual behaviour appropriate to the sex of the percipient and the situation in which it finds itself. For example, the red breast of the robin elicits aggressive behaviour in another male, but sexual response in a female. For animals with small brains, and little capacity for learning or judgment, the possession of these simple labels is obviously of immense value. They are fast, certain, automatic in their function, and require no past experience of the situation in which they operate. The disadvantages of instinctual behaviour are that it is inflexible, undiscerning and by no means foolproof. Even that Lothario of the village pond, the male stickleback, will attempt to mate with a block of wood having a protuberance on its lower side in preference to a pregnant female—however startling her pulchritude—whose distended egg sack (the sign stimulus) happens to be concealed.

Before considering the likely relevance of these phenomena to 'bull', there is one final point. It has been shown with lower species that *supernormal*, or larger than life, sign stimuli have a *greater* capacity for releasing the appropriate instinctual behaviour than have the normal, naturally occurring versions of the same stimuli.

Now if we assume that even humans are not entirely immune to the effects of such sign stimuli—as, say, bared teeth which may release a fear response, or the shape of a baby's face which releases maternal behaviour, or the contours of the female form which release sexual responses in the male—then it is reasonable to regard certain forms of 'bull' as the deliberate setting up of supernormal sign stimuli. The prostitute who pads out her bosom, or applies rouge to her cheeks, has perhaps something in common with soldiers who don tall hats and scarlet jackets. In both cases the little extra aims to elicit desired responses (lust in one case, fear in the other) *of a greater magnitude than would have occurred without these prosthetic extensions.*

Take the case of threatening postures. Like the anthropoid apes, we threaten by rotating our arms inwards and raising our shoulders. With apes this response has the effect of lifting the hair on their shoulders, thereby making the animal look more than usually fearsome. Denied the joy of owning hairy shoulders, modern man discovered 'padding' and the epaulette. But these 'supernormal stimuli' are over-determined. Not only do they intimidate, they also decorate and flatter. Not only do they threaten the enemy, they also boost the ego of the wearer. As well as making him look tough, they make him *feel* tough; and not only do they subserve aggression, they also serve sexual needs. Nelson's immediate dislike of two French naval captains whom he found wearing epaulettes bears witness to the sexual symbolism of these over-determined ornaments. He wrote: 'They wore fine epaulettes for which I think them coxcombs. I shall not court their acquaintance.'[2] His choice of words, coupled with his own predisposition towards making amorous conquests, suggests that he at least saw the epaulettes as a competitive sexual display.

A similar response to an over-determined aggressive-cum-sexual display has been noted by E. S. Turner when writing of British officers in the Peninsular War. 'The regimental officers did not lack subjects for scorn. They despised Spanish officers who made their horses prance and caracole before ladies.'[3]

For a more modern example there is the provocative arrangement of one long cylindrical parachute mine flanked by two red painted spherical sea mines which used to decorate the lawn outside the headquarters of a Royal Engineers Bomb Disposal Unit during the last war.

Another over-determined piece of 'bull' is the military salute. According to one authority, the origin of this gesture was the simple act of lifting the visor of one's helmet, thereby putting oneself at the mercy of an opponent. As such it has much in common with those appeasement responses in lower species which involve presenting vulnerable parts of the anatomy to a potential enemy. But with man, this straightforward communication grew into a highly over-determined piece of ritual. Thus the military salute effectively combines the threat of the raised arm with the reassurance of the open hand in which no weapon could be concealed. A similar admixture of contrary motives occurs in the practice of presenting arms. The trigger is turned towards the 'enemy', but the gun remains firmly in the possession of the presenter.

While the instinctual origins and over-determined nature of these sorts of behaviour undoubtedly contribute towards their tenacity, there are other factors rather more important. These can be subsumed under the general heading of anxiety-reduction.

Perhaps the single most important feature of 'bull' is its capacity to allay anxiety. There are two components to this function, one conscious and rational and the other unconscious and compulsive. Both operate to reduce two sorts of anxiety, the first social and the second instinctual.

Since, at a conscious rational level, orderliness, cleanliness, punctuality and discipline clearly make for efficiency, the knowledge that one belongs to an organization which puts a premium on these laudable traits, that one's rifle will fire and there is a key for the bully beef tin, obviously makes for confidence. At a conscious rational level, therefore, even those aspects of 'bull' which reflect the grossest exaggeration of these traits must seem like steps in the right direction. This confidence may, of course, be misplaced. That a commander insists upon meticulous attention to detail, down to the last shining button, is no guarantee that his strategical thinking is anything other than puerile. Indeed, he could well be unwittingly substituting a lesser for a more important area of generalship. Nevertheless, there are good grounds for believing that those situations in which 'bull' flourishes are ones in which it reduces anxiety *because* orderliness is fairly vital to survival.

Again, the imposed uniformity which is part and parcel of 'bull' obviously makes for group cohesiveness, and that 'we're-all-in-it-together' feeling which combats fear. We must suppose, too, that the heightened conformity which it imposes will, like other forms of perceived conformity, encourage people, through a diffusion of responsibility, to perform acts which they might otherwise avoid.

Yet another useful feature of 'bull', so it has been said, is its role as a distractor and time-filler. According to this theory, a mind preoccupied with buttons and toecaps has little room for gloomy forebodings. The point is well made by A. B. Campbell when writing of naval customs: '... it is the guiding principle of naval service that the ship's company should be constantly employed, and this is the reason—apart from the necessity for scrupulous cleanliness—why there is so much scrubbing of decks and polishing of bright work.'[4]

In the same context this writer compares naval and civilian routine. 'It is safe to say that in many shore jobs routine destroys initiative—this also applies to many factory workers, *but it is not so in the Navy* [where] a routine job builds up his [a bluejacket's] character.'[5] As to why naval and civilian characters should require such diametrically different treatments, Campbell refers to the moments of danger which occur for the former but not the latter. This of course begs several questions. It confuses loss of initiative and blind obedience with the building of character, and makes the unwarranted assumption that naval ratings face greater danger than many civilians, including merchant seamen, steeplejacks, racing motorists, mountain-climbers, single-handed yachtsmen, coal-miners and matadors, not one of whom has to fortify his character by polishing brass or scrubbing wood.

It would perhaps be truer to say that since the imposing of 'bull' upon troops serves to reduce initiative, it will thereby increase the feeling of dependency which they have towards their superiors. This in turn will increase their obedience and loyalty.

Finally, at a conscious rational level there are aspects of 'bull' which may well help to combat social anxieties in military men. Gorgeous uniforms, martial music, prancing horses, and even being saluted, are obviously balm to tender egos and, by promoting soldierly pride, do much to offset the hostility and ridicule to which the military are from time to time subjected by those in other walks of life.

But there is another, less obvious reason for 'bull', namely that it serves to reduce deeper-seated feelings of anxiety which may well have their origins in events, unrelated to the here and now, of which the subject remains blissfully unaware. The response in this case has about it an immediacy which is clearly not the product of conscious deliberation. The most extreme examples of this phenomenon occur in obsessive-compulsive neurosis, a condition in which the patient feels compelled to follow a pattern of ritualistic thoughts and acts. That these often include such bizarre symptoms as compulsive hand-washing, a preoccupation with timing and counting, recurrent ruminative ideas, stereotyped verbal utterings, and *always standing with one's toes absolutely in line*,[6] has obvious significance for more military versions of the malaise.* One underlying feature of such symptoms is that they are repetitive, stereotyped and occur without insight into their origins. Another is that they centre around cleanliness and orderliness. Finally, they are often defences against anxiety or suppressed anger. This is clear from the great distress which may be occasioned by their forcible prevention.

Such symptoms are not, of course, confined to the chronic sick. Milder forms may well occur in the normal population during times of stress. Bead-counting, foot-tapping and the mouthing of dogma are, like the compulsion to make things clean and tidy during periods of menstruation, well-known palliatives for the stressed psyche.[7]

But why should compulsive ritual reduce anxiety, and what are these deeper anxieties?

Let us not beat about the bush. At the risk of offending those with delicate susceptibilities, or who themselves have problems in these areas, it must be said that they involve four matters of primary importance in every human life: sex, elimination, eating and death.

Take sex. There seem to be two main worries here, that of not being what one wants to be, and that of not being sure that one is what one is. As to the first, a psychiatrist colleague once remarked: 'All my patients have the same basic problem: the men all want to be women, and the women want to be men.' It seems that being potentially hermaphrodite leaves some people chronically dissatisfied with their particular position on the sexual continuum—a sort of 'grass is greener on the other side of the fence' feeling.

For those who do not nurse an unconscious urge to be of the opposite sex there is the other problem, equally worrisome, that of not being absolutely certain that they are what they think they are. Thus many men have serious doubts about their masculinity.

These 'triumphs' of our culture and methods of child-rearing are something to which we shall return in due course. Suffice it to say that they constitute just one of the major sources of anxiety which men carry to their graves. They may, moreover, constitute one of the factors which may motivate people towards perpetuating those

* Research has shown a correlation of 0·8 between obsessional traits and symptoms in neurotic soldiers.

aspects of 'bull' that could help to still these nagging doubts. Such trappings of aggressive masculinity as a three-foot sword or pair of pearl-handled revolvers dangling from the region of the crutch can hardly help but be reassuring for sufferers of this ilk. The fact that men with high levels of castration-anxiety (men whose early years were perhaps enlivened from time to time by the maternal threat: 'If you don't leave it alone, I'll chop it off') have been shown to entertain an exaggerated fear of death adds point to this conclusion for the military scene.[8]

But clearly the unique features of 'bull' reflect something more fundamental than defences against mere sexual anxieties. The greatest anxieties concern death and unconstrained disorder. Since the two are inextricably related, a defence against one is a defence against the other also. This is perhaps the crux of the origins of 'bull'.

Let us approach this from another standpoint. Whatever its particular form, 'bull' results in a state of affairs which is opposed to what many people would regard as a primary source of delight: the natural diversity of nature. Towards such diversity it is implacably hostile. It is no exaggeration to say that this aspect of militarism is dedicated to the ironing out of differences. The efficiency with which it destroys variety and imposes uniformity is matched only by its demand for conformity.

To many people such ends are anathema. Indeed, most civilized cultures put a premium upon their opposites. With an insouciance bordering on the reprehensible, they actually applaud individual differences. Whether in people, animals or plants, the variety of nature is regarded as a spiritual bonus. The inanimate and artificial are similarly regarded. In clothes, cars and houses, uniqueness has market value. Not for nothing does current advertising for 'the best car in the world' make only one specific claim—that no two Rolls Royces are alike. As for sex and food, only the most puritanical would deny that variety is the very essence of enjoyment.

But 'bull' inverts these values. It worships homogeneity and frowns on deviance. Whether it is toecaps, buttons or dressing by the left, hair-length, kit inspection or marching feet, the quintessence of perfection resides in conformity to a regulation pattern. This conformity is the product of constraint. Thus even conversation in an officers' mess was confined for many years to topics other than women, religion, politics, sex or talking shop, a state of affairs which once drew from *Punch* the acid comment that 'it would be very dreary indeed if officers were ever thrown back upon their conversation!'

It seems then that since 'bull' is primarily concerned with substituting pattern for randomness, it evidently reduces anxiety by the reduction of uncertainty. But why should the removal of uncertainty in trivial matters assuage anxiety for more important issues?

How is it that uniformity of dress, cleaning rituals and the predictability of exchanged compliments restore peace of mind in the enormous uncertainties of war? Why, in one of the most incompetent navies this world has ever known, when about to embark on a voyage to its ultimate destruction in the battle of Tsu-Shima, was it the case that 'Again and again we washed the gangways with soap and water, we scrubbed the bridges, touched up the paint, scoured the brass work. Engines and stokeholds were not forgotten ... cleanliness became a mania!'[9] (a sailor in the Russo-Japanese War of 1904)?

Anyone who doubts these soothing effects of 'bull' has only to consider two other situations of frightening uncertainty—marriage and death. Few who have played even a minor role in these events would deny the emotional support that comes from the time-honoured ritual of weddings and funerals.

By way of trying to explain these effects, two overlapping theories can be invoked. The first is that of entropy-reduction. This maintains that 'bull' exemplifies a general principle common to all organisms, that of combating randomness.

The argument is simple—living organisms are complex patterns which persist for a time within the essential disorder from which they came and to which they will, with equal certainty, return. To the notion that conception, life and death represent stages in this process we must add one other, namely that living is the process whereby the pattern endeavours to maintain itself in being. It is a truism that this applies equally at microcosmic and macrocosmic levels. Whether it is the single cell, the integrated systems of the total organism, or the external social order, there exist regulators, controls and constraints whose function it is to preserve the pattern, to keep this from that, to maintain purity and separateness. This holds as true for the biological processes as it does for the construction of an urban sewage system. By these lights, social and environmental stress, extremes of heat or cold, viruses, drugs or direct physical injury may be regarded as forces which, because they threaten the organism with a dissolution of its pattern, with mixing up of its constituent parts, call forth one or other of these self-preserving tendencies. Indeed, life can be construed as a fight for orderliness in the course of which much behaviour, both voluntary and involuntary, both external and internal, is directed to this end. The law, and rules for hygiene, prophylaxis, antibodies, rejection-mechanisms, adrenalin secretion and New Year resolutions are just some of the devices which aim to stem the perpetual drift towards disorder.

It is, of course, a losing battle. As Oscar Wilde said: 'Good intentions are useless attempts to meddle with the laws of nature.' One of these 'laws' is that ultimately the forces of dissolution increase beyond the capabilities of adaptive mechanisms to hold them in check, because the very processes whereby order is maintained (of which one is compulsive ritual) may themselves assume destructive proportions. The notion has great generality. Just one special case of this destructive outcome is embodied in Selye's concept of the General Adaptation Syndrome.[10] This refers to the fact that such internal bodily responses to stress as raised blood-pressure may, in the end, precipitate irreversible tissue-damage and death.

According then to this theory of entropy-reduction, 'bull' represents an extreme manifestation of a general and necessary propensity on the part of living systems to resist randomness. This would account for the fact that the sartorial aspects of the syndrome are concerned with removing dirt (matter in the wrong place), with maintaining separateness, with keeping green green and white white; with preserving the status quo—keeping hair short, brass shiny and rifles clean; and with maintaining uniformity by written orders, shouted commands and other behavioural constraints. But, like waking consciousness in contrast to the dream, and normality in contrast to psychosis, 'bull' makes its effect by constraint upon the 'creativity' of thought.

Obviously, the constricting, information-reducing aspects of 'bull' extend beyond the individual and his immediate possessions to embrace the total social scene, thereby preserving the hierarchy of rank and status, separating high from low, and delineating what is, from what is not, appropriate behaviour for every situation.

Just as the General Adaptation Syndrome is the body's response to the internal effects of stress, so 'bull' may be regarded as an organization's response to the threat of *its* disintegration. In the military this threat has two sources: the external enemy, and the aggressive impulses of its own members.* In either case, the greater the threat, the greater the constraints.

Thus the aforementioned etiquette of the officers' mess, which confined conversation to the utmost trivia, had its origins in the wholly rational avoidance of all topics which, in a profession liable to aggressive outbursts on the flimsiest pretext, might result in much wastage of life through duelling.

So much for a general theory. We come now to the second, more specific theory, which aims to explain individual differences in a propensity for 'bull', and the relationship between compulsive cleanliness and a particular sort of personality.

* Josephson has drawn attention to the fact that the supposed hostility of an apparently submissive subordinate in an authoritarian organization may result, perhaps without conscious intent, in behaviour that is actually destructive of the goals of the organization in question.[11]

17

Socialization and the
Anal Character

*'... a form of adaptation is thus achieved by narrowing and distorting
the environment until one's conduct appears adequate to it, rather
than by altering one's conduct and enlarging one's knowledge till
one can cope with the larger, real environment.'*

K. J. W. CRAIK, *The Nature of Explanation*

Line-shooting, deceiving with false appearances, covering up, compulsive cleaning and other mindless rituals are to be found the world over, and have presumably been so since Adam donned a fig leaf and Eve gave a first sly polish to the apple.

But people are not born that way. For several years they show appalling sincerity and an impressive disregard for all forms of cleanliness. They do not know the meaning of disgust, and are unmoved by disorder. On the contrary, they do their very best to exemplify the second law of thermodynamics that 'entropy always increases'.

Evidently, then, adult behaviour and its accompanying attitudes come through socialization. Let us look at this process in connection with the development of the so-called anal character.

The precise details of the process whereby babies—disorderly, demanding, self-indulgent and incontinent—are turned into clean, dry, socially responsible adults is still a matter for debate. According to psycho-analytic theory, between the ages of one and five years the fear of losing parental affection, together with the threat of other dire consequences, moves the child towards renunciation of old habits for some rather bleak new ones. The latter reflect the standards of the society into which he has been born.

Now those needs of a baby which result in the sorts of behaviour for which socialization is required are in fact, as Freud has pointed out, centred on three erogenous zones of the human body: the lips and mouth, the genitals and the anus. In babies, as in adults, stimulation of these areas is evidently pleasurable. This relationship between need and pleasure is hardly to be wondered at, in as much as it provides the essential motivation for the three vital activities of eating, elimination and reproduction. Species which did not enjoy these things, like people with anorexia nervosa, would have a poor chance of survival. In its raw form, however, such enjoyment is hardly compatible with the ethos of adult society, which, in demanding some control of basic drives, attempts to curb their free expression. So begins the slow process of socialization in early childhood. It is a wearing time for one and all. Normally, however, and against apparently formidable odds, this hard-fought campaign draws to a satisfactory conclusion, with parent and child winning a harmonious victory over the dark forces of disorder.

I say 'normally', for sometimes, so it has been suggested, there occurs a concatenation of factors which results in lasting effects of great relevance to the subject-matter of this chapter. They include an unduly strong attachment by the child to the pleasure it derives from its erogenous zones, an unduly strong distaste on the parent's part for manifestations of the child's underlying drives, and, as a consequence, the implementing of an unduly strict training programme. When these three factors are operating, the resulting situation, which approximates to that of an irresistible force pitted against an immovable object, probably reaches its climax in the period of pot (or, as some prefer to call it, 'toilet') training. The nature and outcome of this process may be summarized as follows: the small child obtains considerable pleasure from its bowel movements, but when this pleasure is tempered by anxiety as a result of a harsh training schedule, the usual result is reaction-formation ... 'In extreme cases he becomes parsimonious, stingy, meticulous, punctual, tied down with petty self-restraints. Everything that is free, uncontrolled, spontaneous is dangerous.'[1]*

* According to Kline's excellent review of research in this area,[2] there is considerable support for this constellation of personality traits, and some for their origin in anal eroticism. Methodological difficulties have so far precluded any clear picture of their precise relationship to toilet-training. See also Beloff,[3] and Beech.[4]

In other words, the individual resolves his conflict by developing character-traits which are the exact opposite of those he has had to renounce.

Now it does not need any vast stretch of the imagination to see more than a passing similarity between these obsessive traits and the practice of 'bull'. Both are ritualistic, concerned with cleanliness and orderliness, and designed to hold down, and then cover up, impulses of a totally opposite kind. It scarcely needs adding that the latter half of the word 'bullshit' takes on a new significance in the light of this comparison. The closeness of the relationship between these events of early childhood and aspects of militarism is, moreover, conveyed by two other considerations.

The first concerns the matter of aggression. Socialization necessarily involves frustration, and frustration is, as we know, one of the main instigators of aggression; were it not, the species would once again have a poor chance of survival. But the sort of parents who cause their offspring to develop obsessive rituals against dirt are also likely to be those who cannot tolerate any show of aggression. So this too has to be suppressed or rather replaced by a safer, symbolic, outlet. One such, in humans as in lower animals, is ritualistic behaviour. Just as the male stickleback who encounters a threatening opponent on the borders of his territory 'displaces' his aggression into the ritualistic punching of holes in the bed of the stream, so some humans, made anxious by their own aggressive impulses, find relief in such ritualistic acts as say drumming with the fingers, counting or putting things in order. Hence, so-called obsessive traits may be regarded as defences not only against dirt, but also against aggression — the aggression which originally arose through frustration of infantile desires.* But 'bull' also has a two-pronged purpose: to combat dirt and to prevent illegitimate outbursts of aggression (aggression, that is, towards superiors — the frustrating and potentially dangerous 'parent-figures').

One last connection between 'bull' and obsessional-compulsive symptoms is their tenacity and proliferation. Thus the individual whose compulsive hand-washing increases from ten to fifty times a day has something in common with the devotee of 'bull' whose life becomes increasingly occupied with unrealistic and anachronistic

* Further support for this contention comes from the finding[5] that people with high scores on a test of anality also show high levels of political aggression.

extensions of what were originally quite rational pieces of behaviour.

At this stage in the argument it is necessary to issue a caution. We are not saying that military organizations are hotbeds of obsessional neurosis, nor that those given to 'bull' are necessarily manifesting compulsive symptoms. On the contrary, all that we have tried to show is that the anxiety-reducing, aggression-controlling and tenacious nature of 'bull' becomes at least partly explicable in terms of two non-mutually exclusive theories. As to the second theory, the ontogeny of socialization is no more than a special, learned instance of the first, more general, principle that life depends upon the preservation of a minimum level of orderliness.*

There are four corollaries. Firstly, we would expect a complementary relationship between perceived threat of destruction and the occurrence of compensatory devices to preserve orderliness. In this connection research has not only shown that physiological arousal is *decreased* by ritual, but also that, under threatening conditions, normal individuals behave like compulsive neurotics.[7] Secondly, these compensatory devices or patterns of behaviour might be expected to involve compulsive cleanliness and strict observance of a dominance–submission relationship because the threat of disorder which they are (unconsciously) designed to meet activates a much earlier threat of being overtaken by the forces of disorder and aggression, a threat which is overcome by cleanliness and obedience. Thirdly, since the original causes of these reactions to threat are lost to consciousness, the resulting behaviour tends to resist rational modification. Fourthly, since military organizations represent, par excellence, outlets for and consequently defences against aggression and disorder, they will tend to attract people who have some difficulty in reconciling these conflicting needs, people who overvalue aggression, order and obedience. This conclusion is supported by the finding that patients suffering from obsessional neurosis show *improvement* during military service.[8]

* The nature of these tendencies and their potential usefulness can be illustrated in many ways. The following instance exemplifies an extreme form of the sorts of behaviour to which they might give rise.

It concerns a military commander whose martial interests centred round the design and siting of latrines. This necessary, albeit tiny, segment of warfare absorbed his mind to the exclusion of all else. He lived, worked and slept latrines, and when a new one had been built, insisted on 'christening' his brain-child under the steady gaze of his troops.[6]

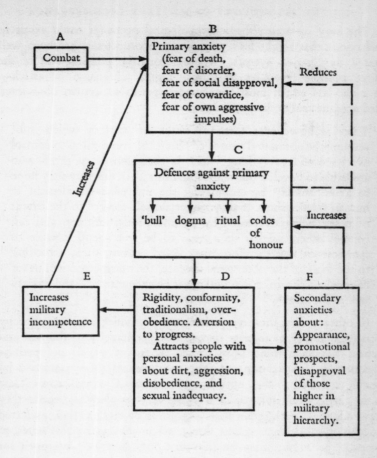

Figure 3. *The role of 'bull' in military incompetence*

Combat (A) produces several sorts of anxiety (B). To reduce these anxieties (and increase efficiency), aspects of militarism (C) are developed. These reduce primary anxiety (the 'dashed' arrow). But defences against primary anxiety (C) necessarily make for rigidity of thinking, etc. (D). They will also tend to attract individuals with personal anxieties about dirt and aggression.

Both aspects of D may be expected to have two adverse effects. Firstly, they will directly reduce military competence (E), thereby increasing primary anxiety in the combat situation. Secondly, they will evoke a number of secondary (social) anxieties (F). Both these effects will tend to increase D, thereby constituting a vicious circle of cause and effect.

The way in which these psychological processes could result in behaviour that might lead to military incompetence has been well stated by Charles Rycroft. Arguing from the position that of the three possible responses to threat—flight, submission or attack—it is the last which most closely corresponds to human obsessional defences, he makes the point that

> There is however a form of attack or mastery which must certainly be accounted neurotic. This is the compulsion to control everyone and everything which is characteristic of those who are liable to develop obsessional neurosis ... In this way they hope to avoid anxiety by eliminating the unpredictable element in human relationships. If they can attain self-control to the extent of never being overcome by an unexpected emotion and can control others so that they cease to be free agents capable of spontaneous and therefore unexpected actions, then, according to the logic of the obsessional defence, the unexpected will never happen and the unknown will never be encountered—and anxiety will never arise.[9]

In other words, those very characteristics which are demanded by war—the ability to tolerate uncertainty, spontaneity of thought and action, having a mind open to the receipt of novel, and perhaps threatening, information—are the antitheses of those possessed by people attracted to the controls, and orderliness, of militarism. Here is the germ of a terrible paradox. Those very people who, because they have adopted attack rather than submission or flight as their preferred psychological defence against threat, are in theory the best suited to warring behaviour, may be the very ones *least* well equipped for other components of successful fighting.

Considered in this light it is a remarkable testimony to biological efficiency that so many military leaders have performed so well.

There is another and related matter to which we must turn our attention—the vexed question of military honour. As a device for maintaining orderliness, quelling anxiety and directing aggression into appropriate channels, honour is to officers what bullshit is to non-commissioned ranks. Whereas 'bull' comprises an array of relatively mindless acts, honour is more concerned with a system of ideas, a code of thought and a set of inhibitions. As can be seen from Figure 3, both are apparently inescapable products of large-scale organized aggression. However, to appreciate the role that 'character' and one of its offshoots, military codes of honour, have played in certain sorts of military incompetence we must first examine their more general psychological significance.

Note on Chapter 17

The author is only too well aware that to suggest that a general's personality may (like anyone else's) bear the hallmark of his 'potty-training' reduces some people to nervous giggles. This being so, it may be helpful to consider the following propositions:

1. A person's psychological make-up is the resultant of only two factors—his genetic inheritance and his life experiences.

2. Life experiences are *most* influential during periods of *greatest* plasticity in development, i.e., before the age of five.

3. Those experiences which are likely to have the greatest lasting effect will be those which make the biggest impact at the time of their occurrence.

4. Before the age of five the most important of these are concerned with socialization.

5. A large part of socialization during this time is concerned with toilet-training.*

To the author, the logic of these simple propositions seems unassailable!

* Notwithstanding, Orlansky has produced some cogent arguments that development of the obsessive personality depends less upon the vicissitudes of toilet training than upon a wide variety of cultural situations. Likewise, Vowles cites evidence for the view that aggressive behaviour is *reactive* rather than appetitive.

18

Character and Honour

'Why should a man be in love with his fetters, though of gold.'
BACON, *Essays: Of Death*

'Moderation in war is imbecility.' MACAULAY

Whatever their profession, most people would subscribe quite happily
to the notion that there are three components to the human psyche—
the *instinctual*, the *intellectual* and the *moral*—and that they develop in
that order. Some physiologically minded psychologists would even go
so far as to relate this tripartite organization to three general regions of
the brain, with instinct being rooted in the activities of the old brain
(suitably sited in the so-called basal areas), intellect a product of the
cerebral cortex, and moral qualities depending upon a proper contribution from higher centres in the frontal forebrain. This crude
topography accords with the effect of lesions in the various different
regions (e.g., damage to the frontal cortex predisposes towards a loss
of moral values) and also with the facts of evolution. Thus, lower
animals who may be strong on aggression and sexually competent,
but devoid of intellect and conscience, have lower brain areas much
like ours but lack the massive human forebrain with its convoluted
cortex.

For those who shrink from the idea that their strengths and weaknesses can be reduced to such anatomical proportions a more acceptable
description can be found in the concepts of psycho-analytic theory.
According to *this* tripartite view we start life with an id—importunate,
randy and aggressive; acquire an ego—intelligent, perceptive and
diplomatic; and then, if all goes well, become blessed (or saddled)
with a superego—the source of conscience and moral imperatives. The

functional relationships between these three components suggest a view of man which, according to one description, seems particularly appropriate in the context of this book. 'Man is basically a battlefield ... a dark cellar in which a well-bred spinster lady and a sex-crazed monkey are for ever engaged in mortal combat, the struggle being refereed by a rather nervous bank clerk.'[1]

These preliminaries are not irrelevant to our purpose, for along with all the other psychological problems which beset those whose business is organized violence is that of deciding whether to plump for intellect or character as the means whereby instinct is controlled and discipline maintained. Generally speaking, the older military organizations have opted for character and the younger ones for intellect.

In the context of militarism the forces of conscience and of character manifest themselves in various guises: in medieval notions of chivalry, in codes of honour, such as the duel, and in the belief that officers must of necessity be gentlemen. As Karl Demeter has shown in his history of the German officer corps, these notions of honour and chivalry brought about and were themselves reinforced by a care to select officer material from the aristocracy and rural landowners—a state of affairs reflected in the contrasts of snobbishness, exclusiveness, sense of honour and lack of intellectual ability which obtained between the officer corps, drawn from the aristocratic junker families on the great estates of Prussia, and those more bourgeois elements from the industrialized south-west of Germany.

A code of honour is a set of rules for behaviour. The rules are observed because to break them provokes the distressing emotions of guilt or shame. Whereas guilt is a product of knowing that one has transgressed and therefore might be found out, shame results from actually being found out—in military circles traditionally the greater crime!

It is usually assumed that military codes of honour serve to reduce fear. This may well be so. Their primary object, however, is to combat not so much fear as the sort of behaviour to which fear might otherwise give rise. In other words, they are designed to ensure that threatening situations are met by fight rather than flight. They do this by making the social consequences of flight rather more unpleasant than the physical consequences of fight. Whereas the latter might lead to physical pain, mutilation and death, the former eventuates with far greater certainty in personal guilt and public shame.

When a soldier in action sees his life in immediate danger ... even the bravest will be seized by a moment of fear. Biologically speaking, fear is the natural reflex-sensation of the instinct of self-preservation which dwells within every man, heroes included. If victory is to be won, this elementary physical sensation must somehow be artificially suppressed—over-compensated by a contrary reflex of a psychic and moral kind, converted into action. The negative content of this counter-reflex is the feeling of shame. 'If', it says, 'you don't stand fast now but run away, the others will laugh at you and despise you.' ...

A soldier must [therefore] be provided—unless Nature has done the job already—with a set of automatic inhibitions that will save him in the moment of danger ... from a collapse of his own morale. Discipline, of course, can hold him steady from without; but his one moral defence against internal weakness is the sense of honour. To arouse this sense in the ordinary soldier, cultivate it and, above all, inspire it by his own example is the officer's highest duty; and to fulfil that duty he must himself have a sense of honour that is well developed, active and finely tuned.[2]

From a psychological standpoint these views are unexceptionable, and from a moral point of view highly desirable; yet the matter needs to be pursued further.

To the extent that a code of honour is reflexive, in the sense used by Demeter, so it is inflexible, thereby leading on occasions to behaviour that is so irrational as to border on the absurd. Just how absurd can be seen from the needless waste of good officer material that occurred through the custom of duelling. So damaging was this practice in the German corps of officers that the 1688 Edict of Elector Frederick II made duelling punishable by death.* In the British Army challenges to duels were still being issued as late as 1880.

While duelling was responsible for eliminating just those officers whose sense of personal honour and physical courage was of the highest order, other aspects of military honour could be just as irrational in their destruction of those who might well not be such whole-hearted parties to the code. The following example also illustrates the juxtaposition of 'bull' and 'honour'. The year was 1755.

General Braddock set out with his two regiments—the 44th and

* The first Prussian law against duelling was dated 1652.

the 48th—and 600 irregulars on a march to Fort Duquesne. About nine miles from it he was ambushed by Indians led by French officers. The result was disastrous. The men in their scarlet uniforms and white spatterdashes, marching in columns, were the sort of target an ambush force dreams of. Helpless because they could not see their enemies, some of the British troops broke for cover and fired from behind trees. This appalled Braddock and his officers; they considered skulking behind trees both undisciplined and unsoldierly. So they drove the Tommies back into columns, where of course they were butchered ... The whole episode was glaring proof that neither leaders nor the system under which they operated were worthy of the troops they used.[3]

But there are other more insidious stresses and strains to which honour can give rise. Since honour may be both personal and collective, the two codes may be in conflict. Thus the pagan origins of the medieval code of chivalry* which sets the ethic of killing and battle above all else gave rise to a collective honour in direct opposition to personal creeds based upon the teaching of Christ. In other words, the knowledge that one is a Christian, and therefore bound to the injunction 'thou shalt not kill', and the knowledge that one is a member of a group that is even more forcibly bound to the injunction 'thou shalt kill' are, to say the least, dissonant cognitions, and therefore productive of stress. For every conscientious objector there must be many whose participation in lethal activities cannot be quite so wholehearted as some would wish. Another difficulty with behaviour directed solely by a sense of honour is that, if its incentive is no more than an avoidance of shame, the resultant behaviour may be irrational, and the very strictness of the code have quite unforeseen consequences for the military way. So unthinkable was it that Japanese soldiers would ever surrender to the enemy that they were not instructed as to how they should comport themselves if they did. As a consequence Japanese P.O.W.s were a relatively fruitful source of information for Allied interrogators.[4]

A code of honour may be likened to an endlessly prolonged initiation rite. So long as the individual accepts its demands he is proving himself

* It could be argued that those features of medieval chivalry described by Huizinga, such as the concern to rescue virgins while they were still intact, were reactions against the violence and uncertainties of those days—yet another example of those responses to threatened disorder mentioned in Chapter 16.

manly and brave, a rightful member of the elite. Moreover, the tougher the initiation, the greater will be his liking for the group,[5] and the more will his fears regarding personal adequacy, virility and courage be stilled. It is yet another aspect of militarism which will attract those who seek assurance on these counts.

Furthermore, since a high code of honour has tended to be associated with wealth and position, belonging to a group which sets great store by honour confers the label of social superiority. Just how important this can be is attested by the vigour with which military castes have resisted entry by individuals from humbler backgrounds. Indeed, from the sometimes erroneous belief that a capacity for honour characterizes gentlemen, it is only too easy to draw the unjustified conclusion that anyone who is not a gentleman must lack this capacity — hence the view, noted earlier, that honour is to commissioned ranks what bullshit and punishment are to other ranks. In fact, there is a considerable overlap. Privates are not without honour, and field-marshals are no strangers to 'bull'. Moreover, since 'bull' can be a concrete manifestation of honour, so honour may beget 'bull' which, in turn, reinforces honour. An ingredient of this escalation is vanity. Thus in the Victorian Army it was a point of honour for commanding officers to try and outdo each other, not only in the splendour of their own accoutrements, but also in those of their troops:

> The 11th Hussars were superb. They wore overalls [trousers] of cherry colour, jackets of royal blue edged with gold, furred pelisses, short coats worn as capes, glittering with bullion braid and gold lace, high fur hats adorned with brilliant plumes … this gorgeousness was largely achieved at Lord Cardigan's expense. It is estimated he spent £10,000 a year on the 11th out of his private income.[6]

Lord Cardigan was not unique in lavishing an attention upon the appearance of his troops which, in a more enlightened age, might be reserved for the ladies of the Miss World competition. Nor was he unique in being a senior soldier who combined an exquisite sense of honour with overpowering vanity, and a renown for skill in duelling with an almost total lack of intellect.

Reference to Cardigan naturally brings to mind another facet of some military organizations which is closely related to honour—

their unrepentant snobbishness. According to the dictionary, a snob is 'a vulgar person who apes gentility or truckles to those of higher rank and position, or regards the claims of wealth and position with an exaggerated and contemptible respect'. Very simply, a snob is one who is impressed by, and therefore tries to identify with, those who are higher up the socio-economic scale, while straining to dissociate himself from those lower down. By these lights, such everyday affectations as name-dropping and paying society magazines to publish photographs of oneself or one's nearest family are obvious examples of snobbishness. Less immediately explicable, however, is much apparently snobbish behaviour in military circles. Why, for example, should Lord Cardigan, who had no reason to be snobbish, behave in such a way? Though extremely rich, with blood sufficiently blue by any standards, his notorious hounding of those 'socially inferior' Indian Army officers who had the misfortune to be in his regiment betrayed a streak of snobbishness bordering on the vulgar. Here was a man, with neither of those inferiorities which are traditionally supposed to underlie this unpleasant trait, displaying a form of it as virulent as that of any jumped-up nouveau riche.

The first and most immediate reason for this kind of behaviour would seem to be that some military organizations even to the present day actually cultivate the psychology of snobbishness as a substitute for merit. Higher ranks are encouraged to regard lower ranks as socially inferior.

To aver that such archaic phenomena as the social exclusiveness of particular regiments, and the tendency for even the wives of higher-ranking soldiers to feel uncomfortable if forced to share a position with those whose husbands are of lowlier military standing, are the result of ancient and once useful practices hardly explains their tenacity.*

One has only to imagine the chaos in society which would result if all great organizations confused merit with class, and tried to reinforce positions on the working hierarchy by a system of snobbery

* An amusing example of the relationship between anxiety and military status has been recounted by Noel Barber. It seems that during the fall of Singapore Rob Scott, a civilian member of the Government War Council, took to attending council meetings in his Local Defence Corps uniform. This, however, proved too much for the general, admiral and air marshal members of the council, who were so embarrassed at having to share a table with a 'corporal' that they felt moved to issue 'a mild tick off'.⁷

to realize that there must be something quite special about a military milieu which actually encourages such ways of thinking. To find out what this something may be, let us go back to first principles.

As a general rule, snobbish behaviour betokens some underlying feeling of inferiority. It is a common characteristic of the social climber, of the individual with low self-esteem, of the person who feels threatened or persecuted because of some real or imagined inadequacy. That there *is* an underlying pathology to the condition seems fairly obvious for two reasons. Firstly, those who are emotionally secure are rarely snobbish. Secondly, the behaviour is itself irrational, compulsive and self-defeating. After all, even the most hardened snob must know that other people are adept at seeing through his affectations. There is nothing, for example, quite so transparent as name-dropping or displaying invitations. He must know at some level that his behaviour provokes at best amusement, at worst ridicule, contempt or even dislike, but he is none the less powerless to curb his snobbishness. Something drives him on.

But why should the military be snobbish?

There are several obvious reasons and some not so obvious. Firstly, because, traditionally, top levels of the military hierarchy were occupied by the rich and highly born, the notions of socio-economic and military status became indissolubly related. Since social status determined military status, so, in time, military status became spuriously equated with social status. While this undoubtedly provided the structure for snobbish behaviour, it does not, of course, account for the underlying motivation. For the latter we have to examine a second reason: the anomalous position which the military hold in society, the plain fact that they are both loved and hated, admired and despised. This ambivalence, which ranges from awed fascination to cold dislike and which, as Kipling pointed out, fluctuates wildly from peace to war, has both conscious and unconscious components. At a *conscious* level society admires bravery, enjoys pomp, is grateful for protection and proud of conquest, while at the same time disliking the authoritarianism, potential threat and enormous cost of military organizations. At an *unconscious* level many people undoubtedly project on to military organizations their own internalized conflicts over aggression, for it is at once fascinating and abhorrent to see others indulging in (and getting away with) behaviour tabooed within oneself. The popularity of books and films dealing with war and violence (particularly evident after a prolonged period of peace), like that for pornography following an age of sexual repression, attests to the pleasure provided by vicarious satisfaction of hitherto frustrated drives. But the breaking of taboos is also a threat to those internalized defences against one's own instinctual impulses—hence the ambivalence.

To this one must add that since civilian populations pay a stiff price for their military organizations, they will quite naturally expect value for their money and be critical of incompetence. As General Gordon Bennett wrote: 'Civilians provide the manpower for our huge armies. Parents provide sons who fight. They make sacrifices, enormous sacrifices for the cause. Wives lose their husbands, children lose their fathers, families lose their breadwinners. They go short of food, clothing and the comforts they are used to. They pay the heavy taxes required to finance our war effort ... [hence] when they know that serious mistakes have been made they want to know why. After all they pay the cost of these mistakes.'[8]

Under the circumstances it would be very surprising if some awareness of these truths did not influence the military, leaving them, to say the least, vaguely apprehensive if not downright defensive.

But they are also stressed from within. To know that they have wedded their lives to essentially destructive ends, that they shoulder great responsibilities, that they may be called upon to carry out tasks far beyond their capabilities, and that the price of failure is enormous, is quite sufficient to initiate feelings of uneasiness. Even notions of retirement are fraught with stress. The knowledge that most ex-officers have little value on the civilian labour market, that their lot is the total obscurity of genteel poverty, that only the very best and very worst of full generals and above are likely to achieve immortality, and that none of them will ever again command the absolute obedience to which they have grown accustomed, can hardly be described as reassuring.*

To the factors underlying the self-protective and compensating aspects of snobbishness must be added what is perhaps the most important one of all: pre-existing doubts about the self. Since, as we noted earlier, there is much in militarism to attract those with doubts about their masculinity and intellectual capacity, it would not be

* According to G. M. Carstairs, retired army officers are among those professional groups with the highest rates of suicide.[9]

surprising to find that a number of men with problems over self-esteem will be discovered at all levels of the military hierarchy. Moreover, since, as we shall see, the desire to bolster sagging self-esteem is a great motivator, we might expect this percentage to work its way into the higher echelons.

One piece of corroborative evidence for these views comes from yet another characteristic of many military organizations: their notorious sensitivity to criticism. A word, therefore, about this curious phenomenon.

In discussing the origins of snobbishness, the point was made that it usually betokens some underlying sense of inferiority, that only the socially insecure need to be snobbish.

Applied to the military, this may well seem difficult to swallow. Snobbish they may be, but insecure—never! This understandable scepticism, by those who perhaps mistakenly equate physical might with emotional stamina, and a gorgeous exterior with spiritual tranquillity, does, however, ignore one salient feature of military organizations: their very great sensitivity to criticism and the fact that this becomes acute precisely at those periods when, following upon a major war, the popularity of their calling seems on the wane.

In Britain this sensitivity was activated by the blunders of the Crimean War, re-emerged with the disasters of the Boer War, and reached a peak between the two world wars. Not very surprisingly, it seemed related to the numbers killed in each preceding conflict. Judging from the abysmally low regard in which the Army was held after Marlborough's campaigns, these antipathetic feelings may well relate more to losses than to incompetence, and to the feeling that even for great commanders, victory at *any* cost is hardly admissible.

In that tremendous combat [Malplaquet] near upon two hundred and fifty thousand men were engaged, more than thirty thousand of whom were slain or wounded (the Allies lost twice as many men as they killed of the French, whom they conquered): and this dreadful slaughter very likely took place because a great General's credit was shaken at home, and he thought to restore it by a victory. If such were the motives which induced the Duke of Marlborough to venture that prodigious stake, and desperately sacrifice thirty thousand brave lives, so that he might figure once more in a *Gazette*, and hold his places and pensions a little longer, the event defeated the dreadful and selfish design, for the victory was purchased at a cost which no nation, greedy of glory as it may be, would willingly pay for any triumph.[10]

Whatever Marlborough's motives, it was during the period following these events that officers were depicted, depending on their age and seniority, as either 'wenching whipper-snappers' or 'gouty tyrants', men who had been doomed to a military career usually as a result of being caught in *flagrante delicto* with 'one of mother's chambermaids'. It was the same Tory pamphleteer, Ned Ward, who wrote that a captaincy in the Guards could be obtained 'by giving some bodily consolation to an ancient lady'.[11] The Army has also suffered in comparison with the Navy. As Bond has written of Victorian times, 'The Navy was the bulwark of the constitution, and had the advantage of seldom attracting public notice. By contrast, the Army flaunted its unwelcome presence everywhere ... '[12]

Recently another reason for sensitivity to criticism has come to the fore, namely the fact that the carrying of arms has become what Abrams calls a 'receding profession'. He ascribes this recession to a steady loss of the profession's monopoly of knowledge relevant to the service it is supposed to provide.[13] This state of affairs is aggravated by a growing confusion as to its role, coupled with an increasing lack of consensus among its members as to what membership should entail.

All in all, the military has often had good reason to feel sensitive about its image. The fear of possible criticism has taken several forms— the finding of scapegoats for military disasters, the whitewashing of senior commanders after military courts of inquiry, the watering down of bad news in official reports, unnecessary censorship, interference with the activities of war correspondents, the refusal to promulgate the findings of committees which were critical of previous military enterprises, the curbing or compulsory retirement of officers who spoke out against defective practices (and, perhaps even worse, actually suggested some improvements), and the refusal to appoint men to key positions when these same individuals had been favourably compared by outsiders to less competent officers. According to Vagts, even the editing of history is not unusual. 'To meet the requirements of their contemporaries and of posterity it has been a habit of generals and their staffs not only to edit the reports of battles but also to word their orders in such an oracular fashion that victory, if it comes, can be traced to them, while failure, if it befalls, can be excused as a misreading by those lower in command.'[14] And Francis Grose's *Advice to Officers*, published in 1762, included this precept: 'When at any time there is a blundering or confusion in a manœuvre, ride in amongst the soldiers and lay about you from right to left. This will convince people that it was not your fault!'[15]

To these indications of sensitivity to real or implied criticism we must add another of recent origin: the hostile postbags of those who have dared to record the less successful activities of senior commanders, often long since dead. One such was the urgent demand for a public apology which Russell Braddon received from a group of military gentlemen following publication of his book on the siege of Kut; another was the pained response of 'the Friends of French' to Smithers's story of Smith-Dorrien's fall from grace.[16] It is not too difficult to see these hurt feelings as evidence of the familial identifications which obtain in military organizations, a sort of 'don't be beastly about father however much he is at fault'. The most striking examples of this response were in connection with General Sir Redvers Buller after his removal for incompetence during the Boer War, and with Major-General Gordon Bennett, who wrecked his career by abandoning his troops after the fall of Singapore. Both Buller and Gordon Bennett were defended by their protagonists long after their shortcomings had been widely proclaimed. In all these cases the lives and deaths of thousands of ordinary soldiers were evidently considered less important than the reputations of their leaders.

In talking of criticism it might seem that we are making a great deal of fuss about nothing. After all, nobody likes criticism, and as for complaints against military historians, it is only natural, indeed laudable, to show loyalty to one's group. There are, however, some special features of the phenomenon in some military men which deserve attention. In the first place, their sensitivity seems out of all proportion to that of other public figures. In terms of fame or notoriety, well-known generals or admirals are on a level with film stars, politicians and even newsworthy academics; hence one would expect that they might come to accept the possibility of negative publicity as part of the game, a small price to pay for the 'perks' which they otherwise enjoy. This they seem unable to do. In fact, there is a distinctly paranoid element in the way some senior commanders have reacted to even the faintest breath of criticism; to the vaguest and most tactful suspicion of a raised eyebrow or cleared throat—almost as if they were being held personally responsible for everything that might go wrong.*

In the second place, their dislike of criticism has, on occasion, been so intense as to lead to behaviour diametrically opposed to the well-being of the organization which they represent. The refusal by Montgomery-Massingberd to disseminate the findings of the Kirke Committee on the First World War, and the blighting of the careers of progressive soldiers like Hobart and Fuller, fall into this category.

In the third place, the response to criticism has, upon occasions, like the aftermath of the Cambrai tank offensive, been so blatantly self-damaging as to fit the label 'neurotic' (i.e., behaviour which the individual cannot help even though he knows that it will rebound upon himself).

Duff Cooper summed it all up when he described 'the soldiers at the top' as 'shut off' and 'unlike other public men, absurdly sensitive to criticism—so thin-skinned. Instead of realizing the value of criticism and its publicity value to the Army they regarded any suggestion, that there had been some muddle, as a personal insult.'[17]

In touching upon this delicate matter we must not lose sight of its significance and relevance in the present context. Whatever else it may be, sensitivity to criticism is a measure of insecurity. It implies a weak ego which, in turn, and by way of compensation, manifests itself in particular character-traits, one of which is snobbishness. Whether this ego-weakness is due to some early shock to self-esteem, or fear of the breakthrough of unacceptable impulses, or some combination of these two influences, the individual so afflicted develops certain defences which help to minimize his painful feelings. This finds support in yet another feature of military organizations—their cult of anti-effeminacy.

* A very important and exacerbating factor has been the regulation forbidding serving officers to write to the press to defend themselves against unfair and often ill-informed criticism. In this writer's opinion, not being allowed to answer back prevents riddance of those unpleasant feelings which criticism evokes.

19

Anti-Effeminacy

'Let war cease altogether and a nation will become effeminate.'
GENERAL CHAFEE

When discussing the various anxieties which militarism serves to reduce, brief mention was made of the fears which some men entertain about their masculinity. Thus it was pointed out (see page 185) that, though primarily concerned with combating the dread of disorder and dissolution, certain sartorial aspects of 'bull' might also help to reassure those with problems in this area.

In putting together the jigsaw of military incompetence, therefore, we can now take up one piece which clearly has great relevance to this topic: the striking antipathy towards effeminacy which characterizes some military organizations, and this despite the fact that the female is usually regarded as 'more deadly than the male'! Evidence of this antipathy is of necessity circumstantial. It embraces such phenomena as:

1. The importance attached to such outward signs of sex-role identification as hair-length. Since the insistence on 'short back and sides' seems correlated with those periods in history when sexual differentiation was linked to hair-length, we can dismiss excuses of neatness and hygiene as rationalizations. Field-Marshal Lord Wolseley stated the true case when he said: 'It is very difficult to make an Englishman at any time look like a soldier. He is fond of longish hair ... hair is the glory of a woman but the shame of a man.'[1]

2. Traditional taboos on certain topics and pastimes. Thus we find a Captain Foley, R.N., Commander of Britannia Naval Training Establishment for Officer Cadets at Dartmouth, forbidding piano-playing because he considered it effeminate.[2] And in 1973 Ian Carr described a comparable incident in connection with a fellow officer Gerald Laing.

In Northern Ireland Gerald had returned to one of his earlier loves, which was painting, and in Germany he began taking lessons. As this passion grew, the Army's disapproval hardened until there seemed to be a continuous ideological battle going on... For the assistant adjutant, a blond youth with a retroussé nose, the word 'artist' was synonymous with 'homosexual', and he cornered me one day (I was considered to be the evil influence on Gerald) and said desperately: 'Now look here ... I've got nothing against art, but queers are bad ... I mean it! ... disgusting!'[3]

Eighty years previously, Rear-Admiral A. H. Markham, whose fatal obedience and lack of initiative, when confronted with an 'impossible order' by his autocratic superior, Vice-Admiral Tryon, resulted in a collision between two battleships, seems to have been afflicted with a similar prejudice. 'Cigarettes', he once told his officers, 'are only for effeminate weaklings.'[4]

3. A deeply rooted prejudice towards women who try to adopt traditionally male roles. An early example is the hostility encountered by Florence Nightingale in her effort to reduce manpower wastage from disease and malnutrition during the Crimean War. A recent example is the lack of co-operation experienced by Mrs M. Pratt during her attempts to obtain information for her book on V.C.s.* Of this she said, "They [regimental associations] considered that a woman was not a proper person to write such a book. Regiment after regiment I found took this point of view.'[6]

Apropos of these prejudices it can be argued that the real threat of women who do men's jobs is that in effeminizing the role they, by association, emasculate those who normally fulfil it. The differences in outlook between those armies which eschew training women for combatant roles and those, like the Israeli Army, which do not adds strength to this hypothesis, as does also the observation that one of the favourite insults hurled at officer cadets who make a poor showing during ceremonial drill is that they look *pregnant*. One suspects that this form of abuse is, paradoxically, unknown in those armies where, because they contain women, there may for once be some truth in the assertion!

Finally, an equation between *defensive* behaviour and effeminacy—

* Mrs Pratt died before being able to complete her book, *Aristocracy of the Brave*.[5]

the feeling that it is cissy to wear ear protectors or build head covers (see p. 55) — has undoubtedly caused much unnecessary destruction of the human body. Since the male is in many respects less tough, less able to withstand pain, more mortal, and, from a purely anatomical point of view, more vulnerable than the female, this equation, though understandable, is sadly ironic. It is a not unreasonable hypothesis to suggest that its most glaring and costly illustration occurred in connection with the issue of convoys. In the First World War, hundreds of thousands of tons of merchant shipping were lost through the Navy's refusal to adopt the convoy system. When Lloyd George eventually forced convoys upon an unwilling Admiralty, losses fell significantly. The lesson was plain for all to see. But in the years between the wars the same irrational dislike of 'mothering' a flock of ships prevented the development of an efficient escort system. It was not until the spring of 1941, after a period of calamitous losses, that reason triumphed once again and the escorting of merchant convoys was resorted to in anything approaching a wholehearted manner. The benefits were once again immediately apparent. But then America entered the war and, unbelievably, in the face of overwhelming evidence, insisted on trying to defeat U-boats without the use of convoys. Rear-Admiral Sims, America's great fleet commander of the First World War, not only foresaw the error of their ways, but appeared to grasp the underlying motivation when he wrote: 'It therefore seems to go without question that the only course for us to pursue is to revert to the ancient practice of convoy. This will be *purely an offensive action* because if we concentrate our shipping into convoy and protect it with our naval forces we will thereby force the enemy, in order to carry out his mission, to encounter naval forces.'[7] (Italics mine.)

But despite the admiral's gallant attempt to convince his kinsmen that they need not feel ashamed of defending convoys, his words went unheeded. Between December 1941 and the following March American losses of merchant shipping grew to the staggering monthly total of 500,000 tons. Eventually, the price of aggressive masculinity embodied in the so-called patrol and hunting operations of isolated warships proved too costly, and convoys were instituted between Boston and Halifax. Losses on this route promptly dropped to zero. But south of Boston ships still sailed independently until by June the number of ships sunk reached the all-time record of 700,000 tons in a single month! Thenceforth convoys were run to the Caribbean without one ship being lost. At last the lesson had been learned. But, as Macintyre remarks, 'at what a cost'.*

In touching on this topic, Janowitz makes the point that 'the cult of manliness and toughness associated with junior officers [in the American forces] is often a reaction against profound feelings of weakness'. It is hardly coincidental that the same writer endorses the view that 'the most peaceful men are generals'.[8]

The argument is simply that a proportion of those youths who opt for a career in the armed services do so out of an underlying fear of being unmanly. Such individuals will be attracted to organizations which set upon them the seal of masculinity. By being admitted to a society of men bent upon the most primitive manifestations of male-ness—violence and aggression—the individual achieves the reassurance he requires. But to maintain this reassurance he will in turn have to contribute, by word and deed, to the elaborate defences against effeminacy of the citadel which he has entered.

Apropos of military incompetence this 'butch' element in the armed forces, whatever its origins, may well have two disastrous consequences. Firstly, we shall find positions of importance filled by some of the 'peaceful' generals of whom Janowitz writes, men whose style of life, a compensation for feelings of inferiority, took them to the top but then served them ill in their role of generalship. A classic example is that of Sir Redvers Buller, a man whose outward trappings, such as his large size and valorous deeds, proclaimed his 'masculinity', but who concealed beneath them a soft and passive personality.

A second unfortunate consequence is that since military organizations constitute virility-proving grounds they give rise to those excesses of drunkenness and overt sexuality which have from time to time seriously threatened fighting efficiency. According to John Laffin, this outcome is particularly evident in the armed forces of the United States:

American fighting men want sex for breakfast, lunch and dinner. Women and sex make up a large part of the thoughts and speech of any army, but with the Americans they amount

---

* Dislike of convoy escort duties has also been ascribed to a preference amongst some naval officers for membership of a large battle fleet. In the light of evidence to be considered presently this preference may well stem from the fact that ship commanders in a battle fleet have in fact far *less* scope for personal initiative than have the escort commanders of merchant convoys.[9]

to an obsession. Officers particularly have often seemed more interested in women than in getting on with their war job; private soldiers have complained about this since the days of the Revolution ... They [the Japanese] were soon convinced that only two things mattered to American servicemen — sex and liquor.[10]

The third and no less disastrous consequence is that by selecting and promoting on the bases of such 'butch' criteria as size, strength, physical courage and prowess at games, the armed forces tend to ignore other attributes which really may be of even greater importance to a senior commander — intelligence, high educational level, resistance to break-down under stress and substantial reserves of moral courage.

There is of course a counter-argument, namely that generals should be heroic leaders, which would necessitate them having at least some of the aforementioned 'butch' traits. Sheer physical size, the possession of decorations for bravery and a fine rugger record would, according to this argument, confer invaluable leadership-qualities upon top military commanders. Unhappily this theory does not stand up. Firstly, the causal relationship between leadership and 'butch' traits is at best one-way. Natural leaders may well have made good captains of a first XV, but being good at rugger in no way ensures the best qualities of military leadership. Secondly, while physical stature, and so on, are certainly advantageous to a would-be leader, so-called masculine attributes count for very little in comparison with personality and knowing one's job. Good leadership is synonymous with inspiring confidence in those who follow, and confidence is born of results. Thirdly, the most cursory glance at military history suggests that many of the really great military and naval commanders — Napoleon, Nelson, Wolfe, for instance — were men of brain and character, not of huge bodies with dazzling records in the field of sport. In case these views should seem heretical let it be said that they accord with those expressed by at least one military man of some repute. Of generalship, Montgomery says: 'The science and art of command ... [involves] an intimate knowledge of human nature ... a commander must think two stages ahead.' He speaks glowingly of the physically 'frail' Wolfe and Nelson, describing the latter as a 'brilliant seaman and most original, intelligent and courageous fighter'. In the same vein he comments on the 'flexibility' and 'brilliant intellect' of Napoleon. But nowhere does the field-marshal dilate upon the advantages of fine physique, hairy masculinity, and a reputation for long-distance running, polo or boxing.[11]

The fourth point is this: in days gone by, when physical strength counted for more on the battlefield than mental ability, and senior commanders could exercise their heroic powers by leading their troops into action, the physical aspects of heroic leadership were no doubt important. But in modern war generals and admirals are rarely if ever seen by the vast majority of their men. Under such circumstances heroic leadership must count for rather less than managerial and technical ability.

Since the foregoing section has clearly been treading on very sensitive ground, let us retrace the argument to make sure that no unjustified injury has been done.

We are concerned to relate and explain two indisputable phenomena: so-called 'peaceful' generals who in times of stress reveal themselves as passive, dependent and indecisive; and the anti-effeminacy ethos of some military organizations. To handle these facts the following points were made:

1. Some men, for reasons rooted in the early family situation, have serious doubts about their sexual adequacy and/or physical strength and size.

2. Such men may deal with their feelings of inferiority by adopting a compensatory style of life in which they strive for reassurance in some suitably symbolic role.

3. The prevailing ethos of many military organizations provides this reassurance.

4. Hence a percentage of men will seek acceptance by the armed forces simply because such acceptance is a warranty of their masculinity.

5. Once in, their continuing and underlying fear of effeminacy (synonymous with inadequacy in the minds of this particular group) produces that well-known pattern of behaviour which we have termed 'butch'.

6. But this compensatory behaviour is itself highly valued in the armed forces. Hence the individual not only profits by, but also contributes to, the anti-effeminacy of his parent organization. It is in his interest so to do.

7. The significance of all this for military incompetence is that 'butch' characteristics are not perhaps the most important criteria for top-level leadership. Let us therefore examine this last point.

20

Leaders of Men

'How can the ability to lead depend on the ability to follow? You might as well say that the ability to float depends on the ability to sink.' L. J. PETER AND R. HULL, *The Peter Principle*

Whatever its other causes, military incompetence implies a failure in leadership. This is hardly surprising. Of the psychological problems which beset military officers few exceed in severity those associated with leadership. In this respect they are required to fulfil incompatible roles. They are expected to show initiative, yet remain hemmed in by regulations. They must be aggressive, yet never insubordinate. They must be assiduous in caring for their men, yet maintain an enormous social distance. They must know everything about everything, yet never appear intellectual. Finally, as we saw in the last chapter, they may well have been selected for attributes almost totally unrelated to the tasks they are expected to perform.

Discussion of leadership is so often overloaded with vague but emotive ideas that one is hard put to it to nail the concept down. To cut through the panoply of such quasi-moral and unexceptionable associations as 'patriotism', 'play up and play the game', the 'never-asking - your - men - to - do - something - you - wouldn't - do - yourself' formula, 'not giving in (or up)', the 'square-jaw-frank-eyes-steadfast-gaze' formula, and the 'if ... you'll be a man' recipe, one comes to the simple truth that leadership is no more than exercising such an influence upon others that they tend to act in concert towards achieving a goal which they might not have achieved so readily had they been left to their own devices.

The ingredients which bring about this agreeable state of affairs are many and varied. At the most superficial level they are believed to

include such factors as voice, stature and appearance, an impression of
omniscience, trustworthiness, sincerity and bravery. At a deeper and
rather more important level, leadership depends upon a proper understanding of the needs and opinions of those one hopes to lead, and the
context in which the leadership occurs. It also depends on good timing.
Hitler, who was neither omniscient, trustworthy nor sincere, whose
stature was unremarkable and whose appearance verged on the repellent, understood these rules and exploited them to full advantage. The
same may be said of many good comedians.

In short, there is nothing mysterious, romantic or necessarily laudable about leadership. Indeed, some of the most effective leaders have
been those who, merely through having more than their fair share of
psychopathic traits, were able to release antisocial behaviour in others.
Their secret is that by setting an example they release a way of acting
that is normally inhibited. This gives pleasure to their followers, thus
reinforcing their leadership.

In military organizations leaders are usually of a rather different kind.
For a start, they are appointed rather than emergent. That is to say, the
needs of the individual soldier play almost no role in deciding the sort
of leader that he gets. Secondly, the military leader possesses constitutional power of a magnitude which surpasses that of leaders in most
other human groups. If he cannot pull his followers by force of
character, he can at least push them by force of law.

The third and related feature of military leadership is that it is
essentially autocratic and operates in what modern theorists call a
'wheel net' rather than an 'all-channel communication net'. In other
words, the flow of essential information is to and fro between the
leader and his subordinates rather than between all members of the
group. Not very surprisingly, the wheel net, though no doubt gratify-
ing to autocratic leaders, produces more errors, slower solutions to
problems, and reduced gratification to the group than does the more
democratic all-channel net.

In the light of these considerations it is perhaps strange that leadership
in the British armed forces should have been as effective as it has.
Indeed, on the assumption that the primary function of officers is to
get the best out of their men, the curious alchemy wrought by the
gentlemanly amateurs of the Victorian British officer corps, and even
by the still relatively unprofessional officers of the First World War,
deserves considerable respect. Since a salient feature of all the campaigns

so far considered has been a remarkable absence of mutinous tendencies and a quite astonishing degree of tolerance, fortitude and bravery shown by the common soldier, we have to ask: was this despite or because of their leaders? And if the latter, how was it that even the most inept and reactionary of them could so touch the hearts of their men that they would give themselves to the fight with a cheerful and destructive energy that could, on occasions, rise to whirlwind proportions.

Even men like Elphinstone, Townshend and Buller, about whose flagrant incompetence in the role of decision-maker there can be no possible doubt, earned a loyalty and affection, albeit far beyond their deserts, which maintained the morale and fighting spirit of their men almost to the end.

By way of trying to explain these curiosities let us consider a few more findings from the extensive research on leadership. The first point to note is the distinction that has been drawn between two roles of a leader: 'task specialist' and 'social specialist'. As task specialist a leader's prime concern is to achieve the group's ostensible goal; in the case of the military, defeating the enemy. For such a role, being likeable is a rather less important trait than that of being more active, more intelligent and better informed than his followers. In his capacity as 'social specialist', however, a leader's main function is to preserve good personal relations within the group, thereby so maintaining morale as to keep the group in being. In the military milieu the function of a successful social specialist would prevent mutiny and reduce such symptoms of low morale as absenteeism, desertion, sickness and crime. Not very surprisingly, the most important attribute of such a leader is that he should be liked. Efficiency and task-ability are of rather secondary importance. While it is obvious that many leaders in the British armed forces have tended to be social rather than task specialists, we have to ask why this should be.

It is easy to answer one part of this question. They were poor task-specialists because ours is traditionally an amateur army in which professional ability, knowledge and military flair have counted for little. But why good social specialists?

Again modern research has come up with some possible answers. It has been shown that whereas low-stressed groups, operating in situations that are devoid of painful uncertainties, do best under democratic leadership, organizations like the military in times of war that are subject to stressing ambiguities actually *prefer* autocratic leadership.

In other words, the feelings of dependency induced by stress success-fully neutralize a person's normal antipathy towards the autocratic leader.[1] While a man like Townshend would not be likely to survive for very long in a modern civilian firm, his autocratic mien was lovingly accepted by men whose lives were hanging by a thread.

But even if, given the right circumstances, an autocratic mien is no bar to being liked, we still need some more positive reasons for the extraordinary popularity of otherwise incompetent commanders. There are three such: 'riskiness', 'socio-economic' status and the past indulgence of the individuals concerned. Other things being equal, a man who is prepared to take risks makes a more popular leader than one not so inclined.[2] By taking a risk he metaphorically, if not literally, stands out in front of the group and is perhaps, by so doing, shoulder-ing the responsibility for behaviour in which the group needs (and wants) to indulge but for which, if left on their own, they would lack the necessary moral stamina. The vicarious pleasure and feeling of admiration which we derive from contemplating big gamblers in any walk of life are components of this psychological phenomenon.

A less readily explicable factor is that of socio-economic status.[3] There are probably at least three components to the influence of wealth and position. Firstly there is the, sometimes no doubt erroneous, belief that 'he must be better than I am' which gives rise to the 'therefore-I-will-follow-him-to-the-grave-if-necessary' feeling. Secondly the tradi-tional good manners and self-confidence of the financially and socially secure obviously makes for a more kindly and humane paternalism towards the underdog.

That officers of the old school earned the love of their men by behaving towards them as they might towards cherished pets was possible because of the real and enormous social gulf which the rank and file perceived between themselves and their rulers. The time-honoured distaste which other ranks have felt for officers who rose from the ranks is all part of the same picture.

Finally, the fact that their position was assured through their wealth meant a relative absence of those unpleasant traits which are associated with feelings of social inferiority.

Another obvious reason for the likeableness which eventuates in good social leadership has been researched by Greer.[4] This worker showed that successful leadership tended to occur if followers had been indulged by their leader. In this case the tractable nature of the group evidently reflects a wish to return past favours. By these lights it is hardly surprising that even the most incompetent generals were often effective social leaders. No one took greater risks than Townshend, no one was more concerned to indulge his troops than Buller, and few could outdo such notables as Lucan, Cardigan and Raglan when it came to a matter of displaying socio-economic status.

However, to someone who has not had the misfortune of serving under any of these officers it may seem scarcely credible that the riskiness of Townshend, the indulgence of Buller and the socio-economic status of the Crimean generals could have compensated for their other characteristics. How, for example, could troops overlook the palpable egocentricity of Townshend and the total 'unriskiness' and glaring incompetence of Buller; and how could they forgive the apparent negligence of Raglan? There are at least three related reasons. Firstly, in war, as in other situations of mortal threat, there is an understandable urge to clutch at straws—the good aspects of a leader are seized upon, the less good conveniently denied. We would guess that this anxiety-reduction will, moreover, be particularly likely to occur in a situation without degrees of choice. The situation of a soldier, in an organization which allows of no escape, confronted by the threat of imminent destruction, is just one such. To put it very simply, he makes the best of a bad job, and this includes wholeheartedly accepting a leader even when the latter was not of his choosing.

Again, it is the nature of military organizations to recapitulate the psychodynamics of an authoritarian family group, one in which the paterfamilias can do no wrong. It is not necessary to be an ardent believer in psycho-analytic theory to realize that, in times of stress, there is a natural harking back to an earlier source of security.

But there is still one other reason for the extraordinary tolerance shown towards disastrous leaders—their 'invisibility'. The reputations of many bad generals have survived simply because the individuals concerned kept out of the way. Like God, they did not often reveal themselves. This analogy between belief in an earthly leader and belief in God may be carried further. Both are sometimes functions of experienced threat, and both may be enhanced by the surrounding mystery. Whether they are in fact good or bad, 'invisible' leaders like Raglan undoubtedly benefited from not being known and rarely seen by the rank and file.

The phenomenon is perhaps best illustrated by that most controversial of figures, Field-Marshal Haig, of whom it has been written: 'To write him down as a blundering, heartless incompetent in the prevailing fashion calls for considerable hardihood on the part of the critic. One fact remains that cannot be questioned: until the echo of the last shot had died away, no condemnation of Haig was ever voiced by the rank and file of the two-million-strong army under his command.'[5] When it is considered that few of these two million ever saw their commander-in-chief but *were* confronted daily with the immediate, and fearful, consequences of his generalship, the parallel between blind acceptance of an 'invisible' military leader and the strengthening of religious convictions (regarding heavenly competence) which follows monumental natural disasters can hardly fail to be drawn.

The ideal military leader is, of course, one who manages to combine excellence as a task-specialist with an equal flair for the social or heroic aspects of leadership. Since the traits required for these two aspects of leadership are rather different, these so-called 'Great Man' leaders have been comparatively rare.[6] Amongst the best examples were Wellington, Nelson, Lawrence and, in recent years, Field-Marshal Slim. Such leaders managed to combine extreme professionalism in the realizing of military goals with a warm humanity which earned them the lasting affection and loyalty of their men. There have, of course, been other 'Great Men' who, lacking the natural talents of a Nelson or a Slim for the role of social specialist, have deliberately simulated the necessary traits. The prime example of this genre is Field-Marshal Montgomery. By nature a rather cold, introverted and autocratic individual (a side of him seen by many of his officers), he nevertheless had the good sense to apply a somewhat contrived bonhomie, helped out with packets of cigarettes and numerous cap badges, which undoubtedly did much to ensure high morale and group-mindedness in the troops which he commanded. To many people, whether they like him or not, it must seem totally incomprehensible that Montgomery should have been actually criticized for his quite deliberate showmanship, which probably did more for civilian and military morale than any act by any other general since the beginning of warfare. Such jealous and unwarranted sniping exemplifies one of the more basic causes of military incompetence, namely the fatal confusion between the practical and symbolic roles of military organizations which results in the sacrificing of military efficiency for the sake of 'good form'.

The evident success of some British senior commanders in their role as social leaders does not mean that military mishaps have never been due to shortcomings in this respect. Three situations in particular have provided scope for military incompetence. The first hinges upon the fact that though the leadership qualities required at one level of command may result in promotion, they are often not those relevant to a higher level of command. Just as a brilliant general, such as the Australian Sir John Monash, may have been an indifferent brigadier, mediocre battalion commander and third-rate platoon commander, so, more seriously, there have been outstanding platoon and company commanders who, *promoted on the basis of their performance at these levels*, ended up as inept if beloved generals. Such examples of the Peter Principle, wherein people are raised to their own level of inefficiency, was never better illustrated than in the case of Sir Redvers Buller, who has been described as 'a superb major, a mediocre colonel and an abysmal general'. In this case, high-level military incompetence must be laid at the door of heroic leadership, for this was the quality which eventually put him where he could do the most damage to his own side.

The second situation in which the motivational as opposed to the intellectual aspects of leadership may lead to military disaster is where obedience, evoked by hero-worship, blunts reason and moral sensitivity to such an extent that the group may embark on behaviour which is little short of suicidal. A classic example is to be found in the psychopathic behaviour of some German units towards Soviet citizens during the invasion of Russia. It is certain that this behaviour helped seal the fate of Hitler's forces by turning potentially sympathetic Soviet peasants into vengeful saboteurs. It is possible that the discrepancy between 'military' behaviour under Hitler's leadership and the older, Prussian code of chivalry produced a sagging of morale and failure of soldierly pride from which the corps of officers could never totally recover. There is nothing so eroding of morale as to dislike oneself.

Happily, heroic leadership in the British military has usually been confined to younger officers. Senior military commanders, by reason of their unassailable rank and sheltering staff, have often remained so isolated from the rank and file that their possession, or lack, of heroic qualities has passed unnoticed.

There is, however, one further aspect of these more nebulous qualities of leadership which has played a not inconsiderable part in the story of military incompetence. It concerns the position which an individual occupies on two related continua: those of boldness to caution, and impulsiveness to indecision. Over the years military incompetence has resulted more from a dearth of boldness than from a lack of caution, and more from a pall of indecision than from an excess of impulsivity. The pusillanimity of Generals Warren and Buller at Spion Kop, which we looked at in a previous chapter, is a good example of this failure of leadership. Another is that of the Suvla operation in the Gallipoli campaign, where 'the greatest chance of the war was thrown away by the most abject collection of general officers ever congregated in one spot'.[7]

In more recent times, the Norway expedition of 1940 displayed not only similar shortcomings in high-level heroic leadership, but also the not infrequent contrast between the verve and initiative of junior commanders and the cautious indecision of those at higher levels of control. Donald Macintyre, writing of these events, records:

> The Commander in Chief, Home Fleet, after hesitating until nearly noon on 9 April, detached a cruiser squadron and destroyer to attack at Bergen, only to have his order annulled by the Admiralty who feared the shore defences might by then have been taken over by the enemy ... What a different approach had been that of Warburton-Lee [the destroyer flotilla leader whose spirited destruction of German naval forces at Narvik earned him a posthumous V.C.]—'intend attacking at dawn!' His initiative and daring had turned Narvik into a mortal trap for nearly half the total German destroyer strength. Yet even now the harvest which Warburton-Lee's sowing had prepared was nearly lost to the British through indecision and hesitation.[8]

The same lack of forceful and decisive leadership at the highest levels of command was also evident in the fall of Singapore (see page 139), when caution, precipitated partly by the fear of injuring civilian morale, resulted in too little being done too late to avert the worst disaster of the Second World War.

One obvious explanation for the failure of the motivational aspects of leadership, in all these instances, is the relatively advanced age of the individuals concerned. Old men are more cautious than young men, and less able to make quick decisions than those whose arteries have not begun to harden. The refusal by the elderly General Burrard to exploit Wellington's crushing defeat of Junot at Vimeiro in 1808 is a classic example of this sort of failure in leadership.

Thus in a loud voice clearly audible to his staff Wellesley exclaimed: 'Sir Harry, now is your time to advance. The enemy are completely beaten, we shall be in Lisbon in three days.' Sir Harry hesitated and Wellesley pressed him again, adding the bait of Sir Harry himself taking part in the victorious campaign ... The French had in fact fled eastwards, leaving Torres Vedras and the road to Lisbon open. But Sir Harry had said No once and he said it again. Enough was enough. He had been created a baronet for doing nothing much at Copenhagen in 1807. Before Junot's attack he had said to Wellesley, 'Wait for Moore.' He repeated it. It was not a pun but a fatuity. Wellesley turned away in disgust, remarking to his officers that they might as well go and shoot red-legged partridges.[9]

Another possible factor is that many of these instances involved combined operations. Even Buller had a detachment of naval artillery which, with incorrigible and fatal obstinacy, he forbore to use against the enemy positions on Spion Kop. Are these perhaps cases where inter-service jealousy, like sibling rivalry, effectively deflated and used up the motivational energies of both the rivals? This was certainly true of the Singapore disaster, both in the Chiefs of Staff quarrels of 1925 and in the lack of effective liaison between Army, Navy and Air Force commanders during the Japanese invasion of Malaya in 1941 and 1942.

But there are other more fundamental and pervasive reasons for these failures in leadership which can be ascribed to the general psycho-pathology of military organizations. Their common denominator is anxiety. It is a feature of armed services that the penalty for error is very much more substantial than the reward for success. Whereas the naval officer who, through an error of judgment on the part of his subordinates, puts his ship aground will almost certainly be court-martialled and stands a fair chance of being heavily punished, the reward for taking a bold action which pays off may be no more than a mention in dispatches or some decoration with little or no effect upon promotional prospects. The net result of this bias towards *negative* reinforcement will be that *fear of failure* rather than *hope of success* tends to be the dominant motive force in decision-making, and the higher the rank the stronger this motive because there is farther to fall. There are of course other reasons for supposing that the anxiety which tends to curb bold initiative will be stronger in the higher levels of command than lower down the hierarchy. For one thing, responsibility is greater and, for another, perhaps for the first time, there is no one higher up to whom the senior commander can appeal.

Finally, mention must be made of a thesis put forward by Simon Raven which may bode ill for the future. It concerns the role of false premises in the training of officers, false premises that have their origin in a simple and obvious fact: that an expectation of superiority in a leader by those who are led will increase the tendency to follow him. If, on a priori grounds, you believe that someone is better educated and knows more than you do, then you will be more prepared to follow his lead than if you are not party to this belief.

For years this simple truth has been confirmed by the attitudes and behaviour of the ordinary soldier towards his officers. For years it was the case that since they were drawn from a socio-economic class that was vastly inferior to that of their officers, the rank and file took it for granted that their officers knew more than they did and were in a very real sense born to lead, i.e., were born into that class from which traditionally the corps of officers was drawn. Even more important, the officers were gentlemen, that is to say they possessed an effortless and uncontrived capacity for radiating self-assurance, good manners and a courteous if paternalistic mien towards those of inferior station. However mistaken they may have been in individual cases, the rank and file were able to look up to such men as being of a superior caste, omniscient, omnipotent, natural, preordained leaders, and, more often than not, benign father-figures (see page 234).

Since the last war all this has changed. Officers are no longer recruited exclusively from the upper classes. Comparatively few are from landed gentry or aristocratic families, and many have not even attended a public school. By the same token, the rank and file are better educated and more sophisticated than their forebears. At first blush this would seem all to the good, giving promise of a democratization in the profession of arms, a trend that would one day place it on a par with most other vocations in a civilized society.

Unfortunately, according to Simon Raven's thesis, something quite other is happening. Confronted with the necessity of recruiting its officers from a section of society that would have been unthinkable in years gone by, the military has made what it regards as the best of a bad job by insisting that, since officers must still be gentlemen, where no natural gulf exists between those who lead and those who follow this must be artificially inculcated by training.

The following excerpts from Raven's article should make the matter plain. We start with a glimpse of life at the Royal Military Academy, and later at the School of Infantry at Warminster, in the 1950s. According to Raven, the products of this training regimen may be as bizarre as those depicted in the four character-studies which conclude this section.

> ... the saluting at Sandhurst is tremendous. If you walk round Sandhurst looking remotely as if you might be an officer, you will receive an incessant barrage of compliments. The muddy boy in P.T. shorts will stop running, square his shoulders and snap his eyes in your direction like knives. The elegant young gentleman in the brown trilby will lift it from his head with a controlled jerk, to replace it an exact number of seconds later at precisely the same angle. Boys in uniform with sticks, swords, rifles or sub-machine-guns will perform a volume of intricate movements, alone or as a body, for your especial benefit.[10]

To the detached observer these quaint antics may seem ludicrous, boring or even faintly embarrassing. However, there will be others so emotionally incapable of distinguishing between compliments paid to the abstractions of rank and commission and those paid to themselves as people that they will actually enjoy these gesticulations. But such enjoyment of these mandatory conventions, based upon a highly motivated if understandable misinterpretation of their meaning, may, like the effects of even the most transparent flattery, provoke wholly unrealistic feelings of self-importance.

Other significant features of the Sandhurst milieu, according to Raven, are: the prefect system, wherein cadets of higher rank are required to discipline and report upon those of lower status; the mind-blunting drill square upon which three apparent essentials for a career of violence — unthinking obedience, an exquisite capacity for keeping in step and a proper concern for the minutiae of dress — are instilled for hour upon hour 'until fatigue and sweat hang over the massed cadets like brimstone over Sodom'; and finally the total loss of privacy, and lack of leisure for the following of idiosyncratic interests and pastimes. Taken together, these features of the Royal Military Academy are designed to 'build character' and imbue future officers with values proper to their calling.

Any gaps which Sandhurst might leave in a total programme for the inculcation of officer-like qualities are admirably filled, says Raven, by the quasi-moral imperatives of Warminster. These cluster round the concepts of 'guts', enthusiasm, humour, sociability and responsibility — traits which every officer should show.

While much of the training was inevitably designed to promote physical fitness, there was nevertheless a strongly held belief that an officer, whether fit or not, should always have so much in the way of pride (or 'guts') that he would never admit to physical inadequacy until he dropped dead or unconscious. This belief, a very significant one, was mystical both in its nature and intensity. During a crippling exercise at the end of the course two or three officers fell out complaining of blisters or other mild indispositions. The Chief Instructor, himself a civilized and self-indulgent man, denounced them in round terms. An officer, he said, simply could not and did not fall out. Willpower, if nothing else, should keep him going for ever. It was all a matter of 'guts'. There was an unspoken implication that, since other ranks could and did fall out, even though they were often physically tougher, the officer belonged to a superior caste. I found it an accepted belief among officers later on that they could perform physical feats or endure physical discomforts without it being in the least necessary for them to train or prepare for such things in the manner required of the private soldier. Officers, for example, just did not do P.T.: they did not need it; they were officers and would endure to the very end, had they stepped straight on to the field from a sanatorium or a brothel.

How much, in the last resort, was thought to depend on superior qualities of morality and character! The military arts were given precise attention, but it was the catchwords with quasi-moral implications ('guts', 'common-sense') that filled the air as the course went on. Another Warminster virtue was a peculiar brand of humour. This was not the ability to see oneself and one's activities in a detached and ironical spirit — that would have been fatal. Humour meant being cheerful in the face of unpleasant circumstances, rallying the men's spirits by laughing with them over some slapstick incident, submitting 'like a good sport' to an unjust punishment given to oneself by the Adjutant and 'laughing about it afterwards in the Mess'. This conception of humour (an obvious branch of 'guts') was in fact discreetly designed to counteract or totally extinguish any tendencies towards an objective (or intellectual) humour that might contain tinges of satire or cynicism—for such a thing would have been detrimental to another highly prized virtue, that of *enthusiasm*.

About enthusiasm I can hardly trust myself to speak. It seemed to mean a sort of blind, uncritical application to any task, however silly or futile, that the neurosis or panic of a superior might suddenly thrust upon one. Since one of the points of enthusiasm was that you started doing whatever it was straight away and without wasting time on questions, enthusiasm could involve a frantic expense of time and energy on some trifling project, wastefully because uncritically undertaken, abandoned half-way as irrationally as it was commenced. This, of course, was just what great soldiers of the past wished to avoid when they deplored the indiscriminate use of 'zeal'. Why zeal—condemned alike by von Clausewitz and Wellington—should now once again be thought desirable it is interesting to speculate. I suspect it is because a superior and analytical attitude is considered undemocratic; and that the influence of such people as Lord Montgomery has dictated a spirit, for junior officers at least, of 'mucking in' and 'getting on with the job'. The heartiness, not to say hysteria, implied by such expressions was of course distasteful to the more fastidious and sceptical officers, for whose benefit yet another virtue, that of loyalty, had to be invoked. Loyalty meant that you were required, in the name of the Queen and the honour of the Regiment, to conceal any impatience or amusement you might feel when the demands on your enthusiasm became operatic, farcical or just plainly impossible of fulfilment. Loyalty, in fact, was a conception often blatantly used to blackmail you into silence when you were faced with the incompetence, injustice or sheer folly of a superior officer.

Sociability was also highly esteemed at Warminster. This, like loyalty, could mean many good things, such as hospitality and the desire to please in social intercourse, but it also implied an unquestioning deference to the convenience and opinions of one's military superiors. That one should obey the orders of such superiors, or even be loyal to them during displays of professional vapidity, is perfectly reasonable; but I could never discover why one should be expected, in a purely social context, to receive as gospel wisdom their views on anything from body-line bowling to revealed religion ...

'Courage under fire', a sort of distilled essence of 'guts', could not exactly be taught, and so had to be taken for granted in all of us, who were tacitly and grimly assumed to possess it. Hence we can pass to a very much boosted commodity — *initiative*. This meant (subject to orders received and unquestioning enthusiasm in the face of these) that oneself must always be ready to devise and sponsor an original course of action. A valuable military quality, most certainly: but unfortunately in peace-time conditions, and even for the most part in war-time ones, communications are now so good and the opportunities for genuinely individual action so rare, that initiative tends to become a highly contrived thing artificially fostered to impress superiors. It becomes, in fact, mere interference with the existing order of things which, to give an excuse for showing keenness and interest, are made out to be in some respect 'slack' or 'unsatisfactory'. A genuinely adventurous spirit is one thing, arbitrary exhibitions of officiousness are quite another; and the sort of person who was praised at Warminster for initiative generally turned out to be a meddlesome bully of the type who reports his best friend to his housemaster for immoral behaviour — thereby himself becoming head prefect in his friend's place. Such interference, I need hardly say, is also taken to show responsibility, which is the last of the great Warminster virtues.

Responsibility is in a way the pivot of the whole system. Like the Holy Ghost, it is supposed to be everywhere, and anything which is not material for the exercise of guts or enthusiasm (or one of the other Warminster qualities) will certainly turn out to be in the realm of responsibility. It covers everything from making an intelligent assessment of how to move a Division down to being careful not to get drunk in the Sergeant's Mess. It means not gambling for high stakes, not being late for parade, not sending reports to the Press, and not going to bed with your Company Commander's wife — however pressingly invited.[11]

Now, according to Raven, these features of officer training — the monastic segregation and disciplinary methods of Sandhurst followed by the quasi-moral imperatives of Warminster — may, rather surprisingly, eventuate in what he calls 'unlooked-for and immoral results'. He illustrates this outcome by describing four of his erstwhile fellow officers.

Second Lieutenant A. was a Roman Catholic from a professional family of moderate means (his father was a lawyer in a provincial town). He had done badly at Sandhurst, since he had limited intelligence and little application, but he had a family connection with our Regiment, was an agreeable and well-mannered boy, and so was accepted by our Colonel. He was sincere in his religious beliefs, drank too much, was sexually indiscriminate. His other amusements were horses (racing, hunting), field sports in general, gambling. Here are some of the things that A. would not do because he considered them 'inconsistent' with his status. He would not: be punctual with inferior ranks ('*they* wait for *me*'); join the troops in a run ('I don't need that sort of thing'); box with the troops ('they will not respect me if they see me with a bloody nose'); use the troops' lavatories; undertake any menial task (even on manœuvres); or accept a dressing-down from a superior who (a survival from the war) had been commissioned after many years in the ranks. It is only fair to add that A. had a real gift for handling men and was outstanding on forest patrols in Kenya.

A.'s conception of his status had nothing to do with his actual position or obligations as an officer, or with good-form middle-class notions of 'proper conduct' — in fact his ideas if anything compromised the former and ran completely counter to the latter. His conception of his position was definitely based on a type of feudal relation with his men.

Or consider Lieutenant B., also trained at Sandhurst. B. had an analytical intelligence, fair general culture, and a liberal outlook. The laziest man I have ever met, he had obviously only drifted into the Army by sheer chance and the lack of positive interest in any other idea. B.'s amusements were reading, desultory conversations of a mildly philosophic nature, mathematics, bridge. He was utterly indifferent to the Army and everything about it, but his acuteness enabled him to get through his duties without any trouble. Here are some remarks he made to me at various times:

'Sergeant-Major X. took me on one side today and said my untidiness made a bad impression in the Company. I told him that tidiness was only essential to the rank and file, who would otherwise just let themselves go altogether ...

'Driver Y. didn't want me to drive his truck, so I drove it at seventy for five miles flat. It's good for them to see a little officer-type driving ...

'Sergeant Z. is getting familiar. He asked me into his quarter for a cup of tea with his wife ... '

This is the pattern. A kind and intelligent man, B. is nevertheless convinced that he cannot take tea with a sergeant's wife. If this had been mere petty snobbishness, it would have been disagreeable but harmless. It was, however, something far deeper: it was a genuine conviction of rooted and superior status which simply would not brook such a proceeding.

Or again, there was Captain C. Unlike either A. or B., he was interested in the professional aspects of soldiering, and spent much time and effort devising new ways of shortening cumbrous military processes (generally tedious matters of administration). C. was always very concerned with his men's welfare, to which he gave genuine consideration (on the face of it, just the kind of competent, thoughtful and public-spirited young officer which a Labour Government would wish to perpetuate in 'a democratic Army'). But C.'s was scarcely a democratic nature. 'They are rather like pet animals,' he said to me of his men one day. 'One must keep them clean and properly fed, so that they do not get diseased and are in good working order. One must teach them to react swiftly and without thought to certain external stimuli or signals. Just as you whistle for a dog, so there must be certain simple and easily recognizable forms of words for the men. They must be given a certain amount of genuine affection, so that they feel loved and secure. They must expect, and on the whole receive, justice — a lump of sugar when they have done well, a whipping when they have been disobedient. But they must also realize that there are too many of them for justice always to work dead correctly in individual cases, and that occasional lumps of sugar will go to the idle and mischievous, occasional whippings to the industrious and innocent ... And they should be made to recognize the signs one sometimes gives when one simply does not want to be bothered with them ... '

Lastly, take the rather remarkable Captain D. D., though of the same middle-class stock as the others, had been at one of the 'top' public schools, had rather more money than the rest of us, and boasted a father who had served (during the war) in a Regiment distinctly smarter than our own. Thus he had certain social pretensions.

D. was keen on horse-racing (though only in a very broad and ill-informed way), and went one Sunday to a small meeting in Germany, sponsored and organized by the British Army of the Rhine, whose men and their horses were alone eligible to compete. It was thus an affair on the scale of a small point-to-point. The only betting was by means of a very amateur totalisator, which was run by members of the Royal Army Pay Corps and presided over by an elderly and plebeian lieutenant-colonel, himself also of the Pay Corps. Both the colonel and his men had given up their Sunday afternoon to do this, and though the tote was hardly a triumph of speed or efficiency, it was, on the whole, remarkable as a piece of makeshift goodwill. To this tote went Captain D. to bet on the last race of the afternoon. He was, of course, in civilian clothes, but his status if not his identity was easily discernible in the closed circle of attendants. The last race had only three runners, one of which, far superior in its record to the others, was owned and ridden by a well-known gentleman rider (who was also, in the time he could spare, an officer of Hussars): the animal could only fail to win through the direct intervention of God. (It was called Satan's Pride.) The tote dividend would be minute, and D. decided to back the favourite with a very substantial sum, hoping that the sheer size of his wager would bring him a tangible return. It was also a good opportunity to show off, which he always enjoyed. Accordingly, he demanded fifty pounds' worth of tickets on Satan's Pride. The corporal at the window went white and hurried off to fetch the colonel in charge, who appeared (since he was on duty of a kind) in a battered suit of battledress, and took D. on one side. There was no reason, he said in broad Midland tones, why D. should not put £50 on a horse; but in this particular race everyone was backing Satan's Pride, the dividend if it won would in any case be negligible, and D.'s £50 would make it almost non-existent— perhaps a penny half-penny in twenty shillings. Would D. consider lowering his stake and letting other people have a look-in? The bets were very modest, and £50 was not only completely wrecking the market but was—well—rather ostentatious.

Instead of answering the old man in courteous tones, D. completely lost control. With an occasional and insultingly stressed use of the word *Sir*, he demanded to know what an ex-ranker member of the Pay Corps, who came to a race meeting in uniform, could be expected to know about betting or gentlemen's habits in the matter, accused the colonel of snivelling, egalitarian, lower middle-class prejudices, and finally shouted aloud that if Pay Corps officers were too mean or spineless to risk £50 on a horse, then it was time they had lessons from their betters. The old man merely shrugged with good-humoured resignation and let D. have his way. (I should add that Satan's Pride, having led by 200 yards, fell at the last fence, leaving Molly's Son to win and pay a dividend of ten pounds odd for a two-shilling ticket.)

The incident was inspired by D.'s resentment that he, an officer, was being criticized by someone who did not, in a proper analysis, belong in the same category. The colonel, for a start, was in the Pay Corps, the officers of which could hardly aspire to the status of the Infantry: and in any case the old man simply was not 'a gentleman' and, having emerged after years in the ranks, had no claim to possess the qualities with which D. (trained as an officer from the beginning) must inevitably be blessed. So D. could ignore the entire official structure of the Army (not to mention the requirements of mere good manners) and give a blatantly feudal exhibition of hysterical spite. Nor is it insignificant that the colonel accepted the situation and let D. get away both with his insults and his demands.[12]

If we can extrapolate beyond the picture that Simon Raven draws, these character studies pose an alarming and complex problem. Here were crash courses in martial expertise and spartan morality designed to turn ordinary youths from respectable middle-class homes into highly professional officers. But somehow, somewhere along the line, the whole enterprise backfired. The youths emerged as four neo-feudal paternalistic despots, extraordinary anachronisms in the military forces of a modern democracy. It seems that all that remained of his training in the mind of each recipient was a faulty syllogism: Officers should be gentlemen; I am an officer; therefore I am a gentleman. After this he seemed to behave neither as an officer nor as a gentleman in any generally accepted sense of these terms. Or, as Raven puts it:

> Once an officer is established, in his own view, as a member of a superior and order-giving class, he never loses this sense; but he can, and often does, lose all awareness of the moral basis of this superiority and all the qualities which constitute this basis. He just becomes superior, as it were, *in vacuo*. He becomes a 'gentleman' ... when this happens, one gets that product so typical of the British—the Amateur English Officer. Highly trained professionally and morally, he has forgotten his professional techniques and sloughed off his sense of moral obligation; but he has retained an unassailable sense of his own superiority (for is it not innate) and absolute right to give orders.[13]

In the light of our foregoing discussion of military organizations Raven's data are not perhaps surprising. Even if specific to the England of the 1950s, they do, nevertheless, raise an issue which may well apply to many military organizations at different times in history. The argument centres around what has come to be known as a person's 'social reinforcement standard': the set of expectancies which he acquires regarding other people's behaviour towards him. From earliest infancy we all begin to build up a set of such expectancies. Thus the attractive child of doting parents may well develop a very high standard of social reinforcement. He will *expect* a fuss to be made of him. Conversely, the child lacking social graces and from an unloving family will be unlikely to expect much from his future encounters with other people. All this is common sense. But perhaps not so immediately obvious are the effects of discrepancies between a person's expectancies and his subsequent experiences. Two such have been found. The first is that both large negative and positive discrepancies are experienced as unpleasant. People who are *too* nice, *too* pleased to see one, *too* gushing or *too* full of praise may well create as much unease as those who fail to measure up to what one expects in the way of social reinforcement. The second effect follows from the first, namely that a person acts in such a way as to close the gap between what experience has led him to expect and what he now encounters. Or, to quote R. M. Baron: 'An individual does not wait passively to see if the social environment will live up to these expectations. He communicates to others his notion of what an appropriate reward is, for a given class of behaviours, by means of his "self-presentation", that is, by his selective display of certain attributes or behaviour in an attempt to influence the rate, direction or type of social reward that he is to receive.'[14]

How does this work out in the context of military affairs? In the old days when officers and men were drawn from widely different social classes the behaviour of one towards the other was what both expected as a result of previous experiences. Saluting and being saluted were natural successors to forelock-touching on the one hand, and the languid wave of the landed gentry on the other.

Nowadays things are different. The social distance between officers and men is, more often than not, contrived rather than rooted in their ancestry. For officers of humble origins this might well be expected to produce sizable problems of adjustment. The first jolt to their social reinforcement standard will be one of positive discrepancy. From being nobodies they suddenly find themselves elevated on to an institutionalized pedestal of dizzying proportions. Being saluted and called 'sir' at every turn goes to their heads. In most other walks of life such a huge positive discrepancy between new and old levels of social reward may well tend to change behaviour towards reducing the unexpectedly great flow of social rewards. We respond coolly to the gushes. But in the military this adjustment cannot occur. It simply is not done to shrug off the R.S.M.'s quivering salute with a 'Come off it, old chap, you embarrass me!'

Unable to protest, these new young officers have no alternative but to adopt a fresh set of expectancies which are so wildly discrepant with what they have been used to that they begin to overplay their hand. Conscious of the gap between their background and what is *now* expected of them, they over-compensate. What is worse, for the individual whose ego is on the fragile side, this tendency to over-compensate will be exacerbated by the discovery that, whereas ninety per cent of his fellow human beings may be heaping him with ego-boosting forms of address, there still remains another ten per cent, his seniors, to whom, in theory if not in practice, he remains the contemporary equivalent of what used to be called 'a wart'.

Under the circumstances the bizarre behaviour of Raven's fellow officers is hardly to be wondered at. Some reactions to Raven's observations may be found in the following letters which his 'Perish by the Sword' drew from three readers of the article.[15]

I have read Simon Raven's paper with great interest. When you ask for my comments you must bear in mind that in 1898, when I was gazetted, I joined a totally different type of army from that of today. It was an aristocratic army, feudal in the sense that it was grounded on leadership and fellowship, in which, with few exceptions, the leaders were the sons of gentlemen, and more frequently than not eldest sons—the privileged son. When I went to Sandhurst we were not taught to behave like gentlemen, because it never occurred to anyone that we could behave otherwise. We were taught a lot of obsolete tactics, as in every army of that day; did a tremendous lot of useless drill; but never heard a word about 'responsibility', 'loyalty', 'guts', etc., because—so I suppose—these were held to be the natural prerequisites of gentlemen.

The men—followers—of that period were a rough lot, simple, tough, illiterate; largely recruited from down-and-outs, men who had got into trouble, vagabonds, and a sprinkling of the sons and grandsons of N.C.O.s and private soldiers—military families—who generally became N.C.O.s. There were therefore two distinct classes (really castes) by birth. On the whole the men looked up to their officers, whether they were efficient or inefficient, and the officers did not look down on their men—why should they? I cannot recall any like Mr Raven's examples, A, B, C and D. The idea of an officer imposing his will on his men never entered his head, because one class was so superior and the other so inferior that it was unnecessary to do so. The superior could not lose caste should he play or mix with his men. I remember on my first tour as orderly officer the Q.M.S. was late for meat issue, and as I thought he had risked it because I was a novice I put him under arrest. When this was reported to the Adjutant, he took me aside and said: 'Strictly speaking you were right, but actually speaking you were an ass, so don't do it again'—and I didn't ...

Of the present-day democratic army I know next to nothing. Like the old aristocratic one it must have its good points, and is probably more efficient; but to me it is folly to try to mix the two. A gentleman is born and not made. It is probably true that it takes three generations to fashion one, and—so it would appear from Mr Raven's experiences—if you try to make them synthetically, you get neither an aristocracy nor a democracy. One sees this everywhere today—among rich and poor, on the roads, and in the factories. I repeat it again, gentlemen are not turned out like sausages; they are men of engrained honour, of principle and of decent behaviour; and some of the finest I have met in my long life have belonged to the humblest classes. Because this caste, rather than class, is becoming extinct is, in my humble opinion, one of the greatest factors at the bottom of the present world turmoil.

MAJOR-GENERAL J. F. C. FULLER

I read with fascination Simon Raven's test-tube study of the shaping of the military mentality. Never having thought much about the backgrounds of the things I detest, I've been mainly concerned with writing about its end-products—this despite attending two Officer Candidate Schools and having lived for several war years on both sides of the brass curtain. Broadly speaking, his indictment also holds for the United States. Money and position aren't as crucial here, but certain other training factors bring about the same dehumanization. One way or another, both the British and American military factories are mass-producing much the same type of commissioned personnel, except that ours are more cynical and yours more insufferable.

Pragmatically the caste system works. The Russians found that out in the 1920s and 1930s, after trying to abandon it; and Chu Teh never even abandoned it, although he modified it. On the other hand it warps both officers and privates as human beings, and it certainly manufactures personalities who tend to accelerate the thrust towards war. Mr Raven denies the latter, but he is referring to lieutenants and such. I would say that age and brass harden the military arteries. Figures like Patton, Haig and MacArthur are beautiful examples of the gulf between professional warriors and men with normal respect for human life, normal tolerance, normal good nature. Very few warriors ever bridge this gulf, and Mr Raven shows us why. Thus, whenever the twain meet on common ground, it is almost accidental. Today this accident exists, but only at the level of high policy. Socially, morally and intellectually, the indoctrinated officer is a man apart, a man almost to be pitied.

Mr Raven says: 'Nor is there reason to look for change'; but in time all this too will pass, for mass armies are as doomed as the battleship.

LEON WOLFF

I have read Simon Raven's article with very great enjoyment indeed, and I can well believe that all he says is true. It is a great mistake to suppose that one can change the nature of an officer-corps by broadening its social base. When between 1890 and 1914 the expansion of the German Army made it necessary to admit bourgeois officers into the officer corps—what William II euphemistically called the *Adel der Gesinnung*—the new entrants outdid the old nobility in their brash insolence and militarism. In general, conscription has always tended not to civilize the Army but to militarize the population—the effect of national service in Algeria on the French is very much a case in point.

The other important point which Raven makes is that of the absence of professionalism in the British officer. This, again, I endorse, but think he overstates it. This is a quality which tends to develop a little later in their careers, at Staff College level; when they suddenly realize (a) that promotion matters, and (b) that promotion does depend on professional ability. In their thirties, army officers tend to become much more serious and dedicated, and their horizon is no longer bounded by the regimental mess. The emphasis on *morale* which the young infantry officer gets drummed into him is designed to give him the complete self-assurance which is essential on the battlefield, maddening as it is anywhere else. The complexities of administration, strategy, logistics, military law *et hoc genus omne*, come at a rather later stage.

MICHAEL HOWARD

From a general study of leadership, it seems there is much in military organizations to invite incompetence. Officers are selected for the wrong reasons, required to fulfil incompatible roles and expected to function adequately in a communication-system of dubious efficiency.

At higher levels of command they are protected from adverse criticism by their invisibility, and by the plain fact that in times of stress even the poorest leaders, like drunken fathers and rabbits' feet, are clung to with pathetic, if misplaced, dependency.

21

Military Achievement

'How my achievements mock me!'
SHAKESPEARE, Troilus and Cressida, IV.ii

Besides providing legitimate outlets for aggression, the gratification of obsessive tendencies and reassurances about virility, armies and navies also cater for another basic human motive: the need to achieve. They do this in several ways. First, they embody related hierarchies of rank, money and class—with rank depending more (in the old days) upon money and class than upon merit. From the poorest private in the Pioneer Corps to a rich and aristocratic field-marshal, the rungs of the ladder climb ever more steeply upwards—an inviting prospect for the would-be achiever. Second, they accentuate the challenge of the promotional ladder by making certain upward movements very difficult indeed (but never quite impossible, as is shown by the case of William Robertson, who rose from under-footman in an aristocratic house to the rank of field-marshal and a knighthood). Third, the ethos of the armed forces is such as to make advancement laudable and highly rewarding. Generals have every advantage, bar that of age, over those lower down the ladder: they are richer, safer and more comfortable. Their chances of collecting honours, orders and knighthoods are also immeasurably greater than those of more junior ranks.

Finally, even the most modest thirst for achievement is encouraged by training and convention. The tabu on juniors speaking to seniors in officer-training establishments, saluting and being saluted, orders of march, rules as to who says 'sir' to whom, all serve to emphasize the horizontal stratifications of military organizations, besides adding lustre to each new level gained.

At first sight these arguments would seem to suggest that the possibility of promotion in a military organization would attract those with a potential for achievement: go-getters, entrepreneurs, innovators and men with energy and drive—in short, people who should make first-class military commanders. Sometimes it does, as in the case of Wellington, Montgomery, Rommel and Zhukov, men with inordinately strong needs for achievement. Unfortunately, however, there are aspects of a military career which are unlikely to attract people with high achievement-motivation. The fact that, traditionally, promotion depends upon seniority, class, wealth, conformity and obedience may well leave them rather cold. Neither the means nor the ends are sufficiently attractive.

Moreover, the military have never smiled upon entrepreneurs and innovators. The cut and thrust of private enterprise, cleverness and even working too hard have not been deemed 'good form'. There is, however, another class of person for whom the military might well be an attractive proposition. These are people whose achievement-motivation is pathological in origin. The crucial difference between the two sorts of achievement—the healthy and the pathological—may be summarized by saying that whereas the first is buoyed up by hopes of success, the second is driven by fear of failure. Both types of achievement-motivation have their origins in early childhood. The former is associated with the possession of a strong ego and independent attitudes of mind, the latter with a weak ego and feelings of dependency. Whereas the former achieves out of a quest for excellence in his job, the latter achieves *by any means available*, not necessarily because of any sincere devotion to the work, but because of the status, social approval and reduction of doubts about the self that such achievement brings.

Although these two sorts of achievement-motive may bring about rapid, even spectacular, promotion, their nature and effects are very different. The first is healthy and mature, and brings to the fore those skills required by the job in hand; the second is pathological, immature, and developing of traits, such as dishonesty and expediency, which may run counter to those required in positions of high command.

Applying these distinctions to the military, it would seem that senior commanders fall into two groups, those primarily concerned with improving their professional ability and those primarily concerned with self-betterment.

Critics of this theory may well object that professional excellence and the protection of self-esteem are not mutually exclusive incentives; that far from being different in motivational make-up, Montgomery of Alamein, for example, and Townshend of Kut were two of a kind. Both were conceited, vainglorious[1] showmen with an eye to their own personal advancement; both had charismatic personalities and were popular with their men. If there was any difference, according to this argument, it was in their luck. Had Montgomery been at Kut and Townshend at Alamein, their subsequent reputations would have been reversed.

Anticipating these very reasonable objections from those who denigrate Montgomery and nurse a sneaking regard for 'Charlie of Chitral', let us see how they may be answered. In the first place we are concerned with *primary* motivation. No one would dispute that Montgomery was motivated towards, and enjoyed, personal success, nor that Townshend had considerable regard for professional efficiency, but in both cases these were *secondary* factors in their careers. This is shown most clearly in those instances when their primary and secondary motives were in conflict. On several occasions Montgomery risked his own career by sacrificing popularity with those on whom his promotion might depend, for the sake of what he felt was the right course of action in terms of military efficiency. Typical was the occasion at the time of Dunkirk when Montgomery, 'a very junior Major-General', had the temerity to tell the C.I.G.S., Lord Gort, that his decision to appoint Lieutenant-General Barker as the Corps Commander to supervise the last stages of the evacuation was *wrong*. On the grounds that Barker lacked the essential qualities for this crucial role, he persuaded Gort to rescind his decision and nominate Alexander. As Ronald Lewin remarks: 'There can be little doubt that to Montgomery's act of intelligent effrontery a good many men owe, if not their lives, then at least salvation from years in a German prison camp.'[2] This sort of personal risk-taking for the sake of larger issues was not a feature of Townshend's make-up. For him it was always self first and Army second. As for the suggestion that Montgomery was lucky to be at Alamein and Townshend unlucky to be at Kut, one can only opine that the characters of these two men and their previous performances make it highly unlikely that Townshend would have been given command of the Eighth Army, and even more unlikely that Montgomery would have abandoned ten thousand of his men to a lingering death in the desert.

But let us leave consideration of these particular personalities for a wider issue. Research suggests that these two sorts of achievement-motivation go along with certain other personality-traits. Thus need-achievement (motivation towards professional excellence) is accompanied by:

1. Greater occupational and intellectual competence;
2. A better memory for uncompleted tasks and therefore a predisposition to finish something once begun;[3]
3. A preference, when choosing working partners, for successful strangers rather than unsuccessful friends;[4]
4. A greater readiness to volunteer for psychological experiments;[5]
5. Greater activity in the institution or community of which they are a member.[6]

There are grounds for believing that this curious miscellany of traits, which are found in people who score high on tests of achievement-motivation, come closer to describing such unequivocally great commanders as Wellington, Napoleon, Nelson, Shaka, Allenby and Slim than it does those who were inept. In terms of their record, the former certainly showed greater occupational competence. All of them are renowned for choosing their staffs, and other subordinates, for their professional competence rather than for some other reason. Allenby's acceptance of T. E. Lawrence, despite the latter's unorthodox behaviour, unsoldierly appearance and 'bolshie' attitude towards senior officers, is a classic example of this trait, as is also the way in which another great general, Alanbrooke, steadfastly supported Montgomery despite the latter's irritating ways and monumental faux pas.

We do not know whether these men had good memories for uncompleted tasks, but certainly they struck at something once begun with painstaking tenacity.

As to the fourth trait, that of a readiness to partake in psychological experiments, this could be taken to signify an adventurousness that is unrestrained by fear of personal exposure—unshakable self-confidence, perhaps. Interpreted in this light it certainly fits such men as Wellington, Shaka, Rommel, Slim and Zhukov. From the following examples collected by Lewin it was also clearly manifest in Montgomery. Thus R. W. Thompson said: 'Montgomery had the knack of creating oases of serenity around himself'; and Goronwy Rees, from his first meeting with him, remembers that 'that air of calm and peace which he carried with him was so strong that after a moment my panic and alarm began to die away: it was something which one felt to be almost incongruous in a soldier.' Similarly Sir Miles Dempsey described how he never failed, at bad moments, to be invigorated by a visit to Montgomery, and that the latter, with his cheerful smile and his confident air, had a way of turning apparent difficulties into phantoms. As Lewin remarks: 'He was sure of himself.'[7]

Finally we have the trait of great activity, strikingly present in Lawrence, Slim, Napoleon, Kitchener, Allenby and Montgomery, and conspicuously absent in Elphinstone, Haig, Buller and Raglan. In distances covered, units visited and troops spoken to, those generals who, we would suggest, rate highly on need-achievement make commanders like Haig and Raglan, who rarely stirred from their headquarters, look somewhat static, to say the least.

This dimension of activity bears upon two other, at first sight somewhat bizarre, findings from research into achievement-motivation. High scorers on need-achievement tests have been found to prefer sombre to bright colours[8] and to produce doodles which differ in several characteristics from those drawn by people low in achievement-motivation.[9] The self-expressive scribblings of high achievers, in fact, have been found to resemble such decorative designs as occurred in particular cultures during times of great architectural achievement and activity. In other words, latter-day tests of achievement-motivation seem to be measuring something meaningful in terms of human activity.

As for the colour-preference data a parallel has been drawn between the liking of subdued colours shown by high achievers and the equally sombre tastes favoured by puritans: people renowned for their high achievement-motivation and single-minded dedication to hard work in pursuit of self-abnegating goals.

Returning to the military scene, it would be nice to find that these data on puritanism are reflected in the personality characteristics of competent as against incompetent commanders. Certainly there are striking differences in asceticism between some of the best and some of the worst. Compare Montgomery with Buller. Chester Wilmot said of the former: 'He was not as other men. He revealed no trace of ordinary human frailties and foibles. He shunned the company of women; he did not smoke or drink or play poker with "the boys".'[10] Buller's specially constructed and elaborately fitted out cast-iron kitchen and attendant wagons of champagne had to be dragged wherever the general's duties happened to take him. Compare Lawrence — roughing it with his tiny Arab force on the 800-mile trek across the desert to wrest Akaba from the Turks — with Townshend — comfortably ensconced in his villa on the Sea of Marmara while his captured troops died in their thousands from exposure, malnutrition and brutality. The contrast in both cases is between the self-imposed asceticism of high achievement-motivation and the self-indulgence of one less concerned with professional excellence than with personal advantage.

Clearly there are exceptions to this rule. Thus the prodigious appetite of Allenby and the Presbyterian origins of Haig went along with behaviour that does not accord with the predictions of achievement theory. One can only opine that the total behaviour of these officers suggests that there must be other variables which contribute to the obvious differences in professional excellence and egocentric self-betterment. What these might be we shall consider presently.

Notwithstanding these exceptions and at the risk of over-simplifying what are really very complex issues, there *are* grounds for believing that high achievement-motivation characterizes manifestly successful commanders. By itself this is not a very surprising conclusion. Its real importance, however, resides in its antecedents and, even more particularly, in the attention it draws to those people whose motivation took them to the top but was clearly not concerned with professional excellence. In considering this reverse side of the coin we have to ask what was the nature of *their* impulses. And how is it that some of the criteria for promotion in military organizations are evidently such as to favour people with a pathological degree of achievement-motivation?

There are grounds for thinking that incompetent commanders tend to be those in whom the need to avoid failure exceeds the urge to succeed. According to J. W. Atkinson and N. T. Feather, such people tend to eschew activities in which they may show up in a poor light, and, unless forced to do so, refrain from taking on any skilled task where there are any doubts about the outcome. They go on: 'Given an opportunity to quit an activity that entails evaluation of his performance for some other kind of activity he is quick to take it. Often constrained by social pressures and minimally involved, not really achievement-orientated at all, he will display what might be taken for dogged determination in the pursuit of the highly improbable goal. But he will be quickly frightened away by failure at some activity that seemed to him to guarantee success at the outset.'[11]

We have surely all known people of this kind. In terms of the older psychological theory of Alfred Adler, we would say that they have an underlying 'inferiority complex', from which springs their fear of failure. Be that as it may, in terms of their chronic emotional state such people may be thought of as starting with a debit account. They are, so to speak, driven from behind rather than pulled from in front. They have to achieve, not for the satisfaction which achievement brings but because only by so doing can they bolster up their constantly sagging self-regard; a case of running hard to stay in the same place. But herein lies their special dilemma. Though they need to achieve, it is the very act of trying which exposes them to what they fear most — failure. They are like people who try to climb mountains out of an underlying fear of heights.

As R. C. Birney and his colleagues point out,[12] this state of mind leads to a number of compromise solutions. Thus the person who fears failure prefers tasks which are either very easy or very difficult. If they are easy he is unlikely to fail; if very difficult then the disgrace attaching will be small, for no one really expected him to win. He will also tend to choose non-competitive jobs while avoiding complex or unfamiliar ones. He will be conformist rather than prepared to stick his neck out. He may gravitate towards careers which offer order, minimal competition, gradual advancement and diffuse responsibility, and if forced into serious situations of achievement will be most concerned with the social approval that is placed on his behaviour.

It would not be surprising to find that such people are attracted to and prosper in the armed services. For if one plays it carefully, the military, in contrast to the world of commerce, offers achievement without tears. Stick to the rule book, do nothing without explicit approval from the next higher up, always conform, never offend your superiors, and you will float serenely if a trifle slowly upwards— a blimp in both senses of the word. But if the military provides a congenial vocation for those with a fear of failure it is also adept at keeping them that way.

Confidential reports, courts martial, reduction to the ranks, cashiering and, in days gone by, the firing squad, are effective deterrents to straying from the straight and narrow path. The net result would be a bimodal distribution of officers at every grade: those who take risks and get away with it—the Montgomerys and Lawrences of this world —and those who have plodded up the hard but safe way—the 'good' boys who never speak out of turn, who make up in tact and conformity for what they lack in enterprise and initiative.*

Contemplation of inept commanders suggests that they were of the latter genre. In the first place, they were renowned, almost without exception, for being hypersensitive to criticism. This fear of criticism follows directly from their need for social approval, which is itself the child of low self-esteem. The efforts which were made to depose Leslie Hore-Belisha because he had dared to criticize the general staff typifies this response. Secondly, such military commanders were adept at disclaiming responsibility for actions which ended in failure.

The reluctance of Raglan and Buller to issue orders and directives, followed, when things went wrong, by the choosing of scapegoats from among their subordinates, exemplifies two well-known devices for avoiding the unpleasant consequences of failure. The blaming of junior officers and men for the failure to exploit the Cambrai tank attack, and the subsequent suppression of the Kirke Committee Report by Montgomery-Massingberd because it reflected on his contribution to the Somme offensive, fall into the same category of attempted evasion of responsibility.

Then again there is the interesting behaviour of people who try to avoid the unpleasant consequences of failure by not really trying. Percival's extraordinary refusal to prepare defences in Johore and the north of Singapore island answers this description. A closely allied phenomenon is that of attempting tasks so difficult that no one expects one to succeed; hence little disgrace attaches to failure. The retreat from Kabul, the Third Battle of Ypres, the foolhardiness of Admiral Phillips which resulted in the loss of the two capital ships from air attack, and the behaviour of Townshend which culminated in the disastrous siege of Kut, could be taken to exemplify this particular defence against being branded a failure. In this connection mention should be made of two other apparently very different phenomena: deliberately sustained ignorance and compulsive acts of bravery.

The first of these may be a short-term, almost reflexive response to the possibility of bad news, as when General Warren insulated himself from information from the top of Spion Kop (see page 65); or it may

---

* Like Rear-Admiral Markham whose blind obedience was largely responsible for the sinking of H.M. *Victoria* in 1893.[13]

be the avoidance of issues which threaten to reveal one's limitations, as when in the 1930s senior commanders resisted progress in the technology of war. They understood horsed cavalry and battleships, they did not understand tanks and aircraft—hence their denial that the latter were worthy of attention, let alone ownership. Brooke-Popham's professed underestimation of Japanese air ability and belief in his own ancient aeroplanes may well be subsumed under the same heading. If you don't have the tools you can't be blamed for not doing the job.

As for the question of physical bravery, it in no way detracts from feats of courage to note that the fear of being afraid, the fear of social disapproval for cowardice, and, most important, the personal shame attendant upon flinching in the face of danger, could drive a man to perform acts of valour far beyond the normal call of duty. This is not to deny that bravery occurs for other reasons—out of pure altruism or patriotism—but merely that some individuals are so lacking in self-esteem that they will gladly exchange the fear of failure for their own physical destruction. In a very real sense military organizations recognize and trade upon this fact of human nature. 'Death rather than dishonour' is no empty platitude but formulates an essential and ancient feature of the military creed. Whatever else they do, military traditions, battle emblems, regimental standards and decorations for bravery all serve to reinforce the goal of achieving social approval at whatever cost.

In this connection it is noteworthy that many of the less successful generals described in this book had a fine reputation for physical courage. By itself, of course, a record for bravery does not indicate some pathology of achievement-motivation but when it is combined, as in the case of Buller, with indecision, passivity and a shelving of responsibility it is more than likely that his acts of courage might well have sprung from a deep lack of certainty about himself.

It cannot be emphasized too strongly that this suggested relationship between valour and the need to prove oneself in no way debases bravery. On the contrary, if one takes the view that the best measure of courage is the fear that is overcome then these were the bravest of them all, for it was only by conquering rational fear that they could nullify their fear of being afraid. The tragedy of this issue is that if military organizations select their senior commanders for their physical (as opposed to moral) bravery, they not only might ignore other equally important attributes but are bound to select a proportion of individuals whose underlying psychopathology is quite unfitted to positions of high command. In this way they invite incompetence. It is to Buller's credit that he had no illusions about himself in this respect and was prepared to say so. As Pemberton notes: 'It is a pity that six months earlier the Government had paid insufficient attention to Buller's honest admission regarding his fitness for the job. He had then told Lansdowne: "I have always considered that I was better as second in a complex military affair than as an officer in chief command ... I had never been in positions where the whole load of responsibility fell on me."' [14] To Wolseley he was even more un-buttoning, loudly objecting to having such a command 'forced' upon him. All in all, Buller exemplifies the fact that a physically brave man does not by any means make a morally brave commander.

The distinction made between senior commanders who evince all the signs of high-need achievement-motivation and those who appear to have been driven by a fear of failure gains further support from two other suggestions which have emerged from research in this area. The first is that people who fear failure prefer practice and 'games' to the real thing. In other words, and not very surprisingly, they prefer those activities which set a lower premium on success. If you fear failure it is better to be beaten at Monopoly than go bankrupt in the property market. Though standing to gain less you also stand to lose less, and for some people the latter consideration is the more important of the two.

Is it too far-fetched to suppose that this goes some way towards accounting for the otherwise inexplicable behaviour of that incredible figure, General Percival?[15] Here was a man of keen intellect whose performance in the field fell far short of the brilliant promise he had shown when conducting war games on the sand-tables of Staff College. That aggressive and equally controversial senior commander, the Australian Major-General Gordon Bennett, who served under Percival, seems to have sensed this possibility when he drew an angry distinction between 'thinkers' (men who have passed through Staff College) and 'fighters' (men who had not). 'Thinkers' according to Bennett were yes-men who lacked the aggressive spirit so necessary on the battle-field. 'Fighters', with whom he clearly identified himself, were practical men who, unencumbered with theory, didn't give a damn for anyone and got on with the job.

It was this same general, whose actual behaviour often belied his strong views about yes-men (he was always very careful not to criticize any particular person *senior* to himself), who drew attention to another shortcoming of those who fear failure: their love of privacy. Since by concealing his failures a man escapes the possibility of social disapproval, we might expect that the greatest and the more controversial senior commanders would have rather different attitudes towards publicity. In general this seems to be the case.

Thus, while Montgomery made it his business to accommodate the press and to give as wide publicity as possible to his Army and himself, commanders like Buller, Haig, Percival and Brooke-Popham did their utmost to conceal from the public what was going on. Gordon Bennett is typically outspoken on this subject,

> The conservative soldiers and sailors of the old school would prefer to tell the public nothing. They resent the inquisitiveness of war correspondents and the public about naval or military affairs, which they look on as their close preserves. They seem to think that this is their war and that all journalists are insolent 'nosey parkers' ... in the interview they clumsily show their inexperience in such matters by bluffing or attempting to bluff their interviewer, telling them little or nothing.[16]

Montgomery was of course quite exceptional in this respect, though even he, as Lewin points out, could be understandably reticent about his less successful ventures. But this natural reticence is a far cry from the deliberate censorship and/or falsification of news which, rationalized in the name of security or preservation of civilian morale, has characterized some military enterprises.

The last trait of those who harbour a fear of failure concerns the selection of subordinates. It is a common observation that those over-concerned about their image devote considerable attention, energy and time to a continuous self-assessment against some external standard, usually another person. Ostensibly such behaviour may be directed towards promotional prospects. In the words of Liddell Hart: 'It is amazing to find how much time many rising soldiers spend in studying the Army List and its bearing on their own promotional prospects. One prominent general, Ironside, even kept a large ledger in which he entered details of the service record of all the officers above him in the list, with his and other people's views on their performance, health and prospects.'[17]

There are really two components to this process. The first concerns the way an individual sees himself in comparison with his competitors, and the second the way he thinks others will see him in comparison with his contemporaries. In either case he may well try to elevate his own self-estimation by choosing a low standard with which to make comparison. Hence the phenomenon of people who tend to shun the company of individuals more gifted and even to choose workmates or *select as subordinates* people whom they consider inferior to themselves. By so doing, their own position is not threatened by the possibility of being supplanted by a bright underling; they feel themselves superior when in the company of those less able than themselves; and finally they appear to others better than they really are when viewed against a background of individuals duller than themselves.

Clearly, possession of this particular trait of achievement-motivation could have disastrous consequences when a senior decision-maker is dependent upon the competence of his staff. There is, however, another side to the coin. So great may be the driving force of pathological achievement-motivation that it can on occasions bring to the fore an able individual whose ability alone would have been insufficient to guarantee his rising to the top. We may not like such people and some of their characteristics may not constitute perfect ingredients for great generalship, but it could be argued that this hardly matters if their urge for self-advancement makes their talents available on those occasions when there is no one better to fulfil the role.

Of all the senior military commanders whose records illustrate this simple truth, Haig seems the prime example. The acknowledged dunce of the family, Haig's military career seemed directed towards trying to prove otherwise. It is an astonishing tribute to powers of disturbed achievement-motivation that out of a nation of twenty million people fighting for its life, there should have arisen a leader of such *apparently* limited capacity: a man of such *apparently* mediocre intellect that he had the greatest difficulty in passing even the Sandhurst entrance examination and actually failed the Staff College examination, where 'he attracted unfavourable comment' from his examiner, General Plumer; a man who at thirty-eight was still only a captain, a man who had been so completely outmanœuvred in the pre-war training exercises of 1912 that the manœuvres had to be abandoned a day earlier than scheduled; a man whom Lloyd George was to call 'utterly stupid' and Briand '*tête du bois*'. How did it come about?

For a start, it is highly unlikely that Haig was in fact of low intelligence—hence my reiteration of the word 'apparently' in the above paragraph. Had he been as stupid as his detractors maintain, it is unlikely that Lord Haldane would have chosen him as military adviser before the First World War. For reasons that are dealt with in a subsequent chapter, it is far more likely that Haig's educational backwardness, and such other intellectual characteristics as his undoubted administrative ability, were products of a mind constrained and inhibited by the emotional consequences of early damage to his self-esteem.

In a word, he was a victim of the naive belief that backward boys are necessarily dim—a belief that is responsible for one of the most malign of vicious circles to beset the growing child.* For Haig, as the following points suggest, the outcome of this vicious circle was a lasting impairment of his achievement-motivation.

To begin with, there are grounds for supposing that, born as he was into a very successful family of entrepreneurs, and saddled with an ambitious mother, Haig was probably made conscious from a very early age that he did not measure up to his highly competent relations, people who would be likely to score high in need-achievement, and therefore inclined to goad their 'dullard' offspring. Painfully aware of his limitations, he then tried to enter the one profession open to the dunce of the family, the Army, only to find that even here he failed to shine. It is not unreasonable to suppose that when he was passed over when in competition with his fellow officers, these further injuries to an already injured self-esteem would have pushed his fear of failure to breaking-point.

At this stage more might never have been heard of him were it not for his elder sister Henrietta. For Henrietta knew the Duke of Cambridge, who, in his turn, was able to arrange Haig's entry into Staff College. It was but a small step from these felicitous coincidences to a marked change in Haig's fortune. Though made Chief of Staff to French in the Boer War, his part in this unfortunate affair seems not to have slowed up his ascent. On the contrary, he was subsequently

* The invalidity of this belief has been well documented by the Illingworths in *Lessons from Childhood*. I am indebted to Hugh L'Etang for drawing my attention to this work.

appointed A.D.C. to the King and became respected for his conventional opinions, e.g., that 'Cavalry will have a larger sphere of action in future wars' and 'Artillery only seems likely to be really effective against raw troops'.

From that time on, Haig's behaviour helped him along the road. It included marriage to one of Queen Alexandra's maids of honour, a tireless currying of favour with the King, a steady denigration of his competitors, and the removal of his superior commander.

Haig's stated views on his military colleagues bear this out. Of his Chief of Staff, Major-General Sir Archibald Murray, he wrote: 'I had a poor opinion of his qualifications as a general. In some respects he seemed to me to be "an old woman". For example, in his dealings with Sir John [French]. When his own better judgment told him that something which French wished put in Orders was quite unsound, instead of frankly acknowledging his disagreement, he would weakly acquiesce in order to avoid an outbreak of temper and a scene.'[18] In the space of some seventy words, he manages to express contempt not only for his subordinate but also for his superior—the commander-in-chief. Secondly, by criticizing Murray for acquiescence he betrays a remarkable lack of insight into the inconsistency of his own attitudes. For in another entry in his private papers he condemns another of his Chiefs of Staff, Brigadier-General Gough, for *not* 'weakly acquiescing' to the plans of his master: 'After dinner at Mareuil he [Gough], in his impetuous way, grumbled at my going on "retreating and retreating". As a number of the staff were present, I turned on him rather sharply, and said that retreat was the only thing to save the Army, and that it was his duty to support me instead of criticizing. He was very sorry, poor fellow.'[19] In the light of his manifest disloyalty to his own boss, Sir John French, we can only assume that what was sauce for the goose was evidently not sauce for the gander.

To his criticism of Murray and French, Haig added: 'However, I am determined to be thoroughly loyal and do my duty as a subordinate should, trying all the time to see Sir John's good qualities and not his weak ones, though neither of them [French or Murray] is at all fitted for the appointment which he now holds, at this moment of crisis.'[20]

Denigration of his colleagues was not confined to his assessment of military performance. Thus he criticized Major-General Monro for being fat and the French commander D'Urbal for being unpleasantly polite. Lieutenant-General Henry Wilson he described as 'such a terrible intriguer ... Sure to make mischief ... [his face] now looks so deceitful.' In the same generous spirit he described the Military Secretary, Lambton, as 'weak' and 'stupid'.

Haig's talent for finding fault with everyone but himself was particularly keen whenever events had resulted in a military setback. After the defeat at Neuve Chapelle, with the loss of many British lives, Haig tried to fasten the blame on a Major-General Davies: 'He was unfit to command a division at this critical period of the operations in France but should be employed at home.'

When this attempt at finding a scapegoat misfired, as a result of General Rawlinson, Davies's superior, taking the blame, Haig quickly shifted his aim. 'Rawlinson is unsatisfactory—loyalty to his sub-ordinates, but he has many other valuable qualities ...'[21] Haig, the man who set such great store by loyalty to himself, could evidently not tolerate the same sentiment between his fellow soldiers. His double standard on this issue was conspicuous in his behaviour towards Sir John French.

During a review of the B.E.F. at Aldershot he buttonholed the King and told him, 'as I felt it my duty to do, that from my experience with Sir John in the South African War he was certain to do his utmost loyally to carry out any orders which the Government might give him. I had grave doubts, however, whether either his temper was sufficiently even, or his military knowledge sufficiently thorough, to enable him to discharge properly the very difficult duties which would devolve upon him during the coming operations.'[22]

That was on August 11th, 1914. On October 2nd, 1915, after the almost total destruction of the old British Regular Army Haig wrote of his commander-in-chief: 'It seems impossible to discuss military problems with an unreasoning brain of this kind ... the fact is that Sir John seems incapable of realizing the nature of the fighting that has been going on and the difficulties of getting fresh troops and stores forward and adequate communication-trenches dug.'[23] On Saturday, October 9th, he made an impromptu report in private to Lord Haldane. This was highly critical of, and shamelessly disloyal to, his commander Sir John French.

By way of consolidating his position, Haig also made use of his close friendship with General Robertson, French's Chief of Staff. To Robertson, he said:

Up to date I have been more than loyal to French and done my best to stop all criticism of him and his methods. Now, at last, in view of what happened in the recent battle over the reserves ... and of the seriousness of the general military situation, I have come to the conclusion that it is not fair to retain French in command on this, the main battle front. Moreover, none of my officers commanding corps have a high opinion of Sir John's military ability. In fact they have no confidence in him.[24]

Robertson communicated all this to the King, who promptly came out to see for himself. Haig used the opportunity to tell the King, after a further preamble about French's incompetence, that 'for the sake of the Empire, French ought to be removed!' He added the hint that 'I personally am ready to do my duty in any capacity'.

And so Haig got his wish, reaching the pinnacle of

the greatest army that the Empire had ever put in the field in the past, or was ever to amass in the future. A body whose heroism and devotion was such that they could twice in two successive years be ravaged in hopeless offensives, who were in a single day to lose more men than any other army in the history of the world, whom, after twenty-seven months of slaughter and exhaustion, he was to leave so perilously exposed that they were nearly annihilated.[25]

In delineating the characteristics of individuals with a pathological degree of achievement-motivation, Birney and his colleagues make the point that fear of failure predisposes towards secretiveness. So sensitive are such people to criticism that they prefer conducting their affairs in the strictest privacy. Hence it is no surprise to find that Haig nursed a pathological fear of journalists. Not only did he refuse to see them and obstruct their activities but he actually wrote to the C.G.S. 'recommending that no newspaper correspondent be allowed to come close to the front during active operations', i.e., for the duration of the war.

Finally, almost predictably, Haig suffered from one of the commonest of psychosomatic complaints: asthma. Although the specificity of asthma to any given underlying psychological complex is still debatable, there is evidence to support Alexander's hypothesis[26] that the asthmatic patient has a basic unresolved conflict over natural dependency. In trying to defend against this infantile dependency, such patients develop traits of aggressiveness, hostility or over-sensitivity. Other workers have shown the presence of intense hostility in asthmatic subjects,[27] and that such people may be both hostile and constricted.[28]

Since Haig was both hostile and constricted, besides manifesting signs of unresolved dependency, it is perhaps not surprising to find that he had an attack of asthma on the eve of that most violent of outlets for pent-up hostility and frustration, the battle of Loos.

In this chapter we have examined the proposition that one factor which distinguishes the less from the more versatile of military commanders is their underlying achievement-motivation. It has also suggested that there are features of the older military organizations which attract individuals with pathological achievement-motivation. Thus the 'fear of failure' syndrome not only determines vocational selection but by its very nature facilitates acceptance and promotion within the military organization.

In other words, those sorts of behaviour—conformity, obedience and physical bravery—which earn social approval and increased self-esteem are the very ones rewarded by steady advancement in military organizations. Conversely, many of the traits associated with the more entrepreneurial aspects of need-achievement—unconventionality and scant regard for the approval of others—are not welcomed in military circles. As James Grigg once said of Field-Marshal Montgomery: 'There is always "a cold hush" whenever his name is mentioned.'

A number of consequences follow from these differences in achievement-motivation. The first is that while both the drive towards self-betterment and the drive towards professional excellence may take a man to the top, only the latter guarantees that he is fitted for the job of high command, for only in the latter case can we be sure that he has the requisite expertise. (He must have it then, because this was his only qualification for promotion.) Conversely, the man who reaches a position of great power as an outcome of his drive to achieve greater self-esteem may not necessarily have any outstanding military ability, for his ascent did not depend upon professional excellence. More serious is the fact that even if he has the requisite military skills these may be rendered quite nugatory by those other traits that are part and parcel of his underlying personality-structure: moral cowardice, indecisiveness, secretiveness and sensitivity to criticism.

Even if he has, like Haig, considerable ability, and can learn to overcome the more disastrous products of a weak ego, the man who reaches a position of high command out of a compulsive thirst for personal advancement will tend to lack that creative talent and flexibility of mind so necessary in modern warfare. In this respect he is twice cursed—firstly, by his underlying personality, which resulted in his attraction to the military, and secondly by a lifetime of learning to curb initiative and freedom of thought. Freedom of expression and cognitive processes unfettered by inhibitions were not looked upon with favour in military personnel.

It seems then that in the case of achievement-motivation (as with obsessive tendencies), military organizations attract and then reinforce those very characteristics which will prove antithetical to competent military performance.

But these side-effects of trying to professionalize violence are probably less malign in their consequences than yet another by-product of military organizations: their authoritarianism.

Let us, therefore, take a brief respite from the battlefield for an examination of this thesis.

22

Authoritarianism

'It is a strange desire to seek power and to lose liberty.'
BACON, *Essays: Of Great Place*

In discussing military organizations it was suggested that a symbiotic relationship exists between certain characteristics of armed services and the private needs of their individual members. Emphasis was laid upon the central role in this relationship of anxiety, that insidious motivator of much human behaviour. In the military mind, it was pointed out, anxiety has many sources—fear of death and mutilation, fear of supersession, fear of failure and social disapproval, fear of public disgrace, and, underlying all, that fear of total disorder which is an inseparable product of unleashing normally tabooed instinctual forces. Finally it was suggested that a special predisposition towards these several sorts of anxiety may be present in some people as a result of their early childhood. Such people may well be drawn towards military organizations because the latter have, of necessity, perfected devices like 'bull' and discipline, hierarchical command-structures and rigid conventions which not only allow of aggression without anxiety, but actually *reduce anxiety that may have originated at a much earlier period of life.*

In the light of all this, it is encouraging to encounter a substantial body of research which not only provides support for the thesis but also fills in many of the gaps. It is that on the Authoritarian Personality.

For the impetus behind this study of authoritarianism we have to thank the founders and proponents of the Third Reich. They it was who presented to the world a phenomenon the like of which has never been seen before or since—the systematic and bureaucratized murder of six million Jews. To the inquiring mind, anti-Semitism on this scale would seem to demand, at the very least, some explanation. For a group of researchers at Frankfurt and later at Berkeley, in the University of California,* the fact that human prejudice could assume such monstrous proportions suggested the possibility of a particular personality-type being implicated in the perpetration of these dark events.

They were not alone in this supposition. Some ten years earlier, the Nazi psychologist Jaensch reported that he had identified two basic personality-types, 'S' and 'J'. 'S' types were so called because they manifested *synesthesia*, the harmless enough tendency, one might think, to have subjective experiences in one modality when receiving stimulation in another. For Jaensch this artistic gift of being able to experience affinities between, say, colours and sounds amounted to a sort of 'perceptual slovenliness', the careless mixing-up of sensory impressions.[1] Pressing the matter further, he found that this trait went along with other 'regrettable tendencies'. The 'S' type was 'liberal' in his views and eccentric in behaviour. He was also weak, effeminate and prone to the heretical belief that people are largely shaped by their environment and education. All this, Jaensch claimed (largely on the basis of his own prejudices and political leanings), was the result of inter-racial contamination and mixed heredity. 'S', or 'anti-types' as they were also called, included Jews, Orientals and Communists. Fortunately for his racist views, if not for the repute of scientific theorizing, Jaensch also 'discovered' a contrasting class of individual which he modestly labelled the 'J' type. 'J' types were 'good' types and would make good Nazis. Amongst their sterling qualities were purity in perception and the sure knowledge that human behaviour is determined by blood, soil and national tradition. The 'J' type would be a he-man, hard and tough, a man you could rely on. These qualities would, he said, have been handed down by a long line of north German ancestors.

In their investigation of anti-Semitism and ethnic prejudice the American researchers[2] also found two contrasting personality-types, and these were very like those described by Jaensch. Needless to say,

* Towards the end of the Second World War, the American Jewish Committee set up a department of scientific research under Mark Horkheimer, who had previously been director of the Institute for Social Psychology in Frankfurt. When this institute was suppressed by Hitler, Horkheimer and his colleagues Erich Fromm and T. W. Adorno fled to the U.S.A., where they teamed up with N. Sanford for their research at Berkeley into the authoritarian personality.

they evaluated them rather differently. The one that corresponded to the 'J' type they called 'the authoritarian personality'. Such a person was anti-Semitic, rigid, intolerant of ambiguity and hostile to people or groups racially different from himself. By the same token, the polar opposite to this type, Jaensch's contemptible 'anti-type', was individualistic, tolerant, democratic, unprejudiced and egalitarian.*

They arrived at these distinctions by testing the attitudes of over two thousand Americans from many different walks of life. Their tests provided measures of anti-Semitism, ethnocentrism, political and economic conservatism and implicit anti-democratic trends or potentiality for Fascism — the 'F' scale. This, much of which is based on actual utterances by Nazis, measures an individual's predisposition towards:

1. *Conventionalism*, i.e., rigid adherence to conventional middle-class values.

2. *Authoritarian submission*, i.e., a submissive, uncritical attitude towards the idealized moral authorities of the group with which he identifies himself.

3. *Authoritarian aggression*, i.e., a tendency to be on the look-out for and to condemn, reject and punish people who violate conventional values.

4. *Anti-intraception*, i.e., opposition to the subjective, the imaginative and the tender-minded.

5. *Superstition and stereotypy*, i.e., a belief in magical determinants of the individual's fate, and the disposition to think in rigid categories.

6. *Power and 'toughness'*, i.e., a preoccupation with the dominance-submission, strong-weak, leader-follower dimension, identification with power-figures, overemphasis upon the conventionalized attributes of the ego, exaggerated assertion of strength and toughness.

7. *Destructiveness and cynicism*, i.e., generalized hostility, vilification of the human.

8. *Projectivity*, i.e., the belief that wild and dangerous things go on in the world; the projection outwards of unconscious emotional impulses.

9. *'Puritanical' prurience*, i.e., an exaggerated concern with sexual 'goings-on'.

For the final stage in their research the American investigators subjected those who had scored very high or very low marks on ethno-

---

* In many respects, this American research confirmed earlier work in Britain by Eysenck and others.[3]

prejudice to a number of open-ended projective tests and clinical interviews. These were aimed at uncovering such factors in a person's past history, and deeper facets of his personality, which make for authoritarian prejudice. What, they asked, underlies the striking differences which had emerged from the earlier questionnaires?

The results of this study more than justified the energy which had been put into it. At one level they constituted a fitting monument to the six million victims of Fascist prejudice. To the sensitive reader the pages of *The Authoritarian Personality* are haunted by the ghosts of Belsen and Auschwitz. At another level they sound a grim warning for societies beyond that of Nazi Germany. As Roger Brown remarks, some of the data 'are hair-raising. They suggest that we could find in this country [U.S.A.] willing recruits for a Gestapo.'[4]

The results delineated the authoritarian personality. People who were anti-Semitic were also generally ethnocentrically prejudiced and conservative. They also tended to be aggressive, superstitious, punitive, tough-minded and preoccupied with dominance–submission in their personal relationships. That this cluster of traits suggested a unique underlying personality-structure was born out by the clinical interviews. It seems that authoritarians are the product of parents with anxiety about their status in society. From earliest infancy the children of such people are pressed to seek the status after which their parents hanker.

There seem to be two converging reasons why such pressures produce prejudice and the other related traits. In the first place, the values inculcated by status-insecure parents are such that their children learn to put personal success and the acquisition of power above all else. They are taught to judge people for their usefulness rather than their likeableness. Their friends, and even future marriage partners, are selected and used in the service of personal advancement; love and affection take second place to knowing the right people. They are taught to eschew weakness and passivity, to respect authority, and to despise those who have not made the socio-economic grade. Success is equated with social esteem and material advantage, rather than with more spiritual values. Then again, they are imbued by their parents with rigid views regarding sex and aggression. Sex is dirty, and aggression permissible only towards such out-groups as Jews, Negroes and law-breakers. To complete this gloomy pattern, the sex-role stereotypes of an 'upright' middle class are rigidly implanted. Boys must be masculine, tough and strong, and girls (under a respectable cloak of frigid femininity) alive to the possibilities of granting their favours in the service of status-seeking.[5]*

In the second place, the interview data collected by the Berkeley researchers also suggested that the parents of their authoritarian sample imposed these values with a heavy hand. It seems that, for these families, 'the turning of tiny primates into little ladies and gentlemen' was an exercise in punitive repression. It is here that we discover the link between socially insecure parents and the prejudice manifested by their children. The extreme strictness of the parents, coupled with their lack of warmth, necessarily frustrates the child. But frustration engenders aggression which is itself frustrated, for it is part of the training that children never answer back. Hence the aggression has to be discharged elsewhere, and where better than on to those very individuals whom the parents themselves have openly vilified — Jews, Negroes and foreigners — all those, in short, who, being underprivileged, have acquired bad reputations in a status-seeking society?

In other words, albeit quite unwittingly, an authoritarian upbringing kills three birds with a single stone. It produces submission to the authority of the in-group. It arouses aggression, which is displaced on to a carefully defined out-group. By these means the status-seekers achieve their underlying goal, for the relativity of status depends upon the existence of an underprivileged out-group, and how better to ensure this state of underprivilege than by aggressive persecution?

This theory, based upon the psycho-analytic notions of displacement and projection, explains one striking finding of the Berkeley research, namely that authoritarian personalities manifest a monolithic self-satisfaction with themselves and their parents, and this despite the fact that no love was lost between them during the so-called formative years. This apparent paradox is resolved when one considers the dynamics of authoritarian discipline. For the person who is anxious about status, it is imperative that his, and his parents', shortcomings should be strenuously denied. This is achieved in two ways: firstly, as we have seen, by projecting their *undesirable* characteristics on to others; secondly, by nurturing an impeccable and idealized, if wholly false, image of themselves. Like the prejudiced Southern white who projects his repressed sexual wishes on to Negroes, or the latent homosexual or

* For this aspect of American middle-class life the reader is referred to Polly Adler's *A House is not a Home* and the Plainville study.[6]

voyeur who devotes his or her life to advocating harsher punishment for homosexuals or pornographers, the life-style of the authoritarian personality is one of *finding and prosecuting in others what he has come to fear in himself.* This example of attack, being the surest method of defence, would be incomplete, however, if the individual did not also entertain a highly idealized image of himself. (It is this combination of transparent if unconscious hypocrisy and smug self-satisfaction that makes such people particularly insufferable.)

These tortuous machinations of the authoritarian mind ramify yet further. Because he has to deny his own shortcomings, he dare not look inwards. He is fearful of insight, and strenuously avoids questioning his own motives. By the same token, he cannot allow his extra-punitive defences to be threatened by humane considerations for the objects of his hostility. At first sight this may seem a useful adaptation to the tribulations of early childhood. Unfortunately, however, a price is paid—one which can prove crippling to the human mind. In the place of free-ranging, creative and inventive thought, an authoritarian's thinking is confined to rigid formulae and inflexible attitudes. He is intolerant of unusual ideas and unable to cope with contradictions. Recent research[7] confirms the authoritarian's preference for order and simplicity. As Brown has put it: 'If he has a problem the best thing to do is not to think about it and just keep busy.'[8] Similarly, the authoritarian personality is intolerant of ambivalence and ambiguity. Just as he cannot harbour negative and positive feeling for the same person but must dichotomize reality into loved people versus hated people, white versus black* and Jew versus Gentile, so also he cannot tolerate ambiguous situations or conflicting issues. To put it bluntly, he constructs of the world an image as simplistic as it is at variance with reality.

In the years since it was published, the Berkeley research has been both criticized and extended.[10] On points of detail such as the possibility of there being an authoritarian personality of the political Left as well as of the political Right, and whether or not authoritarians are

---

* Shown quite literally in research by Pettigrew et al., wherein pictures of black faces were presented to one eye and pictures of white faces to the other. Whereas English-speaking South Africans and Coloureds could fuse these pairs of faces into a single combined percept, Afrikaans-speaking South Africans invariably manifested binocular rivalry. They either saw the black face or the white face, never a mixture of the two.[9]

necessarily neurotic, there has been considerable argument, but the main findings of the original research stand up. More important for our present purposes are those subsequent studies which shed light on the implications of authoritarianism. One such is the nine-year research by Rokeach, another American, published a decade after the Berkeley research. His *The Open and Closed Mind* centred on the problem of an individual's capacity to absorb fresh information. Humanity, it seems, varies considerably in this respect. At one extreme are 'open' minds, ready and willing to entertain new facts, even if these are incompatible with their previously held attitudes and beliefs; at the other are 'closed' minds, which, as their name suggests, resolutely resist taking in anything that conflicts with their preconceptions and treasured beliefs. Not very surprisingly, the possession of a 'closed' mind turned out to be yet another facet of the authoritarian personality. This finding had great generality. In one experiment, for example, high and low scorers on dogmatism, an aspect of the 'closed' mind, were asked to indicate their liking for different pieces of music. The compositions in question differed in their conventionality of structure. For works of Brahms, composed according to the traditional conventions of the West, no difference was found between the pleasure they afforded 'open' as opposed to 'closed' minds. But for the music of Arnold Schönberg, who used a twelve-tone scale and no key-centre, a marked divergence obtained. Those with 'open' minds evinced an increasing pleasure in the unaccustomed sounds; those with 'closed' minds not only instantly disliked, but also manifested an *increasing* distaste for the strange noises with which their ears were now assaulted.[11]

Before leaving this section on authoritarianism, there are several other findings from recent research which are pertinent to our present thesis. One of these shows the relationship between conformity, authoritarianism and the tendency to yield to group-pressures.

An extreme example of this pattern is the phenomenon of participation in a lynch mob, where the naturally conformist individual happily yields to group-pressure for the perpetration of a criminally aggressive act which, though wholly at variance with the ethos of a wider society, accords with his own narrow self-interest. In this instance this consists of finding an underprivileged, low-status outgroup, the Negro, on to whom he can discharge his aggression and project his sexual anxieties.

Another finding concerns the effect of authoritarianism upon problem-solving in a group-situation. From their research W. Haythorn and his colleagues concluded that 'equalitarian subjects [i.e., those low on authoritarianism] were apparently more effective in dealing with a task and problem than were the authoritarian. This was reflected in higher ratings of effective intelligence, leadership and goal-striving.'[12] On the sorts of leaders who emerged in the group-situation, they had this to say: 'Emergent leaders in the low F groups [equalitarian] were *more* sensitive to others, *more* effective leaders, *more* prone to making suggestions for action subject to group sanction, and *less* likely to give direct orders to others ... ', a conclusion, incidentally, which accords with the observation that authoritarians are less able to appreciate the effect they have upon others, and may well think themselves more liked and popular than they really are! Even people attracted to a career in an authoritarian organization [i.e., naval cadets] have been found to prefer leaders who score *low* on tests of authoritarianism — presumably because authoritarians are less sensitive to the needs of others.'[13]

Finally we come to something touched upon in an earlier section: the psychology of the obsessive personality.[14] It will be recalled that this type of person, orderly, stingy and stubborn, is in reality manifesting the prolonged effects of the early infantile conflict between being dirty and wishing to avoid the wrath of parents who themselves have anxieties about dirt. He resolves this conflict by developing the triad of personality-traits given above. They represent symbolic defences against those tendencies which he has had to renounce or, to be more accurate, keep under strict control. Under the circumstances, it is not so surprising that a positive relationship has been found between obsessionality and authoritarianism.[15]

What is the relevance of all this to military incompetence?

Firstly, it cannot be stressed too strongly that in talking about authoritarianism we have been discussing people towards one end of a continuum. Between this end and the other can be found people with all shades of opinion on the various attitudes measured. The second general point is this: when discussing authoritarianism no value-judgment is intended. Few would dispute that, in moderation, many of the traits which make up the authoritarian personality have a value in society. Civilization requires that there should be some repression of sex and aggression, some exercise of discipline, and a modicum of conformity and orderliness. It is with these caveats in mind that we can take up the matter of military incompetence and advance the thesis that the nature of military organizations is such as to attract, favour and promote people who might be expected to lie towards the authoritarian end of a personality continuum. There is, in fact, considerable evidence for this thesis. It can be concluded from recent research not only that positive attitudes to war, belief in military deterrence and a liking for militarism go along with above-average levels of authoritarianism; but also that military personnel, including officer cadets, tend to score higher on measures of authoritarianism than do people who do not opt for a career in the Army or Navy.[16]

In some respects, assuming that civilization entails the possession of armed forces, these findings are to the good. There are aspects of military behaviour which require the constraints and the discipline of the authoritarian personality. Similarly, it is to the good that soldiers should be anti-intraceptive, unimaginative and fatalistic. Even the tendency to project aggressive feelings on to others has an obvious advantage for intra-species aggression.

In other respects, however, the likelihood of above-average levels of authoritarianism in military personnel may well contribute towards incompetence, particularly when the authoritarian has reached a level of command where flexibility and an open mind are a *sine qua non* of success. To be more specific, the personality-traits of authoritarianism, and the associated characteristics of the closed mind and obsessive character, may contribute to incompetence in the following ways.

1. Since authoritarians have been found to be more dishonest, more irresponsible, more untrustworthy, more socially conforming,[17] and more suspicious[18] than non-authoritarians, they are unlikely to make successful social leaders.

2. Authoritarians will be less likely to understand enemy intentions, and to act upon information regarding such intentions as conflict with the beliefs and preconceptions which the commander might hold. The events following the Cambrai tank offensive in the First World War and, in the Second World War, the repeated inability of senior commanders to accept the possibility of an enemy offensive in the Ardennes are cases in point. As Major-General Strong has written of the First World War: 'It is no exaggeration to suggest that some of the inadequacies seen in the course of World War I, the mistakes of generalship, the poor strategic planning and the many tactical errors, reflected a serious inability to acquire Intelligence or to make effective

and professional use of the Intelligence that *was* available.' And of events in the Second World War: 'I find it difficult to understand how any plan can be made in the absence of a professional assessment of an opponent's strength, capabilities and intentions. The Norwegian campaign was neither the first nor the last example of this extraordinary syndrome in Britain, and other countries have not been free from it.'[19] In 1954 American research showed that people with a high score on tests of authoritarianism had *greater* difficulty than non-authoritarians in recognizing threatening messages when these were presented visually.[20] A year later another study confirmed this finding with threatening words that were heard instead of seen.[21]

3. The inability to sacrifice cherished traditions and accept technical innovations. The history of the machine-gun, the tank and the aeroplane contains striking evidence of this disability.

In his recently published *Modern Warfare*, Shelford Bidwell has drawn attention to a costly facet of this malaise—the failure of the military to carry out experiments. Eschewing the 'uncertainty' of experimentation and suspicious of applying scientific method, the authoritarian mind prefers the cosy if spurious security of belief and dogma. As Bidwell says of armoured fighting vehicles, 'What the early history of the tank provides is a salutary lesson of the merits of numerical and experimental analysis. It *was* possible to arrive at an exchange rate for the tank and that might have led to a different tactical doctrine altogether … In war an ounce of calculation is worth a ton of intuition. It also saves a great many lives.'[22]

4. The underestimation of enemy ability (particularly when the enemy are coloured or considered racially inferior). This ethnocentrism, which cost many lives in the Boer War, was also a feature of the enormous losses at Kut in 1916, Singapore in 1942 and Vietnam in the 1960s. Needless to say, this tendency to underestimate has not been confined to British and American forces:

Von Kluck, the commander of the German First Army at the outset of the war, could not bring himself to believe that the French soldier, after days of exhausting retreat in August and September 1914, was capable of turning at the bugle call and forcing the Germans back to the Marne. Nor did the General Staff believe the British Army was a fighting force to be reckoned with. They failed to appreciate that if the Americans joined the war they would prove effective in combat; a quarter of a century later they made the same error.[23]

It is closely tied up with two other features of authoritarianism: anti-Semitism, and the tendency to categorize people in terms of stereotypes. A word about these psychological phenomena.

When considering the nature of 'bull' (see p. 185ff), the point was made that, by establishing conformity, 'bull' reduces uncertainty in small things, and, to this extent, offsets the greater uncertainties of war.

It is significant, therefore, that authoritarians are *also* concerned to reduce uncertainty. Writing of them, Kelvin says:

> These tendencies reflect on a type of individual who needs to feel that his environment is highly predictable ... he needs to know where he stands; and so he fastens on to norms: he does not 'let himself go', for fear of where this might lead; he looks to authority as a guide ... [He also] relies very heavily on stereotypes in [his] perception of the social environment. Moreover, the stereotypes used by an authoritarian personality tend to be very clear-cut, and the characteristic inflexibility of this kind of person leads to relative inability to modify the stereotype once it has been formed.[24]

These characteristics have a compound significance for the military scene. Firstly, they have undoubtedly been a potent factor in that underestimation of the enemy which has afflicted so many military commanders. The stereotyping of 'wogs', 'wops', 'gyppos', 'huns', 'gooks' and 'japs', though no doubt reassuring, has tended to cloud judgment regarding the real characteristics of the enemy. One cannot know one's enemy by stereotyping him.

Secondly, stereotyping has introduced what is probably a totally irrelevant and misleading factor into the selection and promoting of military personnel. Like most social stereotypes, that of the ideal military man has tended to be based on physical attributes which may have nothing whatever to do with his suitability as a military leader. Just how misleading a physiognomic stereotype can be is shown by B. R. Sappenfield. In his study,[25] people had to rate faces according to the degree to which they thought the owner of the face possessed masculine qualities. Results showed that there was, in fact, not the slightest relationship between the stereotype of masculinity (a very

important stereotype in military organizations) and the actual possession of masculine traits.

Certainly pictures of competent and incompetent senior commanders
suggest that the stereotype of the ideal military face may well have
played a part in the selection and promotion of those who evidently
had little else to recommend them. Needless to say, the selection of
officers because they fit the stereotype only serves to confirm the
stereotype, thereby perpetuating the harmful effects of what is at best
an irrelevant variable.

In view of the significant relationship between authoritarianism and
anti-Semitism, it would be surprising if the stereotype 'Jew' had not
featured in military prejudices. In Britain it almost certainly played a
part in the sacking of War Minister Hore-Belisha. In France, the
Dreyfus case, which did untold damage to the highly authoritarian
French Army, was undoubtedly a result of strong anti-Semitic prejudices in the military elite.[26] In the United States, according to
Janowitz, Jews are unpopular because stereotyped as over-intellectual.
He cites the case of a Jew who stood second, academically, in his class
of officer cadets. When his photograph was included in the class book,
it had to be printed on special perforated paper for easy subsequent
removal![27]

The contribution of authoritarian prejudice to military incompetence
is that it introduces an inappropriate variable into the selection or
sacking of personnel. A telling example was when the C.I.G.S., Field-
Marshal Montgomery-Massingberd, pressed for compulsory retirement of divorced officers because, as he put it, there were plenty of
brilliant men to take their place. Not only was this nice example of
authoritarian concern with 'sexual goings-on' a complete non sequitur,
but it was also untrue. As Duff Cooper observed: 'There were very
few brilliant men in any line.' (Since the field-marshal's sudden concern
for the morals of the Army occurred only a few years before the
outbreak of the Second World War, it is just as well his wish remained
ungranted.)

5. An emphasis upon the importance of blind obedience and loyalty,
at the expense of initiative and innovation, at lower levels of command.
The sinking of the *Victoria* by the *Camperdown* during the peace-time
manœuvre in 1893 when hundreds of lives were lost through blind
obedience to an ambiguous order; what one naval historian has
described as 'the staggering lack of initiative' shown by British ship

commanders during the Battle of Jutland;[28] and the failure of an encircling force to launch a surprise attack upon the Turkish besiegers of Kut—all these are instances of this disability. The *Victoria* disaster, which was even excused on the grounds that orders *must* be obeyed, illustrates more clearly than any other mishap the differences, in nature and origin, between the progressive autocrat and the reactionary authoritarian, and how the clashing of these personality-types can lead to tragic consequences.

Vice-Admiral George Tryon, the man who gave the order to 'form the fleet into columns of two divisions, six cables apart, and then reverse the course by turning inwards' was the product of a happy and secure childhood. He was supremely self-confident, assertive, outspoken, autocratic, a strict disciplinarian, but *not* authoritarian.

The man who failed to seriously question, let alone disobey, the order, Tryon's second-in-command, Rear-Admiral Markham, had emerged from *his* miserably unhappy childhood at the hands of harshly puritanical parents as a 'sensitive, abnormally courteous', prickly, obstinate recluse.

Tryon was an innovator, bent on achieving professional excellence, dedicated to developing efficiency and initiative in his subordinate commanders. Markham was an anxious, conforming, hidebound conventionalist, dedicated to staying out of trouble and not displeasing his superiors.

Tryon was, in many ways, an early version of the great 'Jackie' Fisher—domineering, warm, sociable and outward-going. He was happily married and cared deeply for the welfare of his men. Markham was a bachelor who seems never to have enjoyed anything approaching a physical relationship with a member of the opposite sex. His antihedonism found expression in pep talks to his subordinates on the evils of drinking and smoking. That other characteristic of an authoritarian personality, the repression of his aggressive drive which evidently prevented Markham from asserting himself with his superiors, found an outlet in the slaughter of wild animals.* As Hough remarks: 'Wherever he travelled, Bertie Markham killed.'[30]

* It is reasonable to suppose that this busman's holiday behaviour of killing lower animals when an outbreak of peace precludes the slaughtering of higher ones is symptomatic of those whose otherwise suppressed hostility craves a legitimate outlet. Certainly, such behaviour is a not uncommon feature of some military men; perhaps the single best example being that of Field-Marshal Sir Evelyn Wood of whom it has been written:

Whether or not Tryon confused radii with diameters of turning circles,[30] or intended Markham's ship to pass outside his own, it remains a sad irony that the man who sought to instil initiative in his subordinates should have been killed by the abject lack of that quality in his own most anxious pupil.

Each of these two men was intelligent, dedicated and conscientious, but the authoritarianism of the one collided with the autocracy of the other just as surely as the rigidity of the *Camperdown*'s bows collided with, and ruptured, the thinner plating of the *Victoria*'s hull.

Another way in which excessive obedience may lead to military incompetence has been suggested by the work of Milgram and his colleagues.[31] Milgram asked the question: If a man is ordered by a person in authority to deliver increasingly severe electric shocks to a helpless victim, how far will he go before disobeying the order? The findings were alarming. They indicated that ordinary decent men could be so seduced by the trappings of authority that they would continue delivering shocks up to 450 volts (marked 'DANGER—severe shock' on the control panel), and this even when hearing the agonized shrieks of the 'helpless victim' whom they could dimly see behind a sheet of silvered glass. This readiness to obey, even when the consequence of obedience was an act of outrageous inhumanity, was significantly greater in authoritarian personalities than others.

Besides providing support for one facet of authoritarianism,* this

---

. . . he was always glad to get away from it all to follow the greatest interest and passion of his life next to the army—the hunting and slaughtering of wild animals. Hunting, like sex and eating, is a human appetite becoming in moderation, but unlovely in excess. In Wood's two-volume autobiography *From Midshipman to Field-Marshal* his oft-recurring hunting reminiscences flow through its chapters like a polluted and stinking stream. These become a self-indictment of a 'Christian manliness' which brought so much terror and pain to God's lower creatures with callous indifference to their sensitivity and beauty.'[29]

Doubtless there *are* other reasons for this impulse to kill, though it is noteworthy that an extension in leisure activities of an individual's professional responses does not seem to occur so readily in other walks of life. Do dentists reach for the drill whenever someone smiles or upon opening a piano lid? And do fish-and-chip men hanker for vats of boiling oil whenever they visit an aquarium? One suspects not!

* The prevalence in society of those authoritarian traits which underlie Milgram's data are reflected in the protests which greeted his book. Better, it

finding has a twofold relevance for military incompetence. Firstly, the effect of obeying orders which they knew to be wrong and which conflicted with their normal set of values left the participants in a severe state of mental and physical stress — one hardly conducive to military efficiency! Secondly, it is just this aspect of authoritarian obedience, exemplified in its most extreme form by Adolf Eichmann, which leads to such atrocities as that of the My-Lai massacre — atrocities which, quite apart from their inhumanity, do great harm to the prestige and, therefore, efficiency of the military organization in which they occur.

6. The protection of the reputations of senior commanders, and punishment of those lower in the military hierarchy if they voice any opinion which, however valuable in itself, implies criticism of those higher up. The following conversation cited by Admiral Dewar illustrates this issue:[32]

> *Scene: Captain's cabin ... Captain sitting at kneehole desk. Enter Gunnery Lieutenant with papers.*
>
> LIEUTENANT. I have prepared a report, Sir, on our new fire-control organization with sketches of the voice-pipe arrangements. It may be useful to other ships and I thought you might like to submit it to the Commander-in-Chief.
>
> CAPTAIN. That's good. (*Reads it; and after long pause*) Do you know who is the Controller?
>
> LIEUTENANT. Yes, Sir, Captain Jackson.
>
> CAPTAIN. Do you know that he was President of the Committee that sat on the approval of the existing voice-pipe communications in H.M. ships?
>
> LIEUTENANT. No, Sir, but I suppose he will be interested in reading the report.
>
> CAPTAIN. I am afraid that I cannot forward a report which suggests that the arrangements which he approved are unsatisfactory.
>
> LIEUTENANT. The report shows how they can be improved.
>
> CAPTAIN. Yes, but I am not going to tell him so.
>
> LIEUTENANT. Oh, very good, Sir.

---

seems, that a 'sickness' should remain concealed than that people should have their less pleasant traits exposed!

For another example from the same source, consider the following:

When the *Prince George* returned to Portsmouth the President of the War College, Rear-Admiral Robert S. Lowry, asked me to deliver a lecture on the Japanese Navy to the Senior Officers' Course. It seemed to go quite well, until near the end. I was explaining the different functions of the Ministry of Marine and Naval Staff and had begun to suggest the desirability of separating administration from operations in our own Admiralty, when the President asked me to bring the lecture to an end. 'The War College,' he said, 'is not the place to criticize the Admiralty.'[33]

Such goings-on are not, of course, confined to the Navy. Liddell Hart has related the events which occurred when Wavell gave 'a forward-looking and imaginative' lecture to the Royal United Services Institute. It was on the subject of 'Training for War'. When he had finished, that 'arch-conservative' General Knox rose to say that he strongly disagreed with the lecturer's views. Then Wavell, a man of greater intellect, instead of defending his position, apologized. 'General Knox's knowledge and experience are of course far greater and far wider than my own and if there is any point on which he and I differ he is much more likely to be right.'[34]

In considering these examples it is important to distinguish between the possibly authoritarian personalities of the individuals concerned and the generally authoritarian ideology of the milieu in which they found themselves. Neither Captain Gamble in the first example nor Lord Wavell in the last was necessarily an authoritarian personality, but the anxieties which gripped the former, like the sycophantism of the latter, were a product of a system which punishes those who, even in all innocence, appear to be critical of their superiors. K. G. B. Dewar is explicit on this point. 'Lowry was by no means antagonistic to the discussion of new ideas, but like many of his contemporaries he was a firm believer in the principle of safety first. He knew that those who opposed Fisher [Admiral Lord Fisher, First Sea Lord] were liable to find themselves out of a job.'[35]

This distinction between the authoritarianism of a regime and the authoritarian personality has its counterpart in estimates of an individual's character. Thus, autocratic leaders like Admiral Fisher are by no means necessarily authoritarian personalities in the strictly technical sense of the term. Indeed, the underlying need of the authoritarian personality to be popular with his fellow men (i.e., loved) almost precludes him from being a thoroughgoing autocrat. Conversely, judging from their *total* behaviour, such autocratic leaders as Wellington, Napoleon, Kitchener, Zhukov and Fisher were, as we shall see, the very obverse of authoritarian personalities.

7. Closely related to the foregoing effects of authoritarianism is an individual's propensity to blame others for his own shortcomings. Few better examples are afforded than the 'scapegoating' of lower ranks by Haig, Byng and Smuts after the German counter-offensive at Cambrai. Such a case illustrates how the modus operandi of military organizations reinforces the authoritarian tendencies of its individual members. The basically insecure person who has from childhood elaborated that system of psychological defences which characterizes the authoritarian personality will have his defences strengthened rather than reduced by the sorts of anxieties which service life engenders. That great harm can be done to the service by this cycle of events is exemplified by the notorious case of the *Royal Oak* courts martial, where a commander and his captain were unlawfully punished for writing and forwarding, respectively, a letter of complaint against a superior officer. Since their 'offence' was no offence at all (indeed they had acted in strict accordance with the dictates of the Naval Discipline Act), the subsequent punitive behaviour of their commander-in-chief was both illegal and contrary to the requirements of good naval discipline. Even worse, the Admiralty, faced with the dilemma of a commander-in-chief who had blundered, compounded *his* error by cynically disregarding the law in favour of rank. They backed the commander-in-chief and framed fictitious charges against the captain who had forwarded the original letter of complaint. By subverting the law and encouraging the perpetration of a series of irregularities during the actual court martial, the Admiralty achieved their goal of finding the victim guilty. He was duly sentenced to be severely reprimanded and dismissed from his ship.

But the matter did not end there. An appeal against this miscarriage of justice was dismissed by the Admiralty on the dubious grounds that the trial had been conducted 'with great ability and conspicuous fairness'. Then, to avoid a recurrence of these embarrassing events the Admiralty cancelled the old regulations on the making of complaints, and issued new ones. This *post hoc* attempt to make the regulations fit the crime in fact banned a large class of complaints which had been authorized by the old regulations.

The after-effect of the court martial was that within two years cases of indiscipline in the Navy reached a level unprecedented in the history of the service, and this despite improvements in pay, food, leave and accommodation. This upsurge of mutinous behaviour has been ascribed on the one hand to a growing cynicism among men who had seen their superiors waive the law to suit themselves, and on the other to the new regulations on complaints which effectively destroyed an old safety-valve for feelings of resentment. Making a complaint against a superior officer had always been a risky venture; now it was professional suicide. But whatever the rights and wrongs of the *Royal Oak* case, one thing is certain. The whole miserable affair, which wrecked the careers of three senior officers and did untold damage to the reputation of the Navy, illustrates the sort of price that human beings have to pay for dedication to an authoritarian system.[36]

8. The close relationship between authoritarianism and obsessive traits has also played a significant part in military incompetence. This is a matter which we discussed earlier. Suffice it to say that the worst excesses of 'bull' and the clinging to anachronistic ritual have played a not inconsiderable part in holding back the military machine.

Here again, however, we must be careful to distinguish between the neurotic compulsive behaviour of the anal-obsessive and the deliberate exercise of what appear to be obsessive traits. A distinguishing feature of the former is its gross inconsistency. 'So much attention was devoted to whitening the sepulchre that there was not much left for questions of health and hygiene. On a sunny Mediterranean day the *Hawke* glistened and sparkled in the water of that ancient sea, but she was infested with rats which contaminated the food in the pantries and the messes. They ran over the hammocks and swarmed into the gun room at night. No attempt was made to get rid of them.'[37] One is reminded of the obsessive housewife (described by one psychiatrist) who spent hours polishing her saucepans but whose underwear remained unchanged for weeks!

In contrast, we have the behaviour of such men as Wellington, Zhukov, Kitchener and Montgomery, whose enormous energy and attention to detail might well have appeared obsessive but was in fact part of a deliberate and much-needed policy of arranging to meet every contingency and of leaving nothing to chance—in short, good planning.

9. There is one trait of the authoritarian personality which at first

sight may seem to have nothing whatever to do with military incompetence: belief in supernatural forces. The contrary is in fact the case. As a general issue, since military decisions should be based upon a proper weighing-up of facts, the introduction of metaphysical variables into decision-making necessarily contributes 'noise', which decreases the probability of decisions being correct. Concern with what the stars foretell, or hopes and occasionally fears of divine intervention, constitute prejudices which can bias decisions away from realism and towards wish-fulfilling fantasies.

For an illustration of how fatal decisions can result from an excess of faith in a benign supernature, there is the extraordinary case of Rudolf Hess's flight to Scotland on May 10th, 1941. It seems that this irrational behaviour by an arch-authoritarian was the resultant of two factors—a wish-fulfilling fantasy on the part of Hess and Hitler that they could negotiate a peace with Britain, and a favourable horoscope supplied to Hess by his personal astrologer. According to Richard Deacon, the horoscope had in fact been concocted by British Intelligence![38] There can surely be no clearer illustration of how one side in a war might successfully exploit the authoritarian shortcomings of the other.

There is, of course, a problem here. Few would deny the importance of psychology in warfare, and that belief in a benign deity may sustain troops when all earthly help has failed. What seems to be important, however, is that knowledge of this fact, in the minds of senior decision-makers, should enable them to make a proper assessment of the factor of morale without themselves being misled by their *own* unrealistic fantasies. Research suggests that this objectivity will come more easily to non-authoritarian minds.

10. One of the least attractive aspects of the authoritarian personality is his generalized hostility, what the Berkeley researchers called 'vilification of the human'. This was the trait which was manifested to such an extreme degree by members of the Nazi S.S. that they could commit wholesale murder, not just without guilt or shame but, perhaps more surprisingly, without the slightest evidence of revulsion. This cool detachment and complete incapacity for empathy with other human beings was not only reflected in the bleakly unemotional title for their task—'the final solution'—but was also a sine qua non of its tidy execution. At first sight, this mixture of brutality and bureaucracy is strange, to say the least. After all, it is one thing to shoot help-

less prisoners in the back, to drive old women into gas chambers and to hang your 'enemies' with piano wire and meat-hooks — but quite another to plan such operations down to the minutest detail, to make ledger entries of hair and calcium, wigs and artificial limbs; to stack corpses and extract the gold from their teeth. In fact, of course, this horrific concatenation of traits is an extreme if grotesque example of the relationship touched on earlier — that between authoritarianism and the anal-obsessive personality. It does, however, add a new dimension and meaning to that undervaluing of other human beings which characterizes the authoritarian personality.

Now in the context of purely military behaviour this inhumanity, like other characteristics of the authoritarian personality, could in theory be either useful or disastrous. On the one hand it could be argued that senior commanders should 'hate' the enemy and not be squeamish about sacrificing the lives of their men for the sake of a greater good. Conversely it could be maintained that it is not hatred so much as an understanding of the enemy, and not a conscienceless squandering but a humane conserving of his own forces, which are the hallmarks of an efficient commander.

Perhaps, as with other aspects of authoritarianism, it is just a matter of degree. Certainly such great leaders as Wolfe, Wellington, Shaka, Lawrence, Monash and Montgomery not only displayed a general absence of authoritarian traits but also showed a lively regard for the prime responsibility of a commander: conservation of his force and a concern for the psychological and physical welfare of his troops. As Trevelyan wrote of Wellington: 'It is fortunate for Britain that Wellington was at once a great humanitarian and a great disciplinarian.' In contrast to these highly competent commanders, many less talented military leaders have, along with other authoritarian traits, betrayed a singular disregard for the welfare of their troops and an unnerving capacity to remain apparently unmoved by losses.

However, we must stress that it is not our purpose to level value-judgments about the lack of humanity of particular commanders but rather to point out that *the aspect of authoritarianism which constitutes a lack of humanity makes for military incompetence.* Furthermore, humanitarianism in a senior commander contributes to military success in at least two ways. Firstly, it is a controlling factor in the making of tactical or strategical decisions, for it feeds into the complex process by which such decisions are arrived at two important criteria: economy of force, and the need for safeguards against the possibility of unforeseen disasters. Thus 'Wellington had learnt to keep his eye at certain times on the exit ... As far back as India, when famine threatened his communication, he had rightly dropped any idea of going to the assistance of Colonel Monson against the Mahrattas.'[39] Very different were the policies of Townshend as he pushed up the Tigris towards the 'mirage' of Baghdad.

Secondly, humanitarianism is a prerequisite for those pillars of military success, high morale and physical health. From researches in industrial psychology it has become abundantly clear that, for the workers in any large organization, physical health and mental well-being (and, as a result, productivity) depend rather more upon workers *feeling* that they are being cared for by an interested and benign management than upon such tangibles as large wage-packets. Even such a 'real' benefit as a newly installed air-conditioning system has been found to have more effect upon productivity before being switched on than it had when actually functioning. In other words, it was not fresh air itself but the fact that management had bothered to give them fresh air which counted. If these effects occur in the relatively unstressed milieu of a civilian firm, how much more so will they flourish when the workers are soldiers and the stresses those of war? Commanders like Wellington, Nelson, Montgomery or that paragon of military virtuosity, the Zulu general Shaka, by according to logistical planning the highest priority, achieved not only the affection and loyalty of their men, but also, as a consequence, standards of health out of all proportion to the purely physical benefits which their energy conferred. It should come as no surprise to those versed in the relationship between psychological stress and psychosomatic illness to learn that the sailors serving in Nelson's ships in the then unhealthy environment of the West Indies, like Shaka's Zulu warriors in the unhealthy climes of central Africa,[40] showed a resistance to disease that bordered on the miraculous. These phenomena stand in sharp and significant contrast to the uncaring way in which some generals have let their armies die of cold, misery, disease and neglect. It is not perhaps surprising to discover that, according to modern research, a lack of compassion goes along with moral conformity, having a closed mind and being uncreative.[41] Katherine Whitehorn, writing in the *Observer* (10.11.69), has provided a useful example of this relationship between authoritarian attitudes and lack of compassion.

She 'cites a letter written to the Conservative M.P. Mr Duncan Sandys which, agreeing with his stand on abortion, went on for several pages about the sanctity of human life, and ended "P.S. I'm with you on hanging too".'

A particular area of manpower wastage additional to those considered above is that of sexual casualties. In every war thousands of man-hours are lost through venereal disease.

That authoritarians are not good at dealing with this problem is suggested by the rumpus which greeted Montgomery's order on how to prevent V.D. amongst British troops in France during 1940.[42] The authoritarian attitude towards this problem is that V.D., like other self-inflicted wounds, constitutes a misdemeanour. V.D., they say, should be avoided by chastity. Montgomery took a different and rather more pragmatic view, namely that V.D. should be prevented by rapid medical attention after intercourse.

His order, which made several useful points about prophylaxis, so upset the senior chaplain and the commander-in-chief that 'our greatest general since Wellington' nearly lost his job.

The most surprising aspect of this incident is the unrealism of the authoritarian approach. For active service involves many privations. Good food, shelter, comfort, safety, the presence of one's family are in varying degrees sacrificed to the cause of fighting an enemy towards whom many of the combatants may feel little personal animosity. Within this context they are then required to break the Sixth Commandment. If they obey this injunction towards un-Christian behaviour they are rewarded, if they disobey they are punished. It falls to the lot of chaplains to attempt a reconciliation. Their task is to *reassure* the military flock that, since God is on their side, the Sixth Commandment can be waived for the duration. How they reconcile this with the knowledge that enemy soldiers are in all probability receiving identical advice from *their* chaplains remains one of the mysteries of the ecclesiastical mind.*

Anyway, presumably with the slacking off of one commandment, the Church feels it necessary to tighten up another, rather as one might

---

* According to J. R. Hale, even as long ago as the seventeenth century 'the Church could not preach war without some feelings of unease'. The rationalization of the divines was that war 'could be a social good, a moral cleanser, God's scourge for vice'.[43] It seems that the relationship between religiosity, aggression and authoritarianism is a phenomenon of some antiquity.

adjust different guy ropes on a tent-pole. This is, of course, not only bad luck on soldiers who probably prefer sex to aggression, but in the event a fairly ineffectual attempt to tamper with the laws of nature. For since the inhibition of sexuality, unlike the unleashing of aggression, does not constitute part of their military training, it is hardly surprising that for a proportion of men, torn from their wives and girl friends, promiscuous behaviour, followed in some cases by V.D., becomes one more hazard of war.

This is bad enough for those who glorify aggression and deplore sex, but worse is to follow. For V.D. interferes with the soldier's primary task of aggressing against the enemy.

This sequence of events is no criticism of the Corps of Chaplains, whose record of bravery and selfless devotion to the physical, as well as spiritual, welfare of soldiers remains indisputable. But it is an indictment of an authoritarian ethos which, in trying to deny the undeniable and conform to conventional middle-class values, under wholly inappropriate circumstances, results in military inefficiency and a quite unnecessary extra burden for soldiers who are already suffering more than their fair share of discomfort.

It is perhaps worth noting that Montgomery's enlightened and pragmatic approach to the problem of V.D. is rather more humane than that followed by the American forces with their 1,000-girl 'Willow Run' brothel in Korea, and their regularly medically examined ladies of the official military brothel, 'Sin City', in Vietnam.

Finally, there is the fact that authoritarianism, itself so damaging to military endeavour, *will actually predispose an individual towards entering upon the very career wherein his restricted personality can wreak the most havoc.* For early signs of this predisposition there is the following case study:

Case 19: Cecil R—an obsessive neurotic

His IQ was in the Bright Normal range. Personality-testing indicated that he was very dependent on his parents, but that they were seen as being emotionally remote and extremely demanding. In fantasy, he expressed strong feelings of aggression and anger ... He seemed most interested in the history of wars and in playing war games. He shot darts 'with vigor and delight' in the therapist's playroom and if given a choice would choose war games ... His parents said he refused to play with other children unless the others did exactly what he told them to do. Cecil said he wanted to grow up, and be a general.[44]

This sad state of 'positive feedback',* together with a brief summary of the sort of personality under discussion, is contained in the following excerpt from a recent paper on the military mind.

Reserve Officers' Training Corps students at the University of California, compared with student draft resisters, were found to have experienced strict childhood discipline in relation to a dominant father-figure. They showed a strong concern about proving their masculinity, used more alcohol, felt powerless to influence their country's actions, felt troubled about their sexual inadequacy, defined independence as loss of self-control, preferred a well-ordered and structured environment, admitted being self-centered and egotistical, felt shy with girls but boasted to their fellows of their sexual conquests, claimed little real intimacy and poor relationships with the opposite sex, admitted treating females as objects, tended to seek dominance-submission relationships, and were relatively aggressive, impulsive, irresponsible and non-intellectual, with a poorly developed conscience.[45]

* 'Positive feedback' is the process whereby the characteristics of an outcome serve to accentuate the same characteristics of future outcomes—i.e., a runaway system.

23

Mothers of Incompetence

'... The adult who is under the dominion of unilateral respect for the "Elders" and for tradition is really behaving like a child.'
J. PIAGET, *The Moral Judgement of the Child*

'... there is nothing more common than to hear of men losing their energy on being raised to a higher position, to which they do not feel themselves equal ...'
C. VON CLAUSEWITZ

For the reader not previously versed in the psychology of authoritarianism the preceding chapter may have come as something of a surprise. At first sight the traits of orderliness, tough-mindedness, obedience to authority, punitiveness and the rest may well have seemed the very embodiment of hard-hitting masculinity—ideally suited to the job of being a soldier. Unfortunately, as represented in the authoritarian personality they are only skin deep—a brittle crust of defences against feelings of weakness and inadequacy. The authoritarian keeps up his spirits by whistling in the dark. He is the frightened child who wears the armour of a giant. His mind is a door locked and bolted against that which he fears most: himself.

Since the truth of this may be a hard pill to swallow (for it threatens the whole elaborate structure of personal defences), we shall now describe another piece of independent research which has reached similar conclusions by a different route. Like the research on authoritarianism it started with a quite specific aspect of perception—in this case, the extent to which people can respond analytically rather than globally to the mass of information reaching their senses.[1] A special instance of this ability is the degree to which people can ignore irrelevant *visual* impressions when these conflict with other sources of sensory experience.

In a simple test of this propensity the individual is seated on a tilted chair in a dark room. All he can see is a fluorescent rod, enclosed within an illuminated frame. Both the rod and its frame are tilted out of the vertical. His task is to set the rod into a vertical position.

In theory this should be quite easy, for darkness does not prevent one from knowing one's position (and therefore that of other objects) relative to the true vertical. If it did, walking in the dark or with eyes closed would be quite impossible. The trick is, of course, that we receive a constant stream of information about the gravitational field (and therefore the true vertical) from the balancing mechanisms of the ear. Any departures from the vertical are also signalled by pressure-receptors in the feet and other nerve-cells in the muscles of legs, body and neck. Without this complex system of postural reflexes we should fall about the place like drunks—for alcohol makes the vestibular-kinesthetic-cerebellar system inoperative (as well as unpronounceable).

In theory, then, the 'rod and frame test' should be easy enough for the normal (sober) individual. In practice this is far from being the case. Some people can do it but there are others so dominated by the *visually* perceived frame that they tend to ignore the information from the gravity and postural receptors. For them the rod looks vertical when it is *perpendicular* to the tilted frame. These people have been termed 'field-dependent' because they are dependent on the visual field.

Now obviously, and leaving aside any notions of Freudian symbolism, there is more to life than the erection of rods in a dark room. So what has all this to do with military incompetence? The answer very simply is that field-dependents have other characteristics which differentiate them from those who are adept with rods and frames. They are less mature, more passive and more conforming. They are also *more authoritarian* and tend to value achievement for the social approval which it earns rather than for its own sake. In other words, they are generally restricted in their transactions with the external world and more determined by feelings of dependency than by the realities of a given situation. As we saw when discussing achievement-motivation, authoritarianism and the 'closed' mind, there is plenty here to make for military incompetence.

But the matter may be pressed further. How, for example, do people become field-dependent? A genetic component cannot be ruled out;

thus many more women than men are field-dependent, and Shafer has found that women who lack one of the two X-chromosomes of normal women are *extremely* field-dependent.[2]

Finally, the fact that *some* monkeys share with human field-dependents an inability to extract hidden figures from a complex visual display has suggested an evolutionary factor in this aspect of perception. Certainly a capacity for analysing out what is important must have, and have had, high survival-value. According to this argument, field-dependence could be viewed as an evolutionary older stage of development.

However, there are also grounds for implicating experiential factors. Field-dependents seem to be those who in their early years were cursed with restrictive mothers and then remained psychologically tied to Mother's apron strings—conforming, obedient, good boys and girls, fearful of the dangerous world outside, fearful of Mother's displeasure and for ever watching* Mother's face for the *visual* impressions of approval by which their lives were guided. Conversely, their independent counterparts seem not to have been saddled with such anxious and oppressive parents.

All in all, work on this perceptual typology makes good sense in the light of the foregoing theory of authoritarianism. It supports the belief that the apparently hard, tough, prejudiced, hostile, aggressively masculine exterior of the authoritarian cloaks an individual whose obedience to authority, prejudiced attitudes, closed mind and sexual priggishness reflect an incapacitating legacy of lessons learned at Mother's knee. It supports the belief that the ability to analyse a situation, to extract the essentials from a complex mass of information, and to be 'a bad boy' when the occasion demands—obvious hallmarks of such competent commanders as Wellington, Nelson and Montgomery—are the product of a personality which either did not experience or managed to rise above the more suffocating constraints of early childhood.

It is hardly surprising, then, that authoritarianism and field-dependency should be related. Their common denominator would seem to be a lasting impact upon the adult personality of maternal pressure on the infant mind—relatively malign in the case of authori-

---

* According to research by Konstadt and Forman, field-dependents concerned about their performance when taking a test under stressing conditions look up at the face of their examiner twice as often as do field-independent subjects.[3]

tarianism, relatively benign in the case of field-dependency. If this maternal pressure is towards achieving status, authoritarianism would seem the likely outcome, but if primarily protective then the traits of field-dependency might be more in evidence.*

That the two outcomes are not mutually exclusive but could reflect a shift in emphasis can be illustrated by considering again the characters of Haig and Buller. Both were strongly influenced by their mothers— Haig's the more pushing, Buller's the more protective. Both developed authoritarian traits and those administrative abilities which follow from the need to preserve orderliness. And both were ambitious to the point of being touchy about their status. But, following on these speculations, there their paths divide.

The mother inside Haig drove him to command one of the largest armies the world has ever seen, and to do so with remarkable self-confidence. But the mother inside Buller, the mother to whom he had been devoted,† whose photograph he always carried, kept *her* boy passive and dependent. It is significant that when Buller married he took a mature and motherly woman for his wife. It is significant that when stressed by being placed in top command, with no parental figure to whom he could appeal, he himself took on the traits of his internalized mother, becoming over-protective towards his men. And it is significant that when stressed his mind turned to food and drink. He became obese.‡ Of obesity and over-eating in field-dependent personalities it has been said:

> Their inadequately developed sense of separate identity makes it plausible that under stress they would seek comfort in oral activities that had been an important source of satisfaction in the period of close unity with Mother. As a technique of defense for dealing with anxiety, eating is a non-specialised defense. It is applied indiscriminately in a wide range of stressful situations, and it does not act in a specific, directed fashion upon the source

* Other research has suggested a positive relationship between authoritarianism, perceptual field-dependence and *general persuasibility*.[4]

† She died when he was 16, which perhaps helped to perpetuate his early attachment.

‡ Obesity is probably the commonest psychosomatic disorder. In the U.S.A. alone it accounts for 125 thousand tons of excessive adipose tissue, enough to produce 100 billion calories of heat.[5] Research suggests that the cause is over-eating and the fact that the obese are more influenced in their eating habits by such external stimuli as the sight and smell of food than by their state of need.[6]

of stress. In particular kinds of persons, it may suffuse the organism with an animal pleasure which blurs anxiety.[7]

According to other research, obese individuals tend to be 'excessively passive, dependent, intolerant of responsibility and unable to express aggression.'[8] They tend to be the progeny of 'controlling, over-protective, over-indulgent and cold mothers'[9] who use food to compensate for the emotional deprivation suffered by their children[10] and have a penchant for feminizing their sons.

We cannot, unfortunately, know how Buller would have fared on the 'rod and frame' test, but the following description of this general as he 'directed' the fatal battle of Colenso, taken into account with his other characteristics, suggests that he might not have done too well. 'He ordered his men to make a direct attack upon the Boers. He gobbled sandwiches as he watched the action while shells burst around him, one killing the staff surgeon at his side. Buller was himself severely bruised by fragments of shell— but he just continued observing and eating. As more and more of his officers and men fell dead or wounded, Buller's resolve was strangled by pity. He ordered his troops to abandon their assault and to withdraw ... British casualties in this futile engagement totalled 1,100.'[11]

This excursion into the mysteries of field-dependency illuminates features of militarism discussed earlier. It does so by emphasizing an aspect of aggression described by the American analyst C. M. Thompson: 'Aggression is not necessarily destructive at all. It springs from an innate tendency to grow and master life which seems to be characteristic of all living matter.'[12]

Exploration, independence, self-assertion, the overcoming of obstacles and the domination component of male sexuality depend upon the *positive* aspects of this most fundamental drive.

Thus it is that the *repression* of aggression which results from a restrictive childhood impairs those very traits which are required by the professional fighter. It also conflicts with the original purpose of intra-species fighting; for in lower animals aggression is concerned with capturing and holding territory upon which the species depends for its food supplies. It is also the means whereby the strongest males gain access to breeding females. The vestiges of these instinctual goals are to be found in human fighters: the compulsive preoccupation with holding ground (which proved so costly in the First World

War) and the upsurge of sexual activity which follows victory (an upsurge to which the Duchess of Marlborough drew attention when she confided that upon returning from battle her lord 'pleasured [her] three times with his boots on') have a long and furry history!

Under the circumstances it is hardly surprising that military organizations and military men set such store by the 'butch' trappings of masculinity. Nor is it surprising that they should recoil from the slightest suggestion of effeminacy. With the long dependency of human childhood and an emasculating 'Mummy' hovering in the wings they have much to fight before they ever meet the external enemy.

From the foregoing two chapters, 'authoritarianism', with its associated traits of anal-obsessiveness and the 'closed mind',* emerges as the final product of a massive and largely unconscious conflict between two opposing forces. On one side are ranged the powerful drives of sex and aggression, and on the other the strictures of a bourgeois morality implanted in the child by his status-anxious parents. The conflict is chronic and its effect upon mind and behaviour restrictive to the point of being crippling.

From this we have argued that the psychology of authoritarianism lies at the heart of much military incompetence. Because organizations which are invested with the task of managing a nation's violence develop devices for controlling aggression, they will tend to attract into their ranks people with similar personal problems of control. Such people will tend to be conformist, conventional and over-controlled. They will also tend to seek approval, enjoy occupying a position in a dominance–submission hierarchy, and derive satisfaction from the provision of *legitimate* outlets for their normally repressed aggression. They are, in short, authoritarian. But because the roots of authoritarianism lie far back in childhood such people also tend to manifest those other residues of early socialization: orderliness, parsimony and obstinacy—the so-called anal-obsessive triad. Finally, because such people are threatened by the possible breakthrough of instinctual impulses they tend to be over-controlled, rigid and possessed of 'closed' as opposed to 'open' minds. They like to be governed by rules and abhor what is spontaneous, flexible or unusual.

* Recent research by Kline found significant correlations between authoritarianism, dogmatism, rigidity and anal-obsessionality.

Clearly there is much in military organizations which might be expected to attract such people, and clearly their personality-traits will, because highly consistent with the needs and demands of the group, facilitate their promotion.

Such people may be expected to make a number of contributions to military incompetence. Firstly, they will tend to foster, intensify and perpetuate the more restrictive features of militarism. Secondly, because somewhat lacking in humane feelings towards others, they will tend to be wasteful of human life and make poor social leaders. Thirdly, they will tend to be slow to accept unexpected information and will cling to strongly held convictions. Finally, when reaching the top of the pyramid the anxiety engendered by their unaccustomed lack of a higher authority will eventuate in an even tighter control of their aggressive impulses.

Besides making sense of all those features of militarism—'bull', anti-effeminacy, sensitivity to criticism, 'scapegoating' and pontification—which were mentioned in earlier chapters, the nature of authoritarianism reconciles the old so-called 'bloody fool' theory of military incompetence with the personality-based theory advanced here. As William Eckhardt and Alan Newcombe report in a recent paper:

> Both authoritarianism and dogmatism were negatively correlated with intellectual conviction[13] and with education,[14] so that the authoritarian, dogmatic militarist is anti-intellectual. He already 'knows' all that he wants to know. Knowledge is a threat to his ego-defensive orientation, and is therefore rejected. What he claims to be 'knowledge' is actually a faith, so that the essence of dogmatism is a basic confusion between faith and knowledge.[15]

Needless to say, this attitude towards knowledge brushes off on those whom he sees as purveyors of this dangerous commodity; 'he is frequently hostile towards and suspicious of "intellectuals" whom he accuses of being too clever to see the plain facts.'[16]

In other words, what some writers have taken to be a simple straightforward lack of intellectual ability in some military commanders was perhaps due to the crippling effects of anxiety upon perception, memory and thought. To confuse the second of these explanations with the first would be like confusing the erratic behaviour of an expensive watch that has been dropped with the poor time-keeping of a cheap clock constructed from inferior materials.

This distinction between an inherently efficient mechanism distorted by 'noise' and third-rate mechanism which is doing its best is also implied by contemporary studies of the military mind. Similar adjectives tend to recur in every case — 'over-controlled, aloof, rigid,'[17] 'orderly, frugal, obstinate,'[18] 'predictable, punctual, prompt, decisive, rank-conscious, simplistic.'[19] These occur in statements about personality, *not* intellect, about psychopathology, *not* cognitive disability. As one review of this work has said: 'These "anal" characteristics ... would suggest restricted and rigid childhood training, a child who was expected to be seen and not heard, to conform without rebellion, to fit into the schedule prescribed by authority without question or wonder, in short the same sort of childhood training that has been found for authoritarian and dogmatic personalities.'[20]

If these relationships are valid then we should find them foreshadowed in the childhood experiences of some future military officers.

Before going on, there is one further point. It concerns the distinction that has been drawn between 'irrational' authoritarianism, as dealt with here, and so-called 'rational' authoritarianism.[21] By the latter is meant the readiness to accept and obey the dictates of rational authority. An irrational antipathy towards *all* authority, as evident in some cases of student militancy, may be just as neurotic and non-adaptive as a predisposition towards 'irrational' authoritarianism. The common denominator of 'irrational' authoritarianism and blind anarchy is that both states of mind are compulsive and derive from underlying ego-pathology.

In fact this distinction between 'rational' and 'irrational' authoritarianism has been implied throughout this book. Without the exercise and acceptance of rational authority, without certain minimal levels of discipline, and even without certain features of 'bull', military organizations would cease to function.

It is necessary to labour this point because of some semantic confusion regarding the term 'authoritarian'. Throughout this book it refers to the (irrational) authoritarianism of T. W. Adorno and his colleagues. For so-called rational authoritarianism we prefer the phrase 'autocratic behaviour'. The terms are *not* synonymous. Whereas the autocrat exercises tight control when the situation demands it, the authoritarian *is* himself tightly controlled, *no matter what the external situation*.

24

Education and the Cult of Muscular Christianity

'The cultivation of bold, independent and imaginative thinking is of the greatest importance if the security of the nation is to be advanced.' J. R. MASLAND and L. I. RADWAY, *Soldiers and Scholars*

'The root of the trouble was the low standard of education of the average Army officer—only the Navy and Air Force are worse.'
MAJOR-GENERAL BONHAM CARTER

'These boys were in fact the first future ruling class in British history to be subjected to a powerful and uniform moulding process at all. This in itself was of the utmost significance, dooming the variety, spontaneity and open-mindedness that had hitherto been the saving graces of the British upper classes, while the pattern on which these boys were moulded compounded the harmful consequences of uniform moulding in itself.'
CORRELLI BARNETT, *The Collapse of British Power*

The foregoing account of military organizations has dwelt on those features of militarism—its fundamental authoritarianism and capacity for inculcating a fear of failure—which predispose towards incompetence.

Whether or not they do so will, however, depend upon the natural gifts of intellect and personality which the individual members of a military elite bring to their calling. What a man was *before* he entered the Army or Navy may in theory do a lot to offset the effects of the system which he now enters. Conversely, an intake of people who already possess the ways of thinking and feeling which characterize military organizations might tend to amplify rather than compensate for these limitations. It seems possible that the latter situation has been more usually the case, and this for two reasons. The first we have already noted, namely that people are drawn towards vocations which fit their needs. Even before the days of vocational selection, round pegs tended to gravitate towards circular holes. While this reason is likely to be true for all military organizations, there is a second which may apply only to a few. It concerns the social class and educational background of those who become officers in the Army or Navy.

Since the ending of the practice whereby commissions could be purchased both the strengths and the weaknesses of British commanders must be laid in part at the door of the English upper-middle-class system of education in preparatory and public schools.

There is hardly an element of militarism which cannot be located in the ethos of these schools. Even the obsessive traits which we considered in connection with the phenomenon of 'bull' receive ample encouragement in latter-day versions of Dotheboys Hall. The object of these schools was to turn out gentlemen—traditionally leisured males who, without the necessity for soiling their hands, did not require special skills or technical ability. Hence their curricula devoted many hours to a classical education, concentrating on Latin and Greek, History and English, with an almost total neglect of science and technology.[1]

The reason why these were the subjects taught was that they exercised the mind and stretched the memory much as exertions on the sports-field stretched the muscles. In theory, the supposed intellectual 'fall-out' from a classical education was enough to silence the sternest critic. It was believed that through his study of dead languages the schoolboy would acquire not only the languages, but also the history of ancient cultures *and* the ability to think logically. Unfortunately it is doubtful whether most of the children who are forced to struggle with the fourth declension, or grope their way through the translating of 'Caesar in Britain', ever achieve anything more than a transient ability to decline or translate. Moreover, without instruction in formal logic, the professed benefits for the thinking process is doubtful. Even English and Modern History are often taught in such a way as to guarantee that though the recipient of the instruction acquires an impressive rote-memory for the dates of kings and battles, his understanding of their social context remains negligible. The emphasis is on events rather than ideas.

The reasons for this stultifying educational programme are no doubt many and various, but two deserve particular consideration. The first resides in the belief that enforced application to unpleasant, boring tasks develops 'character', and the second that any truly intellectual exercise, by which is meant the cultivation of independent thinking as opposed to rote-learning, harms that fine sense of loyalty and obedience which such schools strive to inculcate. To think is to question and to question is to have doubts.

In the light of these fears it is hardly surprising that these same schools devote a great deal of time to religion and also to sports, most of which most of the boys will never play again. Since the place of honour which sporting activities hold in private boarding-schools cannot possibly be justified in terms of physical fitness or vocational training, what purposes *do* they serve? Four have been suggested. They dissipate surplus energy and in particular are supposed to sublimate the sex-drive. In this respect they constitute a muscular extension of the compulsory cold shower before breakfast. Secondly, they instil team spirit, competitiveness, obedience and loyalty. Thirdly, they are supposed to inure the participants to fear and physical hardship. Finally, because of the immeasurably greater kudos of sporting, as opposed to more scholarly, pursuits, they help promote an ethos which effectively devalues intellectual curiosity and the products of creative imagination. All in all, the curricula of many private boarding schools are based upon the cult of 'muscular Christianity' advocated with homo-erotic zeal by the Victorian writers Charles Kingsley and Thomas Hughes. As applied to education, this cult of manliness and godliness was admirably designed to fit boys for shouldering the burden of the heterosexually deprived white man in far-flung outposts of the Victorian empire, and for shaping up the sorts of leadership qualities required at Balaclava and on the North-West Frontier.

They are, however, rather less suited to the exigencies of modern war. In his book, *The Prefects*, Rupert Wilkinson, in fact, goes so far as to suggest that the intellectual shortcomings of public schools contributed to the two greatest failures of British national leadership — the First World War, with its appalling and unnecessarily high losses, and the 'appeasement' policy of the 1930s.[2]

For this calamitous period in British history two traits in particular must be blamed: the first an absence of curiosity and dislike of new concepts, and the second such complete self-assurance as to rule out the likelihood of prudent foresight. To these must be added several ancillary traits, including a mystical belief in the virtues of amateurism backed up by the equally optimistic credo that a mind encumbered with little more than a rudimentary knowledge of the humanities will somehow muddle through, provided its owner has irreproachable good manners, unquestioning loyalty, total obedience and a sense of public duty.

Even a propensity for 'bull' finds encouragement in the reinforcement of obsessive traits which many boarding-schools provide. The following account of bowel-training in a typical English preparatory school of the 1930s illustrates this point. It also demonstrates how an authoritarian regime may, by undue emphasis on such activities, foster a propensity for lying as a means of preventing anxiety. It is worth noting that this expensive boarding-school for boys exemplified, down to the smallest detail, a regime made totally authoritarian by the efforts of just two people—the headmaster and his wife. While both were authoritarian personalities, with well-developed obsessive traits, their roles in the school were typically demarcated. He administered the cane and she the purgatives. In this way, both attacked the same area. Like the interlocking of two well-fitting pieces of a jigsaw puzzle, they displayed the two aspects of the anal problem: constraint and performance. Predictably, both were cold, stingy and obstinate.

Immediately after breakfast the headmaster's wife would set herself up with a brown exercise book at the long table in the school hall. Each line of her book recorded the day-by-day bowel behaviour of the child whose name appeared in the left-hand margin. With a crisp briefing—'Smith minor No. 6', or 'Forbes-Hetherington No. 10'—the children were dispatched to the various school lavatories ... His mission completed, each child would return to the command post and report, 'Yes please', or 'No please', depending on how successful he had been. She would then enter the result in the appropriate square, a '1' for 'yes' and a 'o' for 'no'.

The system was hardly a success. If the boys were away too long, or not long enough, there would be a tightening of the lip, and close interrogation. And if they replied 'No', they not only earned her obvious displeasure, but suffered retaliation—either liquid cascara, or, for habitual sins of omission, castor oil. Under these circumstances, it is hardly surprising that the boys learned to say 'Yes' rather than 'No', nor, indeed, that many of them became more or less chronically constipated.

There was one other unfortunate by-product of these events—one which the headmaster's wife liked to call 'lavatory mis-demeanours'. No doubt bored by having to wait the required period behind closed doors, the boys would while away their time with carving designs on the seat, dismantling the plumbing and, to use the headmaster's words, 'desecrating the toilet roll'; this usually took the form of pulling out the centre to make several yards of translucent paper telescope. The results of these peccadilloes were inevitably discovered by the headmaster's wife, whose pleasure it was to inspect the lavatories each day.

Needless to say, retribution was swift and harsh. Indeed, so harsh was the punishment, and so deep the disgrace into which the culprit was plunged, that few, if any, dared own up to their crimes. Under such circumstances the rule was that the whole should suffer for the sins of the individual, and the entire school would be detained.[3]

From the above account, it seems that the link between anality and authoritarianism that was forged in early childhood becomes strengthened in the sort of preparatory schools from which future Army officers are drawn.

Finally, and perhaps most fatal of all, the private school's ethic of honour and fair play, so admirable in itself, leads to disastrous results when mistakenly imputed to those like Hitler who play the game by a different set of rules.*

For further evidence that at least one aspect of military incompetence derives from the Victorian attitudes to which the young have been exposed in English boarding-schools, there is the interesting case of military intelligence.

The lack of adequate military intelligence, which has been a recurring feature of most of the campaigns considered in the preceding

* A plot to assassinate Hitler during the 1930s was turned down as 'not cricket' —the very words used by the Government of the day.

chapters, reached its nadir in the war against Russia. As one writer has put it: 'The military blunders and scandals of omission of the Crimean War revealed the poverty of Britain's military intelligence. Most of the catastrophes of that campaign were due to an almost total lack of information about the enemy.'[4]

This particular weakness of military endeavour continued to feature in many subsequent campaigns. Indeed it is no exaggeration to say that an absence of adequate reconnoitring, the refusal to believe intelligence reports and a general horror of spying have tended to keep our armies wrapped in cocoons of catastrophic ignorance.

This fatal preference for honourable ignorance, rather than useful knowledge gleaned by devious means, was not confined to the soldiers in the field, but, as an attitude of mind, permeated the highest levels of military intelligence. The history of the various departments of espionage and counter-espionage, of 'special operations' and the like, is one of badly staffed, ill-equipped Cinderella organizations struggling to perform their duties in the face of contempt, jealousy and resentment from Army and Navy service chiefs. Society's, and in particular the military's, low opinion of intelligence services set up a vicious circle of third-rate recruitment giving rise to careless work, in turn resulting in deepening distaste. Thus between the wars the secret intelligence service (later MI6) was largely staffed by officers who had been 'axed' from the Navy, presumably therefore men of below-average competence.

By the same token, the recruiting of Kim Philby by Guy Burgess into the Secret Service, and the fact that the Director of Naval Intelligence between 1927 and 1930, Admiral Sir Barry Domville, was subsequently arrested in 1940 for his role as founder and chairman of the pro-Nazi organization, the Link, suggest something quite special in the way of incompetence.

As for the underlying motivation behind this neglect of one of the most important ingredients of warring behaviour, there is the revealing case of Lord Portal, during the Second World War, scotching a plan by Special Operations Executive (S.O.E.) to ambush a particularly troublesome Luftwaffe bomber-crew. Portal's reason was that he could not associate himself with 'assassins'. As Deacon remarks: 'Here was the mid-Victorian attitude towards espionage and sabotage rearing its head again.'

Of course there *are* other motives for denigrating intelligence services as when 'Bomber' Harris called S.O.E. 'amateurish, ignorant, irresponsible and mendacious'. He just could not tolerate a single bomber being diverted for the dropping of S.O.E. agents and *matériel*.

Another clue to the feelings which underlay the time-honoured distaste for gathering intelligence is the studied denial by British military and government authorities of their own intelligence services. In 1910 two junior Marine officers volunteered to carry out a survey of German sea-coast defences. Unfortunately they were captured and sentenced to four years' imprisonment, but, when they were eventually released on the orders of the Kaiser, instead of treating them as heroes for their patriotic venture the Admiralty shudderingly disowned the whole venture and refused to recompense them for the considerable financial losses which they had incurred.

This 'not wanting to know' about shameful goings-on emerged again in the 1930s when William Stephenson, a Canadian businessman, and subsequently 'the outstanding executive on the British side of the Secret Service in World War II', discovered that nearly the entire German output of steel was being turned over to armaments. Vital though this information was, only one man, Churchill, listened to Stephenson and prevailed upon him to discover more.

Even as late as 1963 Lord Denning, in his report on the Profumo scandal, felt moved to note that 'the security service in this country is not established by statute nor is it recognized by Common Law. Even the Official Secrets Acts do not acknowledge its existence.'[5]

It seems, then, that a dislike of spying was a fairly widespread phenomenon in certain sections of society. It had not always been so. Wellington had had his own highly efficient intelligence service employing some of the ablest men in the Army. Nelson's scouting frigates had been a potent weapon in British espionage activities. The Royal Navy had even played a major role in the elaborate plot to have Napoleon murdered by the outraged husband of the French dictator's paramour, Madame Fourès.

But then, in mid-Victorian England, all this changed; espionage became a dirty word, and the Secret Service something not mentioned in polite society. It can be argued that this change in attitudes, which, through the ethos of Victorian society and its schools, came to have such disastrous effects upon some future military leaders, was but a thinly veiled expression of one aspect of Victorian prudery—a defence against that natural curiosity of the child which in the adult is called voyeurism. It is surely no coincidence that the society which frowned on mixed bathing, had to dress its table legs and could close its eyes to the greatest boom in prostitution the world has ever seen should have had to leave its espionage to such non-conforming and eccentric amateurs as Kavanagh and Burton. One does not have to be an ardent believer in Freudian theory to realize that neurotic attitudes which centre around a particular segment of human behaviour quickly generalize to other symbolic versions of this same behaviour. Nor should one be surprised to learn that Burton, one of the greatest Secret Service agents of the Victorian era, combined his espionage activities with an 'obsessive interest in sexual phenomena'. There cannot be many spies who have submitted a report on pederasty following nights concealed in the male brothels of Karachi!

Finally, it is surely not coincidental that a country with our record of incuriousness regarding enemy intentions* should also be the one afflicted with attitudes towards pornography which have made us the laughing-stock of other nations.

To return from this apparent digression to the factor of educational systems in military incompetence, the cost of 'muscular Christianity', the tabus on sneaking and the positively Draconian measures taken against the mildest incidents of voyeurism in upper-class boarding-schools implanted a set of values singularly untuned to that 'nastiest' requirement of war—prying into, and reporting back upon, one's enemy.

The notion that some British military disasters may be laid, at least in part, at the door of the English boarding-school system gains support from one other rather obvious truth, namely that the ethos of military organizations *complements* rather than *compensates for* the educational background of their members. The public-school boy entering the Army is no doubt made to feel at home by the discovery that the values instilled in him at school are scarcely distinguishable from those which greet him in the officers' mess. It follows that the military mind receives its particular form for two reasons: firstly, as a natural reaction to the stresses of the job, and secondly, because the corps of officers tends to be drawn from the products of a particular kind of educational system. Acting together, it is hardly surprising

* As Major-General Strong has pointed out in his *Men of Intelligence*, failures to gather and utilize military intelligence have not been confined to the British and Americans.

that these twin sources of intellectual and moral rigidity should have produced a military organization which over the years made up in stamina what it lacked in panache.

Before leaving the role which English boarding-schools play in the psychology of military incompetence, there are some further points to be considered. The first concerns a relatively new concept in psychological medicine, that of separation-anxiety. This refers to the varying degrees of permanent psychological damage which an individual may suffer as a result of being separated from his or her mother (or mother-figure) during early childhood. While the most serious form of this malaise is found in institutionalized orphans, lesser degrees of damage can result from such temporary separations as a period in hospital[6] or being placed in a boarding-school at too early an age. In the latter case the ill-effects of early separation are greatly exacerbated by a cold affectionless milieu in which the child can find no substitute for the warmth and security of the family life from which he has been torn. Such a milieu is, of course, only too typical of the English preparatory school—typical because it has long been the deliberate policy and avowed purpose of such schools to 'build character' by forcing the child to forego, not only the security of an indulgent mamma, but also all those features of the home with which she has become associated. Not for nothing are the matrons in boys' schools referred to as 'hags'.

There are yet other causes of separation-anxiety. Not the least of these derives from the propensity which English mothers have for handing over the care of their progeny to a succession of paid hands: baby-minders, sitters and, in the old days, nannies.

Now it is a simple fact that, so far as the last two sources of separation-anxiety are concerned, those at special risk will be the children—in the case of boarding-schools, the male children—of upper- and upper-middle-class parents. The parents who send their sons to boarding-schools are the parents who have, from Victorian times up to the last war, regarded nannies as as indispensable to the household as we should regard refrigerators today. The simile is not without significance.

Sometimes of course the nanny, like Churchill's Mrs Everest, almost completely fulfilled the role of the mother who had abdicated. But in other cases, like Lord Curzon's Miss Pam, she would, by deliberate unkindness, exacerbate the effects of the emotional void in the young child's life.

Even in its mildest form, separation-anxiety has the effect of withering a person's capacity for affectional behaviour. Having dealt with the original loss of the loved subject by 'switching off' his feelings, the child finds it hard, or perhaps dangerous, to switch them on again. In contemporary psychological jargon he is the unwitting victim of irreversible traumatic avoidance-conditioning! In less scientific language, he has learned to keep a stiff upper lip, that state of facial composure which signals to a boarding-school headmaster that his character-building programme is running according to plan.

Naturally, the psychology of separation-anxiety plays into the hands of those parents, headmasters and subsequent Army commanders who set store by such leadership qualities as stoicism and an apparently unflinching response to situations which threaten physical and emotional privation.

There are yet other satisfactions for the makers of moral fibre. One is that the product of early separation-anxiety, who is then incarcerated in the all-male environment of a boarding-school, will not only discover a lasting satisfaction in the company of members of his own sex, but will subordinate heterosexual love-life to a position commensurate with his job. As Wilkinson remarks when talking of the personal price paid by those emerging from the rigours of Victorian public schools: 'They make good leaders but poor lovers.' It is not that such men eschew the pleasures of sexual intercourse, but rather that they separate the emotional aspects of this activity from its more athletic components. It becomes a subsidiary pastime, more agricultural than romantic.

The other bonus for those who operate a child-rearing and educational system along Spartan lines is that it may result in what are to them three admirable character-traits: an excess of ambition, a positive preference for being in a dominance-submission relationship, and total obedience.

Against these claims for the system, however, must be laid three disadvantages. The first is the eroded self-esteem of those who have suffered a deep emotional rejection in their earliest years; the second that the military, because of its internal structure, provides ample opportunity for promotion to the highest levels of just those people who have developed compensatory traits for the underlying damage; and the third is that such flexibility of thinking, and otherwise creative energy, which might have survived the limitations of a boarding-school education will in all probability be vitiated by the straitjacket of an early fear—that of putting a foot wrong.

Before leaving this section let us be quite clear about the thesis being advanced. It is *not* that middle-class boys who are reared by nannies, and then packed off to boarding-school at an early age, *necessarily* suffer from separation-anxiety, leading to those ego-deformations which eventuate in psychological characteristics of profound significance for military incompetence. No such generalization is intended. What is being maintained is the purely statistical proposition that, of those people who reach positions of high command in the military, a sizable number will belong to that section of the community with an above-average chance of being exposed to the constellation of factors which predispose towards separation-anxiety.

Obviously not all boys exposed to these conditions manifest the effects of separation-anxiety. There are indeed considerable, and largely unexplained, individual differences. But those boys who are stricken by early infantile traumata will be the very ones who tend to have their resulting psychological infirmity exacerbated rather than reduced by subsequent school experience. These are the children who, being more inclined to homesickness and such behaviour-disorders as weeping, fits of temper and enuresis,* will be more likely than their fellows to be the butt of ridicule and criticism. And they are also the children who, if they are not to go under, will develop defences and compensations for the damaged ego. Two such are ambitious striving and the choice of a career which promises certain satisfactions and psychological security. As to the latter, it may well seem paradoxical that a boy who has suffered the attentions of an authoritarian and unaffectionate nanny, followed by the even bleaker embrace of a repressive boarding-school, will actually *choose* to enter upon a career that is in so many respects an extension of his school life. According to psycho-analytic theory, such curious behaviour may be explained as a form of 'repetition-compulsion', the unconscious urge to repeat and thereby resolve the hitherto disastrous situation. A non-mutually-exclusive hypothesis is that having learned to adapt to a given situation

---

* In the writer's school, the ten per cent of boys who were enuretic were segregated from the remainder in a dormitory on the attic floor. This 'leper' treatment can hardly have helped to ease the feelings of rejection which put them there in the first place!

the discerning individual will tend to choose one sufficiently similar for his particular set of adaptations.

An even simpler truth is that, because of their constraint on thought and behaviour, authoritarian, hierarchical organizations offer a level of personal security far in excess of that associated with many more liberal but more cut-throat civilian enterprises.

Whatever the particular motivations, there is a close parallel between these phenomena and the predisposition of long-term, institutionalized prisoners to seek a return to prison when their current sentence has been completed. Long-enforced dependency, however unpleasant its restriction may have seemed, evokes a chronic fear of freedom.

It would be a mistake, however, to assume from all this that a Victorian educational system is a *necessary* cause of restrictive militarism. From their study of American war colleges for officers, Masland and Radway conclude that 'The tendency to conform to a prevailing pattern of thought is manifested in a number of ways.'[7] Just one of these is the avoidance of outside speakers whose views might be controversial!

Since the principal function of these war colleges is to prepare senior officers for higher command, they 'genuinely strive to cultivate the greatest possible freedom of thought among their students'. But somehow the underlying dynamics of military organizations frustrate their good intentions. According to the same writers: 'Students show an "unconscious" reluctance to express views that are counter to existing doctrine ... positions are put forward rather cautiously: if too far out they are withdrawn and modified.'[8]

And on the subject of military intelligence, they found that 'study of the full relationship of intelligence to security planning is neglected at the war colleges.'[9] As Dr R. V. Jones, one-time member of the intelligence branch of the Air Ministry, wrote of British practices, 'while intelligence was of great significance in the war it was rarely discussed and understood.'[10] Masland and Radway make the further point that 'many officers have frowned upon intelligence as harmful to career advancement'.[11] When it is considered that the failure to utilize or understand military intelligence at Pearl Harbor, before the Ardennes offensive of 1944, or at Arnhem, cost America some 83,000 casualties, these attitudes would seem to suggest that the neglect of information about the enemy is due neither to stupidity nor to lack of experience. And in America as in Britain there has been a tendency to hive off from the military any really serious attempts to garner intelligence. To avoid contamination of the armed services, the 'dirty' task is carried out by some other organization, one whose reputation (fairly or unfairly) has sometimes been compared unfavourably with that of any well-run vice ring. In America this other organization is called the C.I.A.

This chapter, which has attempted to show how the education of youths in what is now called 'the private sector' tended to reinforce rather than compensate for those features of militarism which make for incompetence, concludes those sections of the book devoted to the social psychology of military organizations. People who resist psychological explanations of military disasters may well incline to the view that they can be more gracefully explained in terms of the culture of the day. Within limits they are right. The behaviour of incompetent military commanders is partly symptomatic of those Victorian attitudes which, disseminated through schools and churches, moulded the national character.

These attitudes were themselves a reaction to what had gone before. With Trafalgar and Waterloo safely behind us, having become top nation in conquest, trade and sheer quantity of possessions, we could afford the indulgence of turning from greedy entrepreneurial aggression to the more soothing task of putting our consciences in order.

Like a reformed, because enormously successful, burglar who selfrighteously puts down his jemmy to take up proselytizing on the evils of crime, we took to repudiating those very traits—push, cleverness, ruthlessness and sheer naked aggression—that had put us where we were.

But this change to a belief in what Barnett has called 'all that is noble and good' was actuated by the guilty memory of a violent (and disease-riddled) past coupled with a shuddering half-knowledge of such continuing evils as poverty, slavery, child labour and prostitution. Hence it gave rise to a reaction-formation, a disinfecting admixture of evangelism and cleanliness. Assailed by soap and godliness, instinct succumbed to conservatism, rigidity and moon-faced complacency— 'the classic attributes of an army about to suffer a catastrophic defeat'.[12]

In a word, the cultural explanation which Barnett's *Collapse of British Power* offers for industrial and political incompetence is absolutely consistent with some of the reasons given in this book for military incompetence. Where we depart from Barnett's thesis is in trying to show how the culturally determined reactions to guilt, sex, dirt and aggression stem from and work themselves out in the minds of those for whom honour, fair play and 'love thy neighbour' have to be reconciled with the task of killing thy neighbour; and how the incapacitating features of militarism are a product of this reconciliation.

PART THREE

There are no bad regiments, there are only bad officers.'
                                    FIELD-MARSHAL LORD SLIM

*Too much of history is still written as though men had no feelings,
no childhood, and no bodily senses.'*   PETER LOEWENBERG

25

Individual Differences

For centuries civilization has advanced by the subjugation of instinct. But to do so it has had to sanction, encourage and even, on occasions, compel the very kinds of behaviour which it is trying to stamp out.

As Freud wrote many years ago: 'The warring State permits itself every misdeed, every act of violence, as would disgrace the individual man.'[1] Under these conditions, 'thou shalt not' becomes not only 'thou shalt', but 'woe betide you if you don't'.

The unhappy instruments of this policy are, of course, military organizations; they are expected to achieve on a microcosmic scale what the State does macrocosmically—the management of instinct.

To do so they evolve a two-tier system. At the bottom or business end come the ultimate deliverers of aggressive hardware, by tradition the lowest paid and appropriately least cultured members—the common soldiery. Not for nothing were they called 'brutal and licentious'. Traditionally again, this part of the organization was not required to think, feel or be unduly weighed down by conscience—'theirs but to do or die'. The fact that they often died if they *did* was compensated for by the knowledge that they would surely die if they *did not*. Caught between a high probability and a certainty, they chose the former.*

The management of this potentially aggressive force has traditionally been vested in a 'superior caste' of men: the officers.

---

* Recent research amongst American soldiers in Vietnam shows that new arrivals to this theatre *believed* that they might as well fight because they will be shot by their own side if they don't.

Not very surprisingly, in groups as in individuals the control of instinct may well produce internal stresses and strains which in turn give rise to symptoms. In the case of armies and navies, these symptoms constitute the causes of military incompetence.

In developing this thesis, emphasis was laid upon those devices whereby fear is stilled, aggression evoked and disorder prevented. Military organizations were depicted as sometimes cumbrous and inflexible machines for the harnessing and direction of intra-species hostility beneath whose often brightly decorated exterior the psychological process of 'bull', authoritarianism, codes of honour, anti-intellectualism, anti-effeminacy, sensitivity to criticism and fear of failure have contributed to incompetence, both directly and indirectly.

These processes make for incompetence because, since their primary object is control and constraint, they themselves tend to become inflexible and unmodifiable. They resist change, block progress and hamper thought. Just as once useful but now irrelevant drills rob overt behaviour of any verve or spontaneity, so ancient rules and regulations, precious formulae and prescribed attitudes become an easy substitute for serious cerebration.

But all this brings up another matter which may have troubled the careful reader: namely the apparent contradiction that, so far as this country is concerned, military incompetence in battle appears unrelated to ultimate success in war. If, for example, we mismanaged the Crimean campaign, bungled the Boer War, produced glaring examples of crass stupidity in the First World War, and were nearly annihilated through archaic military practices in the first two years of the Second World War, how is it that we eventually won these conflicts? Has our particular brand of military incompetence long-term advantages; is muddling through, perhaps, the best way to win in the end?

This is a fair question to which there is no simple answer. The fact that we won may well be ascribed to various factors, relatively un-related to military prowess. (On occasions one of these has been our allies.) It might equally well be ascribed to the special fortitude, courage, tenacity, humour and superior fighting qualities of the British soldier, who despite—perhaps in some strange way because of—the generals who led him went on muddling through after the rest had stopped.

Then again, it could be ascribed with even greater certainty, as in the case of the Crimean War, to the even grosser incompetence of the

other side. These two explanations have been admirably expressed by
J. B. Priestley when writing of the latter campaign:

> Two things saved this small, odd, rather absurd British Army,
> challenging so far from home a gigantic empire, from immediate
> defeat and then total disaster. First, the Russians, with more men in
> the field, and immense potential reserves, were even bigger
> muddlers than their invaders, and seemed to move in a vague
> dream of battle. Secondly, and not for the first or the last time, the
> British owed almost everything to the courage, obstinacy and
> superb discipline of the regular infantryman.[2]

Whatever the reasons, one important factor is that of individual
differences between military commanders. Certainly in every war we
have won, the good carried the bad.

Part Three of this book looks at this question and examines the
thesis that good generals differ from bad not in their age, colour or
intelligence, but in the degree to which they are able to resist the
psychopathology of the organizations in which they serve. Conversely,
less than adequate commanders will be those whose minds most closely
fit that of their parent organization. For one approximation to this
concept of the 'military mind', there is C. S. Forester's fictitious
General Curzon[3]—cold, strict, unimaginative, humourless, honourable, brave, stubborn, meticulous, spartan, stoical, loyal, obedient
and patriotic. Forester's hero was also ambitious, arrogant, impatient,
insecure, sensitive to criticism, dull, unintellectual, unscientific,
conservative and a moral coward.

This set of traits accords with S. P. Huntington's finding that descriptions of the military mind have usually emphasized its low calibre
and the fact that 'the intelligence, scope and imagination of the professional soldier [compare] unfavourably to the intelligence, scope and
imagination of the lawyer, the businessman and the politician'. The
same writer opines that 'this presumed inferiority has been variously
attributed to the inherently inferior talents of the persons who become
officers, the organization of the military profession which discourages
intellectual initiative, and the infrequent opportunities which an
officer has to actively apply his skill'.[4] By the same token, there appears
to be a consensus of agreement among both military and civilian
writers that the military mind is rigid, logical, inflexible, unemotional,
disciplined and devoid of intuition.

But has all this psychological validity, and what is its relevance to military incompetence? Do these traits really distinguish between good and bad commanders, and what are their origins in the human psyche? By way of answering these questions the last part of this book examines some test cases; first of the concept of authoritarianism and then of particular commanders— good and bad.

26

Extremes of Authoritarianism

*'If society is in danger, it is not because of man's aggressiveness but
because of the repression of personal aggressiveness in individuals.'*

D. W. WINNICOTT

When considering the validity of a complex theory it is often helpful
to examine extreme manifestations of the phenomenon which the
theory is trying to explain. In the case of authoritarianism this has been
done in several independent investigations. Two will be considered
here. The first involves a series of interviews of ex-S.S. concentration-
camp guards and other war criminals indicted for mass murder,[1] and
the second, perhaps the most revealing of them all, an analysis of the
entries in a diary kept by that prototypical authoritarian personality—
Reichsführer Heinrich Himmler.[2]

In focusing on such people we are *not* suggesting that their like are
to be found amongst British military commanders. Just as an understanding of mild depression might benefit from a study of suicide, so
an insight into the nature of 'normal' authoritarianism is deepened by
contemplation of its most pathological manifestation. In this instance
the results from such a study support the view that authoritarian
traits are the product of an underlying weakness of the ego. Thus, from
the first study it seems that the S.S. guards of the Third Reich were not,
as popularly supposed, ideological fanatics, but inadequate 'little' men
for whom the satisfactions provided by the S.S. organization were
tailor-made—all-powerful father-figures, rigid rules of loyalty and
obedience, and 'legitimate' outlets for their hitherto pent-up and
murderous hostility.

The evidence suggests that these men had been emotionally deprived
and subject to repressive discipline in their childhood. Bowing to the

values of total obedience, sexual priggishness and manly decency, they had come to despise those of tenderness, love and sensitivity.

The price paid for their loveless and restricting childhood was an insecure and weakened ego, fear and dislike of their own passivity, and grovelling dependence upon the ordering and regulation of their lives by powerful authority.

One underlying problem of such personalities is that they are forced to renounce, and then denounce, that very part of themselves most in need. As one of them remarked, it was much better to be thought of as a hardened Nazi and brutal murderer than a cissy.

In a word, they were like people who, when they are dying of starvation, come to despise evidence of hunger, for to do otherwise would be to admit (to themselves) the parlousness of their plight. It was this seemingly impossible state of affairs which found resolution by the process of paranoid projection. They hated in others what they could not tolerate in themselves. Hence it was that the weak, the old, the underprivileged, the cowed, and later the starving millions of the concentration camps suffered their fearful attentions.

By the same token, it was not entirely a conscious rationalization, nor entirely their need for justification, which led them to aver that their helpless victims were dangerous enemies, Jewish terrorists, etc., who had to be eliminated. For in a sense they *were* enemies, not of the State, but of their own precariously poised egos.

Once one has accepted the nature of these stunted personalities, other features of their behaviour fall into place—the unquestioning adulation of their superiors, the enthusiasm shown for the uniformed, all-male society of their fellow guards, and a sexual priggishness which could coexist alongside the most brutal and obscene of sexual crimes against the bodies of their victims. Significant in view of the relationship between anal character-traits and authoritarianism was the response of some of these men to the excremental activities of their prisoners. Thus there was the S.S. sergeant and sick-bay attendant of whom it was written: 'Some of G.N.'s atrocities, verified in the court, were such that I hesitate to record them here. Essentially they showed that his greatest venom went to persons who suffered from diarrhoea and were incontinent ... castor oil was his treatment for all abdominal conditions. He would then forbid his patients to go to the toilet, but beat or kill them if they soiled themselves.'[3] It needs no great stretch of the imagination to appreciate how the sight of his helpless victims must have aroused some deep-seated memory of his own early 'socialization' at the hands of over-demanding parents.

The relationship between anal sadism and early toilet-training has been well stated by Henry V. Dicks:

> Withholding a stool as an act of defiance or passive resistance is a well-known nursery manifestation ... But also, to be dirty and faecally uncontrolled is both disgraceful and weak; and it can become a more 'explosive' act of hate, defiance and rejection in the earliest war against authority ... In my own practice patients who are frustrated and enraged with me or with outside persons have developed immediate diarrhoea or vomiting or both. So far then G.N.'s hate-laden inner world can be seen to contain this primitive theme.[4]

The frequent employment by the S.S. and, indeed, other military organizations of such euphemisms as 'disinfecting', 'cleansing' and 'mopping up' to describe their work is not without significance in the context of the anal origins of organized aggression. The use of such phrases contrives to cloak even the most flagrant act of murderous destruction with the mantle of clinical necessity.

So much for the lesser fry in the S.S. hierarchy. Any further doubts as to the concept of authoritarianism tend to dissipate when we contemplate their leader, Heinrich Himmler. So well does this 'prim and pasty-faced ex-chicken farmer' exemplify the general theory of authoritarianism that it is worth spelling out the details.

The theory supposes that authoritarian personalities are the product of status-anxious and repressive parents. Himmler's father, a small-town pedagogue and 'terrible snob',[5] fits this pattern. The theory maintains that fear of parental retribution for any display of sex or aggression leads to a massive blocking of these drives and their attendant emotions. Himmler's diary, unlike that of most normal adolescents, is emotionally flattened, cold and colourless.

The theory contends that authoritarians help to sustain this suppression of their emotions by a refusal to look inwards. The entries in Himmler's diary are confined to the most mundane details of his daily life, and lack any statement as to how he *felt*, or *thought*, about even those trivial issues. His writing is barren of emotion; devoid of rage or love, or any searching of the soul.

According to the theory, an individual's fear of his own sexuality leads to projection. He imputes to others what he cannot tolerate in himself and then, with smug self-righteousness, condemns them. The opprobrium which Himmler heaps on homosexuals and other sexual deviants leaves little doubt as to the nature of its origin in his own deranged sexuality.

Himmler's diary also gives evidence of another classic means for preventing the breakthrough of unacceptable impulses—a compulsive preoccupation with time. As the following excerpt shows, it reads like a railway timetable. His life seems to have been regulated by the clock, leaving no unfilled moments for dangerous spontaneity of action.

8.00 got up. Ran errands. Newspaper.
9.00 to Lorwitz
11.45 ate at Lorwitz
12.20 I was to meet father at the train station, but I only get there at 12.30 ... Joined father in a parlour car in Dachau.
3.00 arrived in Ingolstadt
4.00 into the centre of the city, Mother and Gerhard met us. Drank coffee.[6]

As Loewenberg points out, this obsessive programming of the day's events is one more example of the relationship between authoritarianism and so-called anal traits, for it could well be that strict toilet-training with bowel activity restricted to a particular time of day forces the child to acquire not only a rigid sense of timing but also the tendency to equate orderliness with temporal regularity.

It has been suggested that the enforced control which is a necessary consequence of strict toilet-training may give rise, on the one hand, to the lasting trait of stinginess and on the other to a predisposition towards psychosomatic disorders of the gastro-intestinal tract. Himmler showed both these effects. Each daily entry in his diary concluded with a careful ledger of the day's expenses, and all his life he was plagued with disorders of the bowel.

It will be recalled that anxiety over status plays a significant role in authoritarianism. Once again the Reichsführer runs true to form. Like his father, he was a terrible snob; so obsessed with status and position that everyone referred to in his diary is meticulously prefixed by his correct rank and title.

In accordance with the theory, Himmler's diary contains striking evidence of strong 'reaction-formations' (i.e., unconsciously determined counter-tendencies) against his suppressed hostility. He describes himself as moved to tears by the sight of a young girl coming into conflict with 'her unyielding and stiff-necked father', and is similarly stricken on another occasion by a poor old woman's poverty and hunger. Any doubts one might have as to the real reason for these sentiments tend to wane when it is remembered what this same individual did to some millions of young girls and elderly women.

And so it goes on. There is hardly a feature of the American theory which is not exemplified by Himmler. In some respects, of course, in the extent of his anti-Semitism and cold hostility, he is a grotesque caricature of the authoritarian personality which they describe.

In yet other aspects he illustrates a point made elsewhere (see p. 278), namely the way in which authoritarian personalities are drawn towards ideologies, organizations and relationships which promise to fulfil their neurotic needs. For Himmler there were three such: rigid self-control, a denial of his underlying passivity, and a need to quell doubts about his masculinity—an anxiety clearly expressed in the equation which he made between weakness and effeminacy.

Under the circumstances it is hardly surprising that he prospered in the tough, aggressively masculine world of the S.S. As uniformed police chief, with power of life and death over millions but always under the shadow of Hitler; in an organization rigidly controlled by rules and regulations; working in the hierarchical and bureaucratic world of military administration with its schedules, registers and dossiers—every facet of his damaged personality found expression and reward.

For such a man, Nazi ideology acted like a magnet. 'For Himmler masculine identity meant fighting, wearing a uniform and being in the military ... the forced quality of his values of strenuousness, hardness and impulsive over-activity indicate that *their underlying purpose was to ward off* feelings of passivity, weakness and non-being.'[7] It is also likely, as Dicks has pointed out, that the cult of 'manliness' which Himmler fostered in the S.S. owed on the one hand quite a lot to his own physical imperfections, the weak chest, narrow shoulders, fatty contours and myopic eyes; and on the other, something to his thinly veiled homo-erotic tendencies. This cult, a grotesque caricature of true manliness, was not dissimilar in form and origin to 'muscular Christianity', which we considered in an earlier chapter. Both were the product of an authoritarian ethos and both served to gratify homosexual inclinations in the name of some 'higher' ideology. It is perhaps worth noting that both these cults, one so malignant, the other relatively benign, exemplify extreme versions of the ubiquitous phenomenon of male bonding. According to Lionel Tiger, the tendency for men (as opposed to women) to form close-knit all-male groups, whether these take the form of men's clubs, lodges, military messes or such larger predominantly male institutions as the stock exchange, has its origin in one very early and indisputable talent of the male: his superiority as a hunter.[8] Under the circumstances, it is perhaps hardly surprising that the cult of manliness and hunting of a human quarry (or a quick 'killing' on the stock market) should be so closely intertwined.

One of the least attractive characteristics of authoritarians is their preoccupation with punishment, and their incapacity to feel concern for the human rights of persecuted minorities. In Himmler these traits were well developed. His lifelong interest in torture chambers and similar correctives went along with a sadism towards his fellow men that has never been surpassed. For him, the cult of manliness cheerfully condoned, indeed encouraged, the utmost brutality in his henchmen, particularly when this, like the notorious hose-pipe treatment, involved destruction of the human body via its gastro-intestinal tract. But even here, in his monstrous dealings, the other side of the authoritarian personality, its restrictive aspect, was clearly evident. Thus his extermination of the Jews had first to be 'legalized', by the simple expedient of changing the law regarding Jewish rights. That this farcical manœuvre was something more than a superficial sop to public opinion was evident from his personal reaction when actually witnessing the less pleasant antics of his followers. Though adept at initiating mass murder, the sight of people actually being shot or clubbed to death was more than he could bear, presumably because too close a confrontation with the 'facts' was too close a confrontation with his own repressed psychopathology.

All in all, the history of Himmler, like that of many of his contemporaries in Nazi Germany, is not only a cautionary tale for all so-called civilizations which aim to achieve their ends by a mixture of punitive morality, militarism and repressive control of their citizens in their earliest years, but is a compelling if unpleasant illustration for the concept and theory of authoritarianism. Such evidence, moreover, sheds light on two other sorts of military incompetence: that which results from the over-control of aggression, and that which arises from delusional thinking.

When discussing military organizations the point was made that one of the greatest problems facing such groups was not so much the *display* as the *control* of aggression. Since their stock in trade is violence, their task, from a decision-making point of view, becomes one of deciding when and where and how much to release. They have also to erect elaborate safeguards against the recoil of their own destructive potential upon themselves and innocent bystanders. This, it will be recalled, is one of the purposes of militarism.

Now what is true of the group is also true of the individual. We all have the problem of exercising appropriate control over our aggressive impulses. The main difference between the group and the individual is that in the latter's case the controls are largely internalized. All this is fairly self-evident, but consider now the plight of an authoritarian. For a start his fund of aggression is likely to be greater than most, and his defences against overt hostility are like an army which, though it has been grossly provoked, and possesses an immense capacity for retaliation, has nevertheless been given the order: 'Hold your fire'. He is subject to extreme internal tension, poised uneasily between force and counter-force. It is in this state of unstable equilibrium that some authoritarians veer between violent hostility and rigid over-control. Himmler certainly did so.

Obviously such instability is hardly conducive to rational leadership. Affected by an unconscious conflict (between the urge to express hostility and the fear of so doing), the decisions which authoritarians make will be influenced less by cool considerations than by moment-to-moment fluctuations of impulse over control.

Incidentally, it is perhaps worth noting that an analogous state of affairs may operate in connection with the other great drive that causes trouble to authoritarians— sex. No better example is afforded than that of the missionary in Somerset Maugham's *Rain*. One moment he is the rigid over-controlled puritan condemning Sadie Thompson to eternal damnation, the next he is in bed with her, and the next he is cutting his throat. Some would say, a symbolic castration.

In practice, military incompetence has more often resulted from over-control than from its converse. Hence the observation: 'In the

army of a political democracy the most peaceful men are the generals.'
Or as Brigadier Bidwell remarks in his book on modern warfare:
'No general ever won a war whose conscience troubled him or who
did not want "to beat his enemy too much".'[9]

Clearly we are dealing with a confluence of three factors—age,
conscience and aggression. Since with age a man's hormonal balance
may change, producing on the one hand such external signs as obesity
and on the other a decline in sexual and aggressive tendencies, the
restricting effects of conscience will be more apparent. One aspect
of this see-saw relationship has been well stated by Janowitz:

> The entire process of training and career development places a
> premium on the ability to curb, or at least repress, the direct
> exercise of aggression. The cult of manliness and toughness
> associated with junior officers is often a reaction against profound
> feelings of weakness. Such aggressive pressure can *diminish* [italic
> mine]: as the officer develops actual competence, and as he
> advances in rank and organizational authority.[10]

Take the case of Redvers Buller. As a child he was renowned for
the extreme violence of his temper but later, as an adult, 'What struck
me most with Buller was ... the perfect control he had over his
temper, so great indeed that I wondered if he had a temper or not.'[11]

It seems that, presumably blocked by the gentle injunctions of his
anxious mother, Buller's natural aggression had to be re-routed, for,
as a young man and junior officer, he became known for the brave and
violent way he set about those various dark-skinned races whom
without distinction he labelled 'savages'. But even then the underlying
conflict showed and 'motherly' forgiveness was sought. As John
Walters has remarked: 'Letters that Redvers wrote to his wife reveal
that for him, fighting brought relief to a terrible lust *for which he
candidly confessed shame* [italics mine]. Like a virgin youth lured into a
brothel, he suffered agonizing remorse after shooting and killing had
apparently given him a form of orgasm, emotional or physical.'[12]

But when, years later, he was sent to command the British Army
in its campaign against the Boers all this was changed. His old lust for
killing mysteriously evaporated. Gone now was his aggressive leadership. The erstwhile 'veritable God of Battle', to use Wolseley's
description, dropped into the role of an over-cautious and vacillating

old woman, as harmless to the enemy as he was embarrassing to his own side. Age and the mother had claimed him in the end!

This pattern of psychological processes is no indictment of Buller. Its best and worst aspects may be found, respectively, in the nicest and nastiest of people. Compare the kindly, lovable Sir Redvers, whose personality, so valuable in the sphere of administration and reform, served him so badly in the role of commander-in-chief, with that arch-authoritarian Himmler. Both had a lust for killing and resources of murderous hostility when set against 'inferior peoples'. Both had a great capacity for efficient administration. Both showed signs of sexual inhibition. Both were prone to fits of compassion and excessively over-protective feelings towards certain other people. In Himmler's case these apparently tender feelings extended to young girls and old women. The objects of Buller's solicitude were his mother and the troops serving under him. Both manifested a compulsive desire for status and social approval. (In this connection, it has been suggested that Buller's ardent dislike of General Gordon* was apparently inspired by his jealousy of Gordon's even greater popularity with the British public.[13]) And both in times of stress used food and drink as an anodyne for anxiety. This regressive habit took the form of a penchant for cream cakes in the case of Himmler, while Buller consumed such gigantic meals of rich fare washed down by quantities of Veuve Cliquot that the wagon-train of food and champagne which followed him on active service became a byword in the Army.

Finally, both these men displayed delusional, unrealistic thinking when things went wrong. In Himmler's case this was precipitated by Hitler's demise and in Buller's by his appointment as senior commander in South Africa. The common denominator of these precipitants is that in both cases these men had suddenly found themselves bereft of that one essential for an authoritarian's peace of mind—someone higher up, an all-powerful 'parent figure'.

All in all, the symptom of over-control in an authoritarian person may be likened to what happens when a nervous learner driver in a powerful car is suddenly deserted by his instructor. He slams on the brakes!

---

* When the idea of rescuing Gordon, besieged in Khartoum, was first mooted, Buller's comment was: 'The man isn't worth the camels.'

27

The Worst and the Best

'In a war from which so much human error had been eliminated by technological advances alone, human error was still the principal factor in determining the war's outcome.'

JOHN STRAWSON, *Hitler as Military Commander*

We have considered the various examples of military incompetence, the nature of military organizations, and the psychopathology of those who are attracted to, prosper in and ultimately disgrace military organizations, and will now consider some test cases. If our views are correct, then specific instances of incompetence and the possession of authoritarian traits should be related. The least competent should have a personality which manifests profound disturbance of the ego, rigidity, dogmatism and fear-of-failure motivation. Conversely, the striking characteristics of highly competent commanders should be their absence of authoritarian psychopathology, enormous self-confidence and general robustness of the ego.

For the first group there is no better example than Adolf Hitler. On the grounds that, under his direction as commander-in-chief, Germany suffered the most devastating defeats of all time, we should be justified in assuming that, though in some respects a brilliant tactician and political strategist, he may also have been on occasions grossly incompetent. Looking more closely at his record, it becomes clear that he was, and that the forms of his military incompetence were precisely those predicted by the theory. By the same token his psychology as studied by William Langer presents a horrifying amalgam of those factors operating in the less infamous military incompetents who were considered earlier.

Incompetent commanders, it has been suggested, are often those who were attracted to the military because it promised gratification of certain neurotic needs. These include a reduction of anxiety regarding real or imagined lack of virility/potency/masculinity; defences against anal tendencies; boosts for sagging self-esteem; the discovering of loving mother-figures and strong father-figures; power, dominance and public acclaim; the finding of relatively powerless out-groups on to whom the individual can project those aspects of himself which he finds distasteful; and legitimate outlets for, and adequate control of, his own aggression.

By these lights, life in the Army and subsequently the role of commanding the most authoritarian military organization this world has ever known must have drawn Hitler like a magnet. According to Langer's researches,[1] Hitler's neurotic needs can be attributed to a concatenation of factors which included an inadequate father whose unstable personality oscillated between that of a pompous, pretentious small-town official and (when *en famille*) that of a drunken bully; an anal-obsessive and over-indulgent mother; the birth of a younger brother; and watching his parents copulate. This latter experience, besides intensifying hatred of his father, left Hitler disgusted at his mother's disloyalty to himself, and mortified by his role of impotent bystander.

As if all this were not enough, Hitler's discovery that he was monorchic* seems to have confirmed his feelings of inadequacy. According to G. L. Waite, Hitler showed all the characteristics usually associated with this condition: 'impatience and hyperactivity, sudden development of learning difficulties and lack of concentration, distinct feelings of social inadequacy; chronic indecision; tendency to exaggerate, to lie and to fantasize, identification of the mother as the person responsible for the defect ... concern about bowel-training and castration-fantasies.'[2] Such people are defensive when criticized, believe they are 'special people with an unusual mission to perform', and are given to fantasies of revenge and megalomanic daydreams. In his early years as a postcard painter and later in his preoccupation with the plans for grandiose buildings, Hitler showed another documented symptom of the monorchic—'a passion for creativity, redesigning and reconstruction'.

With this psychological background it is hardly surprising that the

* According to the historian R. L. Waite, the lack of one testicle was confirmed by an autopsy on Hitler's body which Russian doctors carried out in May 1945.

opportunity for military service was enormously attractive. Of it Hitler wrote: 'To myself those hours came like a redemption from the vexatious experiences of my youth.'

The Army provided him with the strong, masculine father-figures which he craved. He reacted to them with typical authoritarian submissiveness. As Langer notes: 'The one thing that all his comrades commented on was his subservience to superior officers.'

> ... during his career in the Army we have an excellent example of Hitler's willingness to submit to the leadership of strong males who were willing to guide him and protect him. Throughout his Army life there is not a shred of evidence to show that Hitler was anything but the model soldier as far as submissiveness and obedience are concerned. *From a psychological point of view life in the Army was a kind of substitute for the home life he had always wanted but could never find.*[3] [italics mine]

It also equipped him with strong defences against his underlying anality. He welcomed 'bull' and in so doing was transformed.

> It is ... interesting to note a considerable change in his appearance. From the dirty, greasy, cast-off clothes of Jews and other charitable people he was now privileged to wear a uniform. Mend, one of his comrades, tells us that when Hitler came out of the trenches and back from an assignment he spent hours cleaning his uniform and boots until he became the joke of the regiment. Quite a remarkable change for one who for almost seven years refused to exert himself just a little in order to pull himself out of the pitiful conditions in which he lived among the dregs of society.[4]

Though relatively undistinguished as a soldier in the First World War, Hitler's underlying authoritarianism was well satisfied by the new social environment in which he found himself. While subservient and ingratiating to those above him, he, like the cadet under-officers of whom Simon Raven writes (see page 224), was not averse to informing on those below him.

While Himmler illustrates the less pleasant aspects of authoritarianism, Hitler provides the clearest illustration of the relationship between authoritarian psychopathology and military incompetence. Whatever else he may have been (Langer describes him as a neurotic psychopath), Hitler was nothing if not authoritarian. Of the defining traits described by Adorno and his colleagues, he possessed five in an extreme form and the remainder to an extent far in excess of that enjoyed by the average normal person. Ethnocentric, violently anti-Semitic, obsessed with notions of power and dominance, demanding of complete acquiescence and submission by those under him while contemptuous of, and on occasions inordinately hostile towards, out-groups, Hitler was also superstitious (i.e., he believed in mystical determinants of an individual's fate), anti-intraceptive (i.e., opposed to the imaginative and tender-minded), and destructive, cynical and with-out compassion for human suffering. By the same token he was a master of projection, imputing to others the aggressive intentions which in reality were his own, and sexually inadequate.

Hitler also had, in full measure, those other traits which tend to go along with authoritarianism. Rarely was there a mind so 'closed' and rarely a personality so clearly marked by impulses of an anal-obsessive kind. As to the latter, Hitler was obstinate beyond belief and in his particular sexual perversion★ showed the mixture of intense anal interest and grovelling submission which betokens a more than usually disturbed anal neurotic.

In the matter of achievement-motivation Hitler manifested a pro-found fear of failure and the various traits normally associated with this state of mind.

In the light of his personality and underlying psychopathology Hitler's particular brand of military incompetence is precisely what one would expect:

1. He showed a total unconcern for the physical and psychological welfare of the men in his armies. As General Zeitzler, Chief of the General Staff, wrote after Stalingrad: 'Paulus's report affected him not at all. The figures of dead and wounded ... left him totally unmoved. Even the dramatic descriptions by eye-witnesses of the hell that was raging near Stalingrad, that was becoming more atrocious every day, left him quite cold.'[5]

2. This imperviousness to human suffering which resulted in such enormous wastage of his own forces was a contributory factor in his stubborn refusal ever to relinquish ground gained. At Stalingrad, in North Africa and subsequently in North-West Europe, Hitler's philo-sophy of 'Victory or Death', and his insistence that 'there can be no

★ He enjoyed being defecated on by a woman.

other consideration save that of holding fast, of not retreating one step, of throwing every gun and every man into the battle' amounted on many occasions to gross military incompetence.*

3. From his extreme ethnocentrism came another well-known form of military incompetence: that which results from a gross underestimation of the enemy and in particular of the ability of civilian populations to withstand the effects of war.

4. While many of Hitler's decisions were militarily disastrous, his underlying ego-weakness and fear of criticism eventuated in several other traits which are undesirable, to say the least, in a senior military commander. He promoted his aides and advisers for their sycophancy rather than their ability—Jodl and Keitel were two such. He refused to accept, believe or even listen to unpalatable intelligence. And when things went really wrong he was the first to find scapegoats.

As General Zeitzler recalls: 'When Hitler learned that the counter-attack by Panzer Corps II had failed, his fury knew no bounds. Turning to Field-Marshal Keitel, who was in charge of disciplinary procedure within the Army, he shouted: "Send for the corps commander at once, tear off his epaulettes, and throw him into jail. It's all his fault." '[7]

5. Like his henchman Himmler, even Hitler could on occasion show that over-control of aggression, that procrastination which has incapacitated some other authoritarian military commanders. Perhaps his most disastrous decision of the war was when he halted the German advance before Dunkirk, thus allowing the British to escape. Equally inexplicable in the light of the military realities of the situation was his failure to seize the most favourable opportunity for a crossing of the Neva. Here was a chance to join up with the Finns and seal the fate of Leningrad, but once again he held back until it was too late.

In 1944, after the Allied landings in France, Hitler again interfered to curb German aggression, this time by halting the advance on Caen of two Panzer divisions which had been called up by Rundstedt. According to Lieutenant-General Zimmerman, Hitler's rationalization for this particular error of judgment was that the divisions should be retained in reserve, in case the main invasion was yet to come from another quarter. Suffice it to say that by the time he was persuaded to let the advance continue it was already too late.

6. Finally, on April 22nd, 1945, 'Hitler failed as a military Commander in a way that he had never failed before. In abdicating responsi-

* An exception was when Hitler halted the retreat from Moscow.[6]

bility he betrayed his command ... the Führer abandoned leadership and duty alike.'[8]

All in all, then, Hitler fits our theory pretty well. If not the most incompetent of military commanders he certainly approached that distinction (a view which seems to have been held by most of his generals). As Strawson puts it: 'Hitler's achievements as supreme commander in the Second World War were inferior to his achievements as an ordinary soldier in the First.' Though capable of such ruthlessly correct decisions as that of halting the retreat from Moscow, he could also commit enormous blunders, and these, when they occurred, seemed less a product of stupidity than of his total, sustained, all-pervasive authoritarianism. Of his total, sustained, all-pervasive authoritarianism there can be no question.

But what of the other side of the coin? Are the most competent of commanders also the least authoritarian?*

As a first step towards answering this question, those military and naval commanders about whose competence there has been complete agreement were studied for their possession or absence of authoritarian traits. The results of this analysis are outlined in the following notes and sketches.

GENERAL SIR JAMES WOLFE

'Oh! he is mad, is he? Then I wish he would *bite* some other of my generals.' This retort by George II to one who had complained that Wolfe was insane understates the case. General Sir James Wolfe stood far out from his fellow generals for his humanity, open-mindedness and military efficiency.

Two other features of this remarkable general confirm that he, like our other great commanders, did not manifest those defects of personality which characterize authoritarianism. Firstly, he was bitterly (and unfashionably) opposed to what he called those 'spirit-breaking tactics of harsh punishment and drill'. Secondly, as this very remark suggests, he did not court popularity and was quite ready to deviate from the accepted norms of the military elite of his day.

Thirdly, he, like Nelson, was quite prepared to disobey orders if these conflicted with what he knew was right. After Culloden, General Hawley, 'The Hangman' to his troops, found the young Charles

* In the technical sense in which this term has been used throughout this book; not to be confused with 'autocratic'.

Fraser of Inveralochie lying alive among the bodies of his fellow clansmen. He turned to Wolfe and ordered him to shoot 'the Rebel dog'. Wolfe refused, 'offering his commission and Hawley found a soldier who killed Inveralochie without scruple'.[9]

Wolfe's moral courage enabled him to achieve standards of military efficiency which were rare for that era. Thus E. S. Turner has described him as 'the most enlightened regimental officer of his day'.[10] At the risk of making himself unpopular, he forced his officers to attend to the welfare of their men, to visit their living quarters, have regard to their health and generally get to know them as fellow human beings.

These are not the hallmarks of authoritarianism.

WELLINGTON

It is significant that the man whom many would regard as one of the greatest military commanders of all time was totally devoid of those traits which characterize the authoritarian personality, and a complete stranger to those imperfections of character which signify a pathologically weak ego and impaired achievement-motivation.

Wellington did not evince signs of emotional restriction, did not remain unmoved by human suffering, did not seek popularity, was unimpressed by 'bull', and did not seek scapegoats for his military set-backs.

He was a commander who put efficiency and the welfare of his army above all personal considerations, a man of whom Trevelyan could write: 'It was fortunate for Britain that Wellington was at once a great humanitarian and a great disciplinarian.'

Of him Elizabeth Longford writes: 'Self-confidence gave him decision ... [he] felt that the sacrifices he made in popularity were repaid in the ultimate perfection of his army ... [he] spurned the decorations of authority—the large staff, sentries, gold braid, cock's feathers ... When he heard that the Prince would also permit him to bear "a Royal augmentation, in the dexter quarter of the arms of Wellington" he shied away from such ostentation ... nor did he wish his son to become an earl unless it should be necessary.'[11]

Wellington's self-confidence is also reflected in his refusal to make scapegoats of others. Thus of the Burgos fiasco he said: 'I see that a disposition already exists to blame the government for the failure of the Siege of Burgos ... it was entirely my own act.'

As for his humanitarian (anti-authoritarian) tendencies: 'Far from being a "butcher" Wellington stood out among commanders for his repeated refusals to sacrifice lives unnecessarily ... [his] feelings in the hour of victory and the days immediately following were something less than joyful. As usual after a battle his mood was set by the losses not the glory ... On the morning after the siege another Wellington showed himself to his deeply astonished staff. He visited the dead on the *glacis* and seeing so many of his finest men destroyed—he broke down and wept.'[12] That Wellington's humanitarianism was not confined to the battlefield, nor blocked by any ethnocentric sentiments, seems evident from the fact that he was in the forefront of those who pressed for the abolition of slavery.

Six other facets of the Wellington character are worth noting. Firstly, he displayed an 'open' mind to new ideas, was quick to innovate and see the possible advantages of progress in technology. Secondly, while quite prepared to 'lacerate' his officers for inefficiency, he was remarkably laissez-faire regarding their dress, only drawing the line at the carrying of umbrellas on active service. Thirdly, he did not commit that cardinal error of so many military incompetents—under-estimation of the enemy—nor was he prone to that ethnocentrism which characterizes the authoritarian personality.

Fourthly, and this an aspect of high achievement-motivation, he took infinite pains in his military planning, left nothing to chance, selected officers for their efficiency, always reconnoitred the ground, moved among his troops and, as he put it, 'always had to see things for himself'. Fifthly, he displayed a wit and profundity of thinking not readily associated with the restricted confines of an authoritarian mind. Thus, 'I don't know what effect these men [a fresh draft of troops] will have upon the enemy, but, by God, they terrify me'; 'All the business of war, and indeed all the business of life, is to endeavour to find out what you don't know by what you do; that's what I called "guessing what was at the other side of the hill" '; 'There is nothing so stupid as a gallant officer'; 'Nobody in the British Army ever reads a regulation or an order as if it were to be a guide for his conduct, or in any other manner than an amusing novel.'

Finally, Wellington showed no evidence of that sexual repression which is said to characterize the authoritarian personality. On the contrary, he shared with Nelson a predilection for the fair sex which could on occasions invite some fairly adverse comments from his contemporaries. It is reasonably certain that he, at least, did not choose to enter upon a military career for the wrong reasons.

SHAKA

This Zulu King and commander-in-chief, a contemporary of Wellington, was in many ways one of the most remarkable of all the great commanders.

For a brief estimate of his accomplishments one cannot do better than cite an interesting comparison drawn by E. A. Ritter. In 1879, sixty-three years after Shaka, the conquest of Zululand took a British force of 20,000, armed with breech-loading rifles, cannon and rocket batteries a full six months. This achievement also necessitated the use of colonial mounted troops, thousands of Natal Native levies and some thousand ox-drawn wagons. The area conquered amounted to 10,000 square miles.

Shaka, starting with a nucleus of 500 untrained spearsmen and fighting the same hostile tribes as opposed the British, conquered an area *ten times as great* and made his influence felt over an area twelve times greater still. He did it by building up in twelve years one of the most efficient and well-disciplined armies the world has ever known.

Of his generalship it has been written: 'Shaka's particular genius lay in his meticulous personal attention to detail, and sheer hard work. If at all possible he always insisted on inspecting everything himself. In every one of his critical battles he insisted on personally reconnoitring the ground and the disposition of the enemy forces. He invariably checked all reports by procuring collateral evidence. He was a firm believer in the maxim, "it is the man's eye which makes the cow grow fat".'[13]

In a related area of military competence, the obtaining and use of military intelligence, Shaka was no less adroit. Not only did he use spies and run a first-rate intelligence service but he became a great exponent of surprise and deceptive ruses. As a tactician he delighted in using defensive actions to lure the enemy into a position favourable to himself, and would scorn that hallmark of so many incompetent commanders: the frontal assault upon the enemy's strongest position.

Shaka invariably made a close study of enemy dispositions, with the result that his army suffered far fewer casualties than those of his opponents. This conservation of his force, a distinguishing characteristic of the competent commander, coloured all his generalship. On the battlefield, he, apparently quite coincidentally, not only adopted Wellington's famous British Square but in fact went one better by modifying it to a circle, thus avoiding corners, those points of possible weakness in the British version. For a military mind to sacrifice the beauty of straight lines for the greater usefulness of circles could be taken to suggest a remarkable emancipation from the neurotic attraction of 'bull'.* Another example of his flexibility and refusal to be dominated by tradition was Shaka's banning of sandals for his fighting men. By making them run barefoot, a considerable and by no means popular break with tradition, he invested his army with a speed of movement far in excess of that achieved by his enemies. The displeasure he incurred through this innovation was hardly reduced by an order to his warriors that they should harden their feet on a parade-ground strewn with thorns. Those who hesitated to follow his example in this painful initiation were instantly clubbed to death.

On another occasion this same autocratic (but non-authoritarian) military leader, who would gladly sacrifice popularity for military efficiency, led his entire army, including his overweight councillors, on a forced route march of 300 miles in six days. By way of sustaining morale, official 'slayers' were deputed to slaughter any stragglers. It is interesting to note that nearly a century and a half later another great commander, also concerned to improve the health of his army, took a leaf out of Shaka's book. Montgomery's version of the same treatment has been admirably described by Spike Milligan:

In 1941 a new power came on the scene. Montgomery! He was put in charge of Southern Command. He removed all the pink fat-faced, Huntin', Shootin' and Fishin' chota peg-swilling officers who were sittin' round waitin' to 'see off the Bosche'. To date we'd done very little Physical Training. We had done a sort of half-hearted knees-up-Mother-Brown for five minutes in the morning, followed by conducted coughing, but that's all ...

One morning a chill of horror ran through the serried ranks. There in Part Two Orders were the words: 'At 06.00 hours the Battery will assemble for a FIVE MILE RUN!' Strong gunners fell fainting to the floor: some lay weeping on their beds. FIVE

---

* In terms of the entropy-reducing function of 'bull' a straight line represents less uncertainty than a curved line.

327

MILES? There was no such distance! FIVE MILES!?!? That wasn't a run, that was deportation! ... So to the great run. Hundreds of white shivering things were paraded outside Worthingholm. Officers out of uniform seemed stripped of all authority. Lieutenant Walker looked very like a bank clerk who couldn't. Now I, like many others, had no intention of running five miles, oh, no. We would hang behind, fade into the background, find a quiet haystack, wait for the return and rejoin them. Montgomery had thought of that. We were all put on three-ton trucks and driven FIVE MILES into the country and dropped. So it started. Some, already exhausted having to climb off the lorry, were begging for the *coup de grâce*. Off we went, Leather Suitcase in front: in ten seconds he was trailing at the back. 'Rest,' he cried, collapsing in a ditch. We rested five minutes and then he called, 'Right, follow me.' Ten seconds—he collapsed again. We left him expiring by the road.[14]

Like Montgomery, Shaka could also be humane as well as punitive in caring for his army. To ensure that his fighting men were kept warm, well rested and well fed, an orderly was provided for every three soldiers under his command. No battle was fought without adequate supplies of food, water and bark dressings being assembled at strategic points beforehand. And after a battle:

Shaka made another careful round of inspection of the battle lines. All the warriors had to turn about and face him as he strode along the higher ground inside the lines ... From time to time he spoke encouraging words to the lightly wounded in the ranks, and noted with satisfaction that their flesh wounds on arms and thighs were bound up with bark over a dressing of u-joye (Datura Stramonium) leaves, which his forethought had ordered the undibi to bring. For Shaka knew he needed every warrior, and he did everything he could to avoid wastage in the army. Several of the wounded whom he considered too seriously hurt he ordered back to the central depot for those grievously injured. A badly bandaged wound called for instant censure, and remedial measures. So did any defect in armament.[15]

It was this sort of behaviour, the very antithesis of that extended to other soldiers in other campaigns by *their* commanders-in-chief, that accounted for the remarkable health of Zulu forces. When they did get ill through fighting in malarial and dysentery-ridden country their recovery was rapid.

Finally, a word about Shaka's personality. It can be summed up as autocratic, totally non-authoritarian, high in achievement-motivation, and yet capable of great warmth and sympathy. According to Ritter: 'He was highly emotional and sentimental behind a façade of iron self-discipline. The fact that he was the finest composer of songs, the leading dancer and wittiest punster suggests the artist who would naturally have a highly strung nature, and be more sensitive than the common run of the Nguni race.'[16]

In sexual prowess Shaka was athletic rather than authoritarian. With 1,200 concubines, who lived in a kraal called The Place of Love, he outstripped (if that is the word) even Napoleon.

One particularly significant facet was his rejection of superstition. A well-known feature of authoritarianism is a predisposition towards being superstitious—the projection on to supernatural entities of repressed aspects of the individual's psychopathology. Being superstitious is in the nature of a defence against anxiety, hence its prevalence in early primitive or non-scientific cultures where people seek some satisfying anxiety-reducing explanation of natural, but otherwise inexplicable, events.

In the light of these considerations, it is a remarkable testimony to a man's ego-strength and intelligence that he could renounce the superstitious beliefs in which he had been reared since childhood. Shaka did just this; even to the extent of striking at the very heart of the witch-doctor system which until his time had held all Zulus in its often terrifying grip. In this, as in so many other ways, he showed his freedom from the psychopathological aspects of authoritarianism.

NAPOLEON

Of Bonaparte's competence as a general, Pieter Geyl wrote: 'His greatness in this capacity is obvious, from his first amazing successes in Italy to the last wonderful defence on French soil in his adversity. The comprehensive view of positions, the eye for the key point, the capacity to read the mind of his opponent, the ability to take quick decisions, a personality powerful enough to impose obedience, all these qualities Napoleon possessed in their highest form.'[17]

Another writer describes him as 'a man of swift resolve and iron will, a master of the craft of war on its technical side and yet eloquent and imaginative', while yet other historians have commented on a trait which Napoleon shared with Montgomery: 'the ability to pick out relevant facts from the general confusion'. The consensus of opinion would seem to put Napoleon high on the list of 'great captains'.

How then does he stand as regards authoritarianism and achievement-motivation? The evidence suggests that though he was ambitious, ruthless, devious, unscrupulous, grandiose, despotic, Machiavellian, dictatorial and autocratic, he was *not* authoritarian. For instance, as regards ethnocentricity: 'The worst that our generation has had to witness, the persecution of the Jews, had no parallel in Napoleon's system. Indeed that system remained true, from first to last, to conceptions of civil equality and human rights with which the oppression or extermination of a group, not on account of acts or even of opinions, but of birth and blood, would have been utterly incompatible.'[18]

Napoleon's policy, that *all* members of a society should have equal opportunities for education, may be taken as further evidence, of a rather more positive kind, for his lack of ethnocentric tendencies. His desire to break down social barriers is also evident in this passage from the socialist philosopher Pierre Leroux: 'Wherever he ruled or placed his rulers, the Inquisition, feudal rights, all exclusive principles, were abolished, the number of monasteries was reduced, customs barriers between provinces thrown down ... social prejudices which divided humanity into castes, all sorts of inequalities ... '[19]

Yet another historian makes the point that Napoleon 'offered a career open to talent, holding it at once to be the criterion of democracy and one of the prime secrets of statesmanship, so to provide that no citizen, however humble, should be barred by disparagement of birth and connections from the highest office and eminence in the State'.[20] Evidently Napoleon's peace of mind did not depend upon the authoritarian defence of structuring his social environment into in-groups and out-groups.

If sexual repression is taken as a measure of authoritarianism, then Napoleon, like Wellington, Nelson and Shaka, scores very low indeed. There are many components of mature adult heterosexuality: a romantic love, susceptibility, lust, jealousy, compassion and generosity. According to Frédéric Masson, Napoleon was wanting in none of them. Nor was he a slow starter. 'A fortnight after the first

meeting Bonaparte was her [Josephine de Beauharnais's] lover ...
they loved each other passionately.'[21]

His military duties did not deflect his ardour. 'The journey from Paris
to Nice was accomplished in eleven stages: from each of these and
from almost every posting house where the General halted for relays,
a letter was dispatched to the Citoyenne Bonaparte. These letters
breathe nothing but passion ... At Nice, whether issuing his laconic
orders ... or rapidly devising a system whereby his exhausted soldiers
may be fed, equipped and disciplined in preparation for that rush upon
the Alps ... letter after letter flies to Josephine.'[22]

Nor did he fail to derive comfort and military motivation from his
love-life: ' "When I am inclined to curse my fate, I lay my hand on
my heart and feel your portrait there: I look at it and love fills me with
joy unspeakable." Victory was a means of seeing her again, of possessing her, of having her near him, with him always.'[23]

Napoleon's capacity for jealousy, compassion and forgiveness is
illustrated by his response to Josephine's infidelity.

But now Bonaparte's personal stronghold had to be stormed.
After knocking repeatedly at the door, in vain she knelt down
sobbing aloud ... the scene was prolonged for hours, for a whole
day, without any sign from within. Worn out at last Josephine
was just about to retire in despair, but her maid ... led her back to
the door and hastily fetched her children; Eugene and Hortense,
kneeling beside their mother, mingled their supplications with
hers. Thereupon the door opened: speechless, the tears streaming
from his eyes, his face convulsed with the terrible struggle that had
rent his heart, Bonaparte appeared, holding out his arms to his
wife ... it was pardon, no grudging pardon ... but reconciliation
generous and complete, forgiveness, nay oblivion of past errors ...
Not only could he forgive the faulty wife, he showed the rarer
virtue of magnanimity to her accomplices ... he was never known
to deprive them of life or liberty ... he scorned to injure them
even in their fortunes.[24]

Unlike such sexually deranged arch-authoritarians as Hitler and
Himmler, Napoleon could be the reverse of extra-punitive. 'The fault,
he argued, lay not so much with these men as with himself. He should
have kept a stricter watch over his wife. A man had been allowed to

enter the harem obeying the instincts of his sex, he had persuaded, and she had yielded, as it was her nature to yield.'[25]

As to his propensity for more casual adventures, there is the following account of his numerous relationships of a more transient and purely physical nature. Even here, however, Napoleon displayed an intraceptive and imaginative capacity quite inconsistent with authoritarian personality-traits.

> But other actresses were admitted to the secret apartments of the Tuileries, whose visits became more or less habitual. They were young women of easy virtue, for whom it was impossible that Bonaparte should have any serious feeling ... With his passionate admiration for tragedy he naturally addressed himself to the interpreters of tragedy ... Phaedra, Andromache, Iphigenia, Hermione were something more than mortal women; they were supernatural, almost divine beings glorious with all the treasures of poetry and history. His imagination kindled at the thought of them: the actresses who represented them attracted him, not on their own merits, but as embodiments of the characters they personated. There was no sense of degradation in intercourse with them, and thus he veiled a purely sensual satisfaction in a mist of poetic feeling ...
>
> Many others ... climbed the staircase and ... passed along the dim passage ... to the small room ... M. Bernard, the court florist, arranged a bouquet every morning for the secret room. But the flowers thus renewed each day faded less swiftly than the fancies inspired by the visitors ... So numerous did they become ... that it would be a difficult matter to give the names of such a multitude.[26]

Of course, sexual activity on such a scale is itself so abnormal as to suggest either that Napoleon was grossly oversexed, or that he was concerned to prove himself because of some real or imaginary shortcoming in procreative ability. It seems the latter was more likely to have been the case. Not only were his external genitalia reported to be unimpressive, but as A. J. P. Taylor has remarked,* 'Most women spoke slightingly of his sexual performances. He himself doubted his capacity. When he achieved a child from Waleska he was amazed and

* Personal communication.

took up the idea of a second marriage. On the other hand, after he had seduced Marie Louise in her travelling carriage, she said: "Do it again." There is no record whether he did.'

The point to notice is that far from showing suppression of his sexual behaviour and condemnation of sexuality in others, Napoleon went all out to prove his fears were groundless. If he had problems, they seem to have been more anatomical than authoritarian.

Other traits commonly associated with authoritarianism include rigidity, meanness, lack of spontaneity and having a closed mind. Judging from the following excerpts, Napoleon did not manifest these symptoms of 'Himmleresque' constriction:

Stern and imperious in his business hours, Napoleon was all ease and sunshine to his intimates. They admired his pleasant wit, his unaffected gaiety, his rich and brilliant handling of moral and political themes. They found him kindly and not inaccessible to counsel, immensely laborious, but always able to command the precious obedience of sleep. There seemed no limit to the span of his activities and interests. Now he would listen with his staff to Monge discoursing on geometry, now in a lazy interval he would weave dreams and ghost stories. His confidence was boundless, his ascendancy unquestioned.[27]

At Vienna he was struck by the appearance of a young girl who professed an enthusiastic worship for him. By his orders a message was sent to her, summoning her to an interview ... Napoleon soon discovered that her passionate admiration for him was of the most innocent and ingenuous description. He gave orders that she should be at once conducted to her home, undertook to find her a suitable husband, and gave her a dowry of 20,000 florins. Such an instance of respectful chivalry was by no means unique in Napoleon's career.[28]

Finally, like many of the other commanders on our list Napoleon was without that vanity which betokens a weak ego, was notoriously careless about his dress, had a wide range of intellectual interests, and promoted his subordinates on the basis of their efficiency. Nor did he display that debilitating over-control of aggression which has on occasions paralysed the warlike behaviour of less successful commanders. On the contrary, Napoleon, like his fearful predecessor

Genghis Khan, could be ruthlessly destructive, even apparently profligate, with his own troops, but not authoritarian.

NELSON

When Nelson died, a sailor wrote in his letter home: 'I never set eyes on him for which I am both sorry and glad, for to be sure I should like to have seen him, but then, all the men in our ship who have seen him are such soft toads, they have done nothing but Blast their Eyes and cry ever since he was killed. God bless you! Chaps that fought like the Devil sit down and cry like a wench.'[29]

For reasons of expediency some authoritarians may cultivate a false bonhomie, particularly when dealing with those whose approval they seek, but, as we saw earlier (page 263), authoritarians are often sadly deceived regarding their popularity. Nelson, who has been described as 'warm, vital, human', who was unaffectedly loved by his officers and men, who could maintain a sincere and lasting friendship even with the husband of his mistress Emma Hamilton, and who delighted in the company of his midshipmen, was not like this.

Imbued as they are with the materialistic values of status-anxious parents, authoritarians rate their acquaintances in terms of their potential usefulness. Their generosity, their ingratiating behaviour, their hospitality and their self-sacrifice tend to be reserved for the rich and influential. They do not respond to past kindnesses, they cannot afford to waste time upon those who can give them nothing in return. For them people are objects to be used.

Nelson, who, when his brother Maurice died, went out of his way and deprived himself to help Sarah Ford, the blind woman with whom his brother had been living, was not like this either.

Nelson is to the Navy what Wellington is to the Army: great amongst the greatest of commanders and quite unhampered by authoritarian traits. The following quotations and excerpts from Oliver Warner's *A Portrait of Lord Nelson* touch upon those aspects of the man which are of particular relevance to the present thesis. It seems, for instance, that, like Wellington, Napoleon, Montgomery and Slim, Nelson did not display a compulsive concern with the orderliness of his dress: ' ... when Captain Nelson, of the Albemarle, came in his barge alongside ... his dress was worthy of attention. He had on a full laced uniform: his lank unpowdered hair was tied in a stiff Hessian tail, of an extraordinary length; the old-fashioned flaps of his waistcoat added to the general quaintness of his figure, and produced an appearance which particularly attracted my notice; for I had never seen anything like it before, nor could I imagine who he was, nor what he came about.'[30]

Nelson did not feel compelled to fawn upon or adulate his superiors, just because they were his seniors in rank. Of Admiral Sir Richard Hughes, he said: 'He bows and scrapes too much ... the admiral and all about him are great ninnies.'[31] It was the wife of this self-same admiral who commented on an aspect of Nelson which we touched upon earlier, his behaviour towards his midshipmen, whom he invariably referred to as his children.

> Among the number it may reasonably be supposed that there must be timid as well as bold: the timid he never rebuked, but always wished to show them he desired nothing of them that he would not instantly do himself: and I have known him say: 'Well, Sir, I am going a race to the mast-head, and beg I may meet you there.' No denial could be given to such a wish, and the poor fellow instantly began his march. His Lordship never took the least notice with what alacrity it was done, but when he met at the top, began instantly speaking in the most cheerful manner, and saying how much a person was to be pitied who could fancy there was any 'danger, or even anything disagreeable, in the attempt'.[32]

By the same token George Matcham said of Nelson that he was 'anxious to give pleasure to everyone about him, distinguishing each in turn by some act of kindness, and chiefly those who seemed to require it most'.[33]

On the quality of his thinking and open-mindedness, Nelson's secretary John Scott had this to say: 'His political, able and ready decisions astonished me beyond measure, indeed, all his public business is transacted with a degree of correctness peculiar to himself, nor does the most trifling circumstance escape his penetrating eye; from a knowledge of his private and secret correspondence I am led to consider him the greatest character I have met with, in fact he is a wonderful great man, as good as great.'[34]

Any residual doubts one might have regarding Nelson's freedom from the crippling effects of a weak ego should be resolved by considering his most famous characteristic, disobedience. 'Possessing boundless moral courage, he was himself prepared to disobey if he thought it to

the advantage of his country (or Naples) and he was often right ...
Nelson was in fact always urging others, even allies, superiors and
officials of the army to disregard their orders, if necessary, in what he
thought to be the general interest of the cause.'[35]

Nelson's own view of this matter was uncomplicated. As he said to
the Duke of Clarence: 'To serve my king, and to destroy the French,
I consider as the great order of all, from which little ones spring; and
if one of these little ones militate against it (for who can tell exactly
at a distance?), I go back and obey the great order and object.'[36]
Coupled with this healthy disregard for blind obedience went great
physical courage and that aggressive spirit which knows no faltering
or holding back.

It is probable that Nelson's absence of those traits which have led to
much military (and naval) incompetence may be laid at the door of his
secure and happy childhood, his devoted parents, and the remarkable
Miss Blackett who, upon his mother's death when he was nine, became
Nelson's nurse. It can be attributed, in part at least, to the general
milieu of a country rectory, where the cultivation of intellectual
pursuits and a love of reading were considered more important than
the striving for the goals of status-anxious parents. And it doubtless
owes something to the fact that, because his family were hard up, he
and his siblings were encouraged to take up various employments at a
relatively early age instead of being incarcerated in those sorts of
schools in which later generations of status-anxious parents like to
put away their offspring.

FISHER

If a man's influence for good upon the fighting service is taken as a
measure of his greatness, then 'Jackie' Fisher's impact upon the Navy
of his day qualifies him for a high place in British naval history.

In the words of Richard Hough: 'In an age when the grip of
patronage and privilege was still exclusive and tenacious, Fisher
brashly fought his way through to become at the age of sixty-three
the controller of the most powerful single force of destruction in the
world.'[37]

In fact, this Admiral of the Fleet and First Sea Lord during the First
World War not only controlled but was largely responsible for the
very existence of this force.

Like the other great captains considered in this section, he was totally free of those defects of personality which spring from ego-weakness. An ebullient man, with enormous self-confidence, he manifested a host of traits consistent with our thesis. He was a progressive techno-crat, and this in a service with a well-earned reputation for resisting anything that smacked of progress. He may not have been a great strategist but was eager for responsibility, articulate, incisive and, in his search for efficiency, not the least perturbed by the enemies which he made on his way. Apparently devoid of modesty, he showed none of that shrinking from publicity which characterizes pathology of achievement-motivation.

As might be expected of a man with an ego of positively tank-like proportions, Fisher suffered little repression of sex or aggression. He was not a peaceful man but neither was he cold. A great womanizer, and loved by women in return, his warmth and humanity extended far beyond the confines of sex and marriage. In the words of Admiral Bacon: 'He always had at heart the comfort of the officers and men under his command.' Others have written of 'his emotion and kind-liness towards the younger officers and middies'.[38]

Autocratic but non-authoritarian, highly motivated to achieve but not deflected by the fear of losing the approval of others, Fisher quite consistently showed a dearth of such anal traits as meanness over money. Though always relatively poor and totally without private means he had a well-deserved reputation for generosity.

All in all, Fisher, like Nelson, was well equipped to use his intel-ligence and drive in the pursuit of naval efficiency. He was apparently quite unhampered by those feelings of infantile inferiority which have crippled other military and naval leaders. It was not without insight that he wrote: 'I attribute my present vitality to the imbibing of my mother's milk beyond the legal period of nine months.'

T. E. LAWRENCE

Of his generalship Liddell Hart wrote: 'Lawrence can bear comparison with Marlborough or Napoleon in that vital faculty of generalship, the power of grasping instantly the picture of the ground and situation, of relating the one to the other, and the local to the general. Indeed there is much to suggest that his topographical and geographical sense was more remarkable than theirs.'[39]

In likening Lawrence to Marlborough, 'our most representative military genius', Liddell Hart considered that Lawrence also showed the same 'profound understanding of human nature', the same 'power of commanding affection while commanding energy' and the same 'consummate blend of diplomacy with strategy'. 'To Lawrence, by the verdict of those who have seen him in crisis and confusion, may aptly be applied the words with which Voltaire depicted Marlborough: "He had to a degree above all other generals of his time that calm courage in the midst of tumult, that serenity of soul in danger, which the English call a cool head, and it was perhaps this quality, the greatest gift of nature for command which formerly gave the English so many advantages over the French." '[40]

Liddell Hart also considered that Lawrence, 'the most widely read of generals', was 'more steeped in knowledge of war than any of the generals of the last war'. In the light of later evidence Liddell Hart probably overstated the case but even today few would dispute that 'Lawrence's military skill ... earned him a place among the great guerrilla leaders'.[41]

In personality, Lawrence is probably the least authoritarian senior commander the world has ever known. He was totally without personal ambition, refused promotion, honours and awards for himself, and deplored the pomp, vanities and ritualized bowing and scraping which one associates with the power structure of hierarchical command systems. The fact that he could renounce his name for that of Ross, and later Shaw, and happily resume his role of a lowly ranker *after* achieving worldwide fame, indicates a degree of self-effacement quite unique amongst military men.

Contrary to a characteristic predisposition of authoritarian individuals, Lawrence disliked interfering with other men's freedom. He disliked giving orders and in fact exercised effective command largely through the tendering of advice. That this advice was acted upon suggests that, by his personality, he achieved a level of leadership rarely attained by military commanders. He himself was prepared to obey foolish orders but disliked passing these on to others.

As Liddell Hart remarks: 'In war such orders often result in the useless sacrifice of men's lives. In peace they often contribute to the sterilization of men's reason.'

Lawrence was a great respecter of reason and considered that the possession of knowledge was of primary importance for a military leader. In his opinion 'the perfect general would know everything in heaven and earth'. By the same token this most open-minded of men deplored the closed and vacuous minds of his military compatriots, men who displayed 'a fundamental, crippling incuriousness'. For his part, Lawrence made a point of studying and making himself proficient with the technology of war—whether this involved learning how to use different types of automatic weapons, flying, or designing a successful air/sea rescue launch.

In an earlier chapter we examined the relationship between anti-intellectualism and militarism, a relationship which reached its most depressing proportions in the burning of books by the uncultured and anxious leaders of the Third Reich. In Britain we have seen how the fashionable contempt of military men for intellectual pursuits had its origins in the cult of 'muscular Christianity', as fostered by English public schools. And we considered how the psychology of 'butch' behaviour, with its exaggerated and 'beery' masculinity and its equating of intellectualism with effeminacy, sprang from the deep anxieties with which men with weak egos view their own passivity.

It is in this context of what has almost come to be accepted as a fact of nature, a sort of secondary sex-characteristic, that the actual behaviour of Lawrence and many other competent military leaders needs to be viewed. For Lawrence, perhaps even more so than any of the others, demonstrated with shattering finality just how false is the belief that intellectualism necessarily betokens effeminacy, cowardice or weak leadership in military affairs. Suffice it to say that Lawrence, poet and scholar, sometime Fellow of All Souls, Oxford, author of one of the finest books ever written in the English language and able to speak six other languages, lacked nothing in the way of soldierly virtues —nothing, that is, except a taste for alcohol.

Non-authoritarian, non-ethnocentric, an achiever of professional excellence, and totally unseeking of social approval, this most remarkable man showed three other traits consistent with our theory of military incompetence. He was liked by many, loved by some and adulated by others—and this from common soldiers and aircraftsmen at one end to eminent generals and politicians at the other. He had a warm 'gentle leg-pulling' sense of humour, and he was the antithesis of mean. He was without so-called anal-obsessive traits and, presumably as a consequence of his relatively secure childhood, did not conceal beneath his somewhat insignificant exterior a weak or damaged ego. No doubt he was a complicated person and possibly homosexual, but even his fabrications and other shortcomings revealed by Knightley and Simpson in *The Secret Lives of Lawrence of Arabia*[42] can be ascribed to such non-authoritarian traits as an over-fertile imagination and great natural sensitivity.

ALLENBY

This man, of whom T. E. Lawrence wrote: 'Allenby was morally so great that the comprehension of our littleness came slow to him', was outwardly much like any other British general of his day—big, red-faced, choleric and immaculately dressed. Thus did Lawrence see him on their first encounter. In point of fact Allenby's unpromising appearance quite belied his nature.

For a start, his mind was neither 'closed' nor narrow. In military affairs he was flexible and progressive. His readiness to adopt new technology, such as the armoured car for desert warfare, was only less remarkable than the apparent ease with which he incorporated Lawrence into his plans. For Allenby, ends evidently justified means, however scruffy and faintly 'bolshie' these might be. His off-duty activities, which included yachting, naturalizing, collecting and pressing wild flowers, archaeology and reading poetry, covered a range of interests which not only extended beyond that of most military men but included hobbies that other less secure egos might well consider insufficiently masculine. (It has even been suggested that he gave up game-hunting because of his love of wild life.)

Though a strict disciplinarian and given to alarming explosions of rage, Allenby was non-authoritarian. He managed to combine an almost Wellingtonian attention to detail and a concern for military efficiency with great emotional warmth. From his letters to his wife one gets the impression of a virile and affectionate man with a sense of humour quite unrestricted by authoritarian defences.

One good measure of ego-strength and inner confidence is the degree to which a person can risk unpopularity when the occasion demands. Though in fact popular, not only with his men but also with his officers (itself a comparatively rare combination), Allenby, like Wellington before him and Montgomery after him, could be blisteringly out-spoken and showed little evidence of curbing his tongue if the behaviour of others conflicted with what he considered desirable from the point of view of military efficiency. Moreover, unlike those generals who preferred to make scapegoats of men of lowly rank, Allenby was quite prepared to hit high, with scant regard for rank or station.

'I have no use for these modern major-generals,' he said, and, when referring to senior commanders in the Boer War: 'generals with no more brains or backbone than a rag doll.'

According to the original theory, one striking feature of authoritarianism is a chronic state of generalized hostility which may express itself indirectly in the enjoyment of aggressive fantasies (see Himmler's preoccupation with forts and torture chambers described on page 314). Allenby showed no inclinations of this kind and professed a sincere dislike of war.

Finally, on the matter of achievement-motivation, he did not manifest that 'fear of failure' which has hampered weaker men. He did not shun publicity and shelter behind a protective cloak of secrecy. He did not confine his activities to what was either very easy or impossibly difficult, but was prepared to take well-considered risks. And he did not refrain from promoting men for their efficiency, or employing subordinates who might, in theory anyway, challenge his own self-esteem. Though sensitive he did not lack self-confidence. He was, moreover, a competent general.

SLIM

'He inspired us by his simplicity, his own rugged type of down-to-earth approach to men and events, his complete naturalness and his absolutely genuine humour. He was a great leader—true; he was a great commander—true; but to us he was above all, the well-loved friend of the family.'[43]

'Personally, I consider Slim was the finest general the Second World War produced.'[44]

Together these two quotations affirm the thesis of this book.

Except, possibly, by their wives, authoritarians are rarely loved. Field-Marshal Lord Slim was loved by his army perhaps more than any other commander has been loved by his men since Nelson.

Of his complete competence as a general, fighting the most difficult campaign of the war in an impossible terrain and unspeakable climate, with an army ravaged by disease, there can be no possible doubt. When he took command of 14th Army in 1943, the daily sick totalled 12,000, mostly from malaria. Thanks to his efforts and the strict anti-malaria discipline which he initiated, casualties from this scourge were down to one in a thousand by 1945. He also instituted advanced field hospital facilities immediately behind the front lines in order to provide speedy treatment for the wounded, and to save, whenever possible, the long agonizing journey back to rear areas.

On the more spiritual aspects of morale his biographer writes:

> He did not subscribe to the idea that the average soldier's thoughts dwelt merely on the discomfort and unpleasantness of the country to which he had been sent to fight. He believed that be they British, Indian, Gurkha or African, if they were told the reasons for fighting, the justice of the cause and the importance of beating the enemy, and were kept in the picture, within the bounds of security, they would respond with enthusiasm. To this end he spent a great deal of time visiting units, talking to them informally in their lines, and encouraged his subordinates to do the same. Naturally, he did not confine these talks entirely to this theme, but interpolated subjects of a more personal nature such as rations, pay, leave, mails and beer, combining these with the necessary amount of humour. 'There was no "brass hat" about him,' wrote General Messervy.[45]

Any residual doubts one might entertain about the strength of Slim's ego should be removed by that acid test: his response to failure. Writing of his faulty decision to evacuate Gallabat in 1940, Slim said: 'Like so many generals when plans have gone wrong, I could find plenty of excuses, but only one reason—myself. When two courses of action were open to me I had not chosen, as a good commander should, the bolder. I had taken counsel of my fears.'[46] There is no evidence here that tell-tale defence—projection—and this even though he had ample opportunity for making scapegoats of those subordinates who had given the advice which ended in failure.

If further proofs were needed that great generalship depends upon an absence of authoritarianism, Slim provides them. It is no exaggeration to say that he had three traits without which the outcome of the Burma campaign might have been very different. Firstly, he was non-ethnocentric and therefore able to achieve the almost impossible, but vitally necessary, goal of maintaining a good relationship with his Chinese allies, however frustrating they may on occasion have been. By the same token, his brilliantly successful leadership of Gurkhas,

Africans and Indians, as well as Europeans, would have been impossible had there existed a trace of ethnocentrism in his make-up. Even his absolutely essential good relationship with the difficult General Stilwell and even more difficult General Wingate depended upon a lack of that narrow prejudice, towards people who are odd or 'different', which has disfigured so many authoritarian military commanders. Between Slim and the American forces in South-East Asia there existed none of that friction which tainted Anglo-American relations in Europe after D-Day.

Reference to Wingate brings up another index of Slim's inner strength: his absence of petty jealousy. When Wingate was killed, Slim wrote:

> As the hours passed and no news of any sort arrived, gloom descended upon us. The immediate sense of loss that struck, like a blow, even those who had differed most from him—and I was not one of these—was a measure of the impact he made.
>
> There could be no question of the seriousness of our loss. Without his presence to animate it, Special Force would no longer be the same to others or to itself. He had created, inspired, defended it, and given it confidence; it was the offspring of his vivid imagination and ruthless energy. It had no other parent ...

And this of the man whose last words to Slim had been: 'You are the only senior officer in South-East Asia who doesn't wish me dead!'[47]

Of the same genre, and in striking contrast to the miserable relations that had existed between the Air Force and the Army in Singapore (see page 132), was Slim's wholehearted use and enthusiastic support of Allied air forces—an appreciation of a 'rival' service which led Air Marshal Sir John Baldwin to write: 'Slim was quicker to grasp the potentialities and value of air support in the jungles of Burma than most Air Force officers. Particularly did he understand what the air required and was always ready to understand their difficulties and limitations.'[48] Compare this with the childish sibling rivalry and costly bickering that had occurred between the three services between the two world wars.

Earlier in this book we noted that for a non-authoritarian, unconventional soldier to reach the highest ranks he *must* be good. Slim exemplifies this proposition. When he was at Staff College he nearly failed to get a good report because of his lack of aptitude for, and interest in, games. It is to the eternal credit of his commandant that despite these serious shortcomings for a potential staff officer he received the highest grading. In view of his subsequent performance, exceptional powers of leadership and enormous physical resilience, we might well consider this a sorry blow for the proponents of 'muscular Christianity'.

Finally, because weak egos can be likened to fragile balloons which puff up or deflate with every transient change of pressure, there is this tribute by another great and discerning commander, Field-Marshal Sir Claude Auchinleck: 'Success did not inflate him nor misfortune depress him.'

ROMMEL

'Germany produces many ruthlessly efficient generals: Rommel stood out amongst them because he had overcome the innate rigidity of the German military mind and was a master of improvisation.' From this comment by Field-Marshal Auchinleck we might conclude that Rommel, Germany's greatest military commander in the Second World War, did not display those personality-traits which we have associated with military incompetence.

Apropos of the present thesis, the single most significant feature of Rommel's personality lay in his attitude towards Hitler and the Nazi regime. For Hitler, his feelings of admiration and respect changed to disillusionment and distaste. For the Nazi Party he never had much sympathy. Later, when their deeds became known to him, he felt loathing and revulsion for them.

Though a loyal and patriotic soldier, his open-mindedness and absence of authoritarian traits enabled him to repudiate utterly the ideology of the ruling party in the very country which he served. There is no question but that this man put his goal of complete military efficiency above thoughts of personal advantage. He steadfastly refused to sacrifice his own values for expediency, even though this eventually cost him his life.

A feature of authoritarianism is a compulsive urge to submit to higher authority, coupled with a tendency to ingratiate oneself with powerful father-figures. Rommel showed neither of these traits. His well-known and cordial dislike of Himmler, Keitel and Jodl, suggests that he at least was not mesmerized by the power structure on which his own security depended.

Like his equally famous counterpart, Montgomery, Rommel's outspokenness could verge upon the tactless. When, in 1935, the plans for an inspection of his battalion involved a single file of S.S. men, standing out in front of his own troops for Hitler's protection, he flatly refused to turn his battalion out. This threatened 'disobedience' resulted in an interview with Himmler at which Rommel won his point. The S.S. guard did not materialize.

On another occasion Rommel did not come off so lightly. It was at a conference in 1944.

Rommel did not improve the atmosphere by protesting to Hitler against the incident of Oradour-sur-Glade, which had occurred a week before. Here the S.S. Division, Das Reich, had, as a reprisal for the killing of a German officer, driven the women and children into the church and then set the village on fire. As the men and boys emerged from the flames, they mowed them down with machine-guns. Afterwards they blew up the church and some six hundred women and children with it. It was unfortunate, they admitted, that there were two villages named Oradour and that they had inadvertently picked the wrong one. Still, reprisals had been carried out. Rommel demanded to be allowed to punish the Division. 'Such things bring disgrace on the German uniform,' he said. 'How can you wonder at the strength of the French Resistance behind us when the S.S. drive every decent Frenchman into joining it?'

'That has nothing to do with you,' snapped Hitler. 'It is outside your area. Your business is to resist the invasion.'[49]

Such temerity on Rommel's part, particularly at a time when Hitler's nerves were becoming increasingly frayed, betokens an ego unimpaired by authoritarian weakness.

Taken into account with his other traits—his warmth towards his family, his absence of rigidity, his parsimony with the lives of his men, his ability to improvise, his popularity with his troops and relative lack of concern regarding his popularity with his equals, it should come as no surprise to learn that this chivalrous, autocratic and most efficient of generals, enjoyed a happy childhood apparently unmarred by those stresses and strains which may weaken the ego and stunt the personality.

MARSHAL ZHUKOV

Like Rommel, Georgi Zhukov, Russia's greatest general, was that remarkable phenomenon: the non-authoritarian who for a while prospered in an authoritarian regime. And like Rommel, he eventually came to grief through his refusal to accept the limitations of his particular regime.

As Marshal Konev said at the time of Zhukov's dismissal from the Ministry of Defence: 'Zhukov's mistakes were made worse by certain statements he had made about Soviet military science and the development of the armed forces: Zhukov had stated that Soviet military regulations played a negative role in the education of commanders and did not help them develop creative initiative.'50

Even before becoming Minister of Defence, Zhukov had fallen out with his boss, Stalin. To begin with, he did not shrink from the most reprehensible of 'crimes' in an authoritarian organization, that of answering back. 'On July 29, Zhukov told Stalin that Kiev would have to be surrendered ... Stalin asked: "What kind of nonsense is this? How could you surrender Kiev to the enemy?" Unable to restrain himself, Zhukov retorted: "If you think that I as the Chief of Staff can only talk nonsense, then I have no business here. I ask to be relieved and sent to the front." Stalin replied that if Zhukov felt that way, "We can do without you." '

Authoritarians do not argue with their father-figures, particularly when by so doing they jeopardize some position of power which the weak ego craves.

The second reason for Zhukov's fall from grace was no less significant—he was far too popular. As a fellow soldier, Colonel Antonov, remarked: 'Stalin never tolerated around him people in the Party ranks who were very popular, as Zhukov had undoubtedly become.'51

Since Zhukov was a harsh disciplinarian and aggressively autocratic, yet withal extremely popular, we can safely conclude that he did not suffer from those afflictions of the ego which underlie authoritarianism.

Other features of this remarkable man support this conclusion. His character-traits were the exact opposite of those which define the authoritarian personality. He was unconventional, unorthodox, flexible rather than rigid, warm, impetuous and unreactionary (e.g., he was an enthusiastic proponent of tanks), concerned for the welfare of his troops and against taking unnecessary risks of heavy casualties. He was uninhibited in sex and aggression, unpuritanical, creative and intellectual. Perhaps most important of all, he, like Rommel, Montgomery and Slim, radiated self-confidence; and, like Kitchener, Fisher and Trenchard, was prepared to drive his fist through the sacred webs of protocol and hierarchical administration.

So much for a highly representative set of 'great captains'. The list is not exhaustive. Indeed of all the commanders who exemplify the principle that 'Competence is the free exercise of dexterity and intelligence in the completion of tasks unimpaired by infantile inferiority', none do so better than Field-Marshal Earl Alexander of Tunis. The product of a happy childhood, free from the curbs of oppressive parents,[52] he was compassionate, versatile, sweet-natured, courageous and temperate, he was the perfect social leader and a highly competent supreme commander.

And there were Guderian, General Sir Richard O'Connor and Field-Marshal Auchinleck, and on the other side of the world their psychological counterparts—the Japanese admiral Yamamoto, victor of Pearl Harbor, another unconventional, non-authoritarian, deep-thinking and humane warrior whose reputation as a trouble-maker in high circles rivalled that of Montgomery; and Douglas MacArthur, who, with all his faults and, to some people, obnoxious megalomanic flamboyance, remains a great, albeit grandiose, impossibly autocratic, yet totally non-authoritarian military commander. This most colourful of generals established his reputation as an individualist in the First World War. '... even then his costume was as notorious as his tactical skill: a floppy cap, a riding crop and often a sweater with a huge wool muffler around his neck—all unorthodox but attention-producing.' It is noteworthy that in 1935, while the British High Command were increasing the allocation of horses to Army officers (see page 116), MacArthur's report at the end of his term of office as Chief of Staff expressed a keen interest in armoured warfare. 'Any army', he wrote, 'that fails to keep in step with this trend is far from making necessary progress towards modernization, going steadily and irrevocably backwards.'[53]

Of course there may well be non-authoritarian generals who were *also* hopelessly incompetent—such as Wellington's drunken and dull-witted General Erskine—but one is hard put to it to find a great general or admiral who was conforming, submissive to authority, punitive, sexually inhibited, over-controlled, ethnocentric, anti-intellectual, assailed by doubts as to his virility, anal-obsessive, superstitious, status-hungry, rigid, possessed of a 'closed' mind, and, as Fisher said of Jellicoe, 'saturated in discipline'—in short, authoritarian.

On the grounds that criticism should start at home, most examples of military incompetence have been drawn from the British scene. It will be noticed, however, that for our common denominators of competence we have looked at senior commanders from widely differing backgrounds and countries. The discovery of similar personality-characteristics among great military leaders, whether they be British, Zulu, Japanese, Russian or French suggests that the central thesis of this book has wide generality. This view is strengthened by another list of competent commanders—American this time—compiled by Janowitz. Writing of what he calls 'the elite nucleus' in the armed forces he provides evidence for the proposition that the effectiveness of military leaders tends to vary inversely with their exposure to a routinized military career, and in so doing makes the further point that among those 'rule-breaking military leaders' who comprised the top one half of one per cent of the U.S. armed-forces hierarchy the men who made the larger contribution 'are characterized by even more pronounced unconventionality in their career lines'.[54]

The nearest we have come to Janowitz's list of highly competent, *civilianized* generals is a man described by Taylor as 'the only general of creative originality produced by the First World War': the Australian Jew, General Sir John Monash—lawyer, civil engineer, archaeologist, botanist and part-time soldier. We can only assume that he, like Janowitz's great Americans, was lucky to have escaped the mind-blunting, routinized career of a large mercenary military organization, where the real skills demanded by the complex task of generalship are gradually expunged by orthodox militarism.

Before concluding this section on the personality-characteristics of great commanders, there is one final point. If our theory of military incompetence is correct, then efficiency at high levels of command should be a function not only of individual personalities, but of the ethos of their parent organization. Not only do highly authoritarian armies tend to attract equally authoritarian officers; they will also serve to cramp the style and hinder the promotion of unconventional commanders.

It follows, therefore, that military efficiency should be relatively higher in organizations which are not suffused with the inhibiting values of traditional militarism. Specific evidence in favour of these propositions is not hard to find. We have already considered one example in Chapter 4, that of the Boer War, where the large Regular Army of Great Britain, highly regimented, steeped in protocol with traditions going back a thousand years, was made to look utterly ridiculous in a succession of resounding defeats at the hands of a band of ill-dressed civilian farmers who had just come together for the purpose of defending their homeland.

In contrast to the aristocratic, stiff-necked and immaculately turned-out British generals, the Boers were served by quick-thinking, resourceful but untidy men whose far-seeing minds and concern for the welfare of their people enabled them to run rings around their adversaries. And when the Boers were eventually put in their place, it was only through the efforts of two atypical British commanders: a commoner who had served his time in the unprestigious Indian Army — Lord Roberts — and a man notorious for a flaunting of military conventions — Lord Kitchener.

In a second and contemporary illustration, there is the Israeli Army, the David of two and a half million Jews who in six days defeated the Goliath of 100 million Arabs. By its competence and vastly superior direction this miniscule army, drawn from a country poor in resources and gravely disadvantaged by its geographical position, managed to defeat an enemy from countries possessing inexhaustible reserves of natural wealth (including one half of the world's hydrocarbon reserves).

Developed from a small group of watchmen at the turn of the century, the Israeli Army is an infant by European standards. Like its parent military organization, Haganah, the Israeli Army is civilianized and free of obsolete army tradition. It had, until recently, no time for 'bull', retaining only such military conventions as are minimally necessary for discipline and efficiency. According to Yigal Allon: 'Its attitude to these matters, was (and remains) strictly functional.'[55] In the eyes of military traditionalists the Israelis, like the Boers, present an incomprehensible paradox of efficiency without authoritarianism. As Robert Henriques (quoted by Allon) wrote in his *A Hundred Years to Suez* (1957):

Although Israeli units can be extremely smart on a ceremonial parade, there is very little discipline in the normal sense. Officers are often called by their first names amongst their men, as amongst their colleagues; there is very little saluting; there are a lot of unshaven chins; there are no outward signs of respect for superiors; there is no word in Hebrew for 'Sir'. A soldier genuinely feels himself to be the equal of his officer—indeed of any officer—yet in battle he accepts military authority without question. I cannot explain, I cannot begin to understand, how or why it works. All my own military experience in the British and American Armies has taught me that first-class discipline in battle depends on good discipline in the barracks. Israel's Army seems to refute that lesson.[56]

Yisrael Galili (1947) put it this way: 'It is said that our men are somewhat deficient in the usual forms of discipline. But against that they have the virtues of responsibility and courage. Any loss due to lack of military discipline is more than made up for by the self-reliance, the initiative and the spirit of our men.'[57]

Of the several issues raised by contemplation of the Israeli Army, two are of particular relevance to the present discussion. Firstly, like guerrilla forces in various parts of the world, and *unlike* the armed forces of the great democracies, the Israeli Army is itself a democratic institution. As Yigal Allon put it: 'Since it is the product of a popular movement in national liberation, directed by democratically elected civil institutions, it is perhaps not surprising that the new Army should have inherited from the Haganah its democratic values.'[58] Since it is itself democratic, as opposed, say, to the British Army, which though theoretically elected by democratic institutions is in fact controlled by an authoritarian elite, the Israeli armed forces can dispense with those artificial devices whereby authoritarianism and mediocrity are maintained. Moreover, since promotion is based on merit rather than class or money, its senior commanders, being actually of superior ability, do not require that their positions have to be shored up by the myths of infallibility and the professed virtues of blind obedience.*

* An apparent exception to these characteristics of the Israeli Army occurred with the appointment of General Tal to command of the Armoured Corps. He announced his arrival with a hitherto unprecedented imposing of discipline upon armoured personnel.

In fact, Tal's 'tightening up' was essentially practical and not motivated by a

If there is saluting and respect, it is for the deserving individual, not some such vague concept as the King's Commission. By the same token, with their minds free from the stultifying effects of ancient traditions, these same commanders achieve a level of flexibility and expertise conspicuously lacking in the high command of many other armies.

A second factor which seems to have contributed to Israeli military competence is, paradoxically, the enormous odds ranged against them. Unlike the armies of Britain and France, they have had to maximize their potential in waging war and have therefore not been able to afford the luxury of honourable but outmoded strategy, and a glorious but improvident attachment to horses and ceremonial parades. Hence their emphasis on the strategy of the indirect approach, on achieving surprise and on technical mastery. Regarding this latter point it could be argued that since, from the highest to the lowest, they are imbued with patriotic fervour and are fighting for survival they are without guilt, and therefore mercifully free from mock heroics and those aggression-inhibitors such as notions of fair play which have so incapacitated the British Army. Not for them the sort of thinking which led us to forego use of machine-guns as weapons only for use against savages and other 'inferior peoples'. Nor have they, for the same reason, had to invoke the Almighty with such simple faith as have the generals of other armies. The Israelis, as indeed many guerrilla fighters, regard war as rather like surgery—an unpleasant but sometimes necessary business which, if it *has* to be waged, should be swift, efficient and precise. Again, because their leadership is superb but unassuming, they have

---

compulsive urge towards 'bull'. The point is illustrated by some remarks which he made to members of a Kibbutz who had criticized his new approach.

'Last week I attended the funeral of a tanker at one of the Kibbutzim. Your people eulogized him as a hero, but I could only feel the pity of it all. That boy did not die a hero. He was killed accidently during a simple training exercise, merely because he was not brought up on the principle that an order should be carried out because it is an order. The Armoured Corps rules that shells must not be stored in tanks without their safety clips, for the simple reason that the static electricity inside the tank might set the shell off. Therefore the order is quite categorical. Shells are to be stored in tanks in no other way, and this order must be carried out regardless of whether or not a soldier is convinced of its logic. This particular soldier did not do this, and was killed as a result.'[59]

Perhaps the most significant feature of this episode is that such a dialogue could and did take place between a general and his civilian compatriots.

had no need to glorify war for the purpose of urging on those who do the fighting. All in all, their attitude seems remarkably mature.

Finally, the forces of Israel are also free of that crippling anti-intellectualism and curious anti-feminism which have been such a feature of other military organizations. They are not a refuge for the dim nor a place where 'butch' young men can 'prove' their virility. Needless to say, according to anecdotal evidence, even the Israeli Army is not without its problems. Two are particularly pertinent to our central thesis.

The first concerns administration. Since the Israeli Army puts a low premium on obsessive behaviour, its administration tends to be disorderly: letters get lost, forms are wrongly filed, and Moshe gets the call-up papers meant for Josef. In other words, the flexibility and battle discipline of a near-perfect fighting machine are evidently bought at the price of minor breakdowns on the bureaucratic front. It seems that you cannot have everything.

The other, related problem is that from time to time officers and N.C.O.s are charged with the 'scandalous offence' of imposing 'bull' upon the men. In this 'Alice Through the Looking Glass' army, 'bull', it seems, is rather like V.D. in less enlightened services— something that soldiers are liable to catch, unless due precautions are taken. Even in Israel, it seems, the management of intra-species aggression tends to excite anal-obsessive defences in people who are perhaps naturally inclined that way. The difference between our army and theirs is that an inhibitory trait which we applaud is regarded by them as little short of a disaster for military efficiency.

A comparison of the best with the worst of military commanders supports the view that military incompetence results from those defects of personality associated with authoritarian and disordered achievement-motivation. When all that is natural, creative, flexible, warm and outgoing in the human spirit becomes crushed and constricted, such qualities of leadership as compassion, bold decisions and military flair give way to conformity, sycophantism, indecision and fear of failure.

While the relationship between competence and an absence of authoritarianism appears to hold across cultures and to be connected in a complex way with the parent organization, our study of great commanders emphasizes that competence depends upon emancipation from the restrictions of militarism. Like children who can free themselves from familial pressures, great commanders are those who can rise above and even criticize their parent organization.*

It cannot therefore be maintained, as some would have it, that military disasters are explicable solely in terms of the culture of the day, of—in our case—the stranglehold of Victorian morality. The set of values and attitudes which this morality comprised, inculcated through school and religion, may have provided the *content* for some disastrous thinking. But whether or not they did so depended upon the robustness of individual personalities.

So far, then, a study of great commanders makes the theory of incompetence seem pretty watertight. But are there exceptions to the rule? The next chapter examines three such possibilities.

---

* As exemplified by General Chaim Herzog's analysis of what went wrong in the Yom Kippur War of 1973. (Far from weakening the central thesis of this book, the general's indictment shows how, given time, even the best of armies may succumb to the hazards of professionalizing violence. The syndrome he describes—underestimation of the enemy, a clinging to once useful practices, indecision and the misreading of intelligence—recapitulates a pattern seen so often in the past. But two curious facts remain: firstly, despite a bad start and the enormity of the odds ranged against them,[60] the Israelis were *not* defeated; secondly, a man who was until recently a senior member of their military hierarchy could then set about *exposing* what went wrong. It is perhaps the second of these phenomena which most clearly helps to explain the first.)

28

Exceptions to the Rule?

'One must not judge everyone in the world by his qualities as a soldier: otherwise we should have no civilization.'

FIELD-MARSHAL ROMMEL TO HIS SON

There are grounds for adding to our list of highly competent commanders three who might at first sight suggest a total negation of the proposition that authoritarianism and competence are negatively correlated. These are Montgomery, Kitchener and Haig.

As popularly used the epithet 'authoritarian' might seem applicable to all three; for all tended to be cold, ruthless, dictatorial and ambitious— as different, one might think, from men like Lawrence, Nelson and Slim as it is possible to be.

As to their military competence, they have the distinction of being controversial figures. Unlike the others, all three have had their detractors as well as devotees.

The question then arises: Were they in fact authoritarian in the technical sense (i.e., did they betray an underlying pathology of the ego) and, if so, did this lead to incompetence in their military careers?

In answer I propose to show that in fact these men fit our theory very well, and this in three respects:

Firstly, they suggest that the relationship between personality and incompetence is not confined to those who were extreme in their ineptitude but may operate along a continuum of military excellence from the worst to the best of senior commanders.

Secondly, the military shortcomings of Montgomery, Kitchener and Haig and their positions along a dimension of authoritarianism are perfectly correlated.

Thirdly, the precise nature of their particular lapses from competence

354

are exactly what one would predict from features of their underlying psychopathology.

Let us consider them in descending order of achievement.

MONTGOMERY

There is one great benefit, which generals confer upon mankind, that is rarely touched upon by military historians—their entertainment-value. Even generals under whom it may have been far from jolly to serve can arouse considerable interest when viewed from afar, and this whether they elicit morbid fascination or merely provide comic relief.

High on the list of these absorbing characters stand those like Kitchener and Haig, who have engendered not only entertainment (in the form of films, shows, anecdotes and verse) but also serious controversy over the precise value of their military performance. It could be argued that Montgomery is well to the forefront of this group on both counts.

Firstly, there must be few generals who have done more to relieve the tedium of war than did Montgomery with his now famous order on the Prevention of Venereal Disease (see page 277). For what other order, in the history of warfare, has inspired such elevating doggerel as this?

Mars Amatoria

The General was worried and was very ill at ease,
He was haunted by the subject of venereal disease;
For four and forty soldiers was the tale he had to tell
Had lain among the beets and loved not wisely but too well.
It was plain that copulation was a tonic for the bored,
But the gallant British Soldier was an Innocent Abroad;
So ere he takes his pleasure with an amateur or whore,
He must learn the way from officers who've trod that path before.
No kind of doubt existed in the Major-General's head
That the men who really knew the game of Love from A to Z
Were his Colonels and his Adjutants and those above the ruck,
For the higher up an officer the better he can f—k.
The Colonels and the Majors were not a bit dismayed,
They gave orders for the holding of a Unit Love Parade,
And the Adjutants by numbers showed exactly how it's done,

How not to be a casualty and still have lots of fun.
The Adjutants explained that 'capote' did not mean a cup,
That refreshment horizontal must be taken standing up,
They told the troops to work at Love according to the rules
And after digging in to take precautions with their tools.
Now the General is happy and perfectly at ease,
No longer is he troubled with venereal disease,
His problem solved, his soldiers clean (their badge is now a dove),
He has earned the cross of Venus, our General of Love.

'Cupid' (Royal Corps of Signals)[1]

Secondly, there must be few generals who have inspired such controversial comments on their competence as these: 'Our best general since Wellington' (Lewin); 'The most overrated general of World War II' (Blumenson); 'A great soldier, great in my opinion, not only by the standards of his generation but in the eye of history' (Attlee); 'Over-cautious, habit-ridden and systematic' (Von Rundstedt).

His biographer R. W. Thompson, in his book *Montgomery, The Field-Marshal*, is also critical. 'The Field-Marshal's failure to seize Antwerp entire, and to advance across the Albert canal to cut off the Bevelands—and the German 15th Army—is his most agonizing failure. In battle Montgomery had his feet firmly on the ground, too often perhaps rooted to the ground. But in his hour of victory, in his sense of elation, a new and daring Montgomery was manifest—in words.'

It is not my purpose to debate Montgomery's greatness. Suffice it to say that while not without blemishes he was in the main a highly competent commander and, as such, needs to be considered in the present context. Does he or does he not support the hypothesis that competence depends upon an absence of authoritarianism and its associated traits?

In fact, the victor of Alamein presents the interesting possibility that when achievement-need is sufficiently high it may nullify those authoritarian propensities which, as we have seen, interfere with the role of generalship. Ronald Lewin has drawn attention to this process when referring to Montgomery's reticence in his memoirs about the period between 1931 and 1934 when he was a battalion commander in Alexandria. 'Montgomery ran his battalion as if Alexandria was a city lying somewhere between Sodom and Gomorrah. Certainly in the early 'thirties that age-old siren offered every possible temptation to both officer and man. Conscious of this, and with a harsh self-righteousness, Montgomery used too firm a grip—and used it tactlessly.'[2]

At first sight this apparent concern with 'sexual goings-on', one of the distinguishing characteristics of the authoritarian personality, coupled with the evident *need for approval* suggested by its omission from his memoirs, might seem to imply an authoritarian cast of mind. But in his memoirs we find him cheerfully admitting that in 1939 he got into serious trouble, including 'a proper backhander' from his corps commander, for issuing the order, already referred to, on the subject of venereal disease. In this order, according to its author: 'I analysed the problem very frankly and gave my ideas about how to solve it.'[3]

Thanks to his brother Brian we know what the order said, and that, unlike his superiors, the future field-marshal did *not* set himself up as the moral arbiter of his men but applied himself to the primary task of a commander—conservation of his force!

The order is neither prurient, condemnatory nor incomprehensible, but lucid, memorable and demanding of attention. It establishes him as a leader who cared about the welfare of his men, who was human and possessed of a sense of humour. It also suggests that he possessed that much-lauded feature of leadership: initiative. He may have been unbearably autocratic but he was not authoritarian.

Montgomery's subsequent career confirms the reality of these traits.

1. When commanding the south-eastern (anti-invasion) army in 1941 he ordered that all officers' families should leave the area of divisions which had an operational role in repelling invasion. Of this episode he wrote: 'I was told that a good officer would never give a single thought to his wife and family in such conditions; his whole mind would be on the battle. I said that I did not believe it.'[4]

It has been suggested by Brian Montgomery that this apparently harsh order stemmed from jealousy and a sort of anger against fate consequent upon the loss of his wife. But the fact remains that this same order implies three other characteristics of Montgomery: his grasp of reality, his preparedness to accept unpopularity, and the implication that he, at least, understood what heterosexual love and loyalty are all about. These are not the characteristics of an authoritarian personality.

2. By the same token, the following excerpts from his memoirs say all that needs to be said about Montgomery's capacity for those traits so conspicuously absent in the extreme authoritarian personality—love and humanity:

> Here I must turn aside to deal with something much more important than my military career, the ten short years of my married life ...
>
> I met Mrs Carver and her two boys aged eleven and twelve. I have always been devoted to young people and I like helping them: possibly because of my own unhappy childhood ...
>
> A time of great happiness then began; it had never before seemed possible that such love and affection could exist. We went everywhere, and did everything, together.[5]
>
> The doctors did everything that was possible; the nurses were splendid; but the septicaemia had got a firm hold. Betty died on the 19th October, 1937, in my arms ... I would not let David attend the funeral and, indeed, would never let him come and see his mother at any time when she was in great pain and slowly dying. I could not bring myself to let him see her suffering. He was only nine years old and was happy at school; after the funeral I went to the school and told him myself.[6]

These are not the words of an authoritarian.

3. Montgomery's regard for human life, a potent factor in the over-cautiousness of which some historians have accused him, is reflected in his attitude to generals of the First World War. 'The frightful casualties appalled me. The so-called "Good Fighting Generals" of the war appeared to me to be those who had a complete disregard for human life.' In this context he recounts the story (see page 374) of Haig's Chief of Staff who, prior to returning to England, decided he would like, for the first time, to visit the Passchendaele front. That the Chief of Staff of the British Armies in Europe should be ignorant of the conditions in which these armies were fighting clearly appalled the young Montgomery.

4. Montgomery's 'illegal' hiring out of W.D. land to a fair promoter in order to raise funds for the garrison welfare services could have cost him his career. An officer prepared to jeopardize his career for the sake of the welfare of his troops is unlikely to have an authoritarian personality.

In 1943, during the desert campaign, Montgomery broke new ground by arranging with the R.A.M.C. that *female* nurses should be employed at the casualty clearing stations in forward areas. As his brother records: 'This innovation was very much appreciated by all ranks and was yet further evidence of [his] determination to maintain morale in all circumstances.'[7]

In an army whose prejudices had changed comparatively little since the days when Miss Nightingale encountered such a chilly reception from high-ranking officers during the Crimean War, this was a considerable step forward.

5. Montgomery also lacked those obsessive traits which tend to accompany authoritarianism. He was not particularly mean or particularly obstinate and, judging from his own dress and lenient attitude towards that of his troops, did not harbour any compulsive urge for 'bull'. In this, as in other matters, his approach was essentially realistic. As his brother remarks: 'He simply could not equate proficiency in the formalities of the parade-ground with the skills required of the infantryman on the battlefield.'

6. A likely feature of authoritarianism is a narrow religiosity and intolerance towards sects other than one's own. Hence it is significant to find Montgomery risking his career by banning formal church parades. On the second point it is illuminating to discover that, despite the strict Protestantism of his parents, who nursed 'a very strong prejudice against the Roman Catholic Church', their field-marshal son sleeps nightly, and presumably soundly, under a portrait of Pope Pius XII, a man for whom he had a very high regard.

It seems, then, that whatever else he may be Montgomery does not evince the well-documented signs of authoritarianism. And yet, even in his case, there remains the undisputable fact that for all his greatness as a military commander Montgomery did have serious shortcomings which *could not be attributed to a lack of professional ability*. A headmaster's report might well have read: 'Though hardworking, energetic, knowledgeable, intelligent, and able to think big, Montgomery does on occasions show surprising lapses in judgment and behaviour.'

These lapses were:

1. An inability to get along with many of his military colleagues. Like Kitchener, he had the knack of making himself enormously unpopular with his contemporaries and preferred the company of younger and more junior officers. As his biographers have pointed

out, this failure to achieve rapport, even in a situation which demanded perfect teamwork, reached catastrophic proportions in his relations with the Americans after D-Day. According to Lewin: 'Montgomery lived inside a cocoon, and this accounts for much of the personal animosity he aroused, animosity of which he was simply not aware and of which he was incredulous when it struck him.'[8]

2. Montgomery's second shortcoming was that he sometimes allowed his own desire for personal glory to influence his planning. 'A military plan tainted by an attempt to satisfy the Commander's ego is unlikely to be the best plan: an irrelevant factor has been introduced into the calculation.'[9]

Some of Montgomery's plans fell into this category. At Enfidaville, in the North African campaign, his ego-needs threatened to destroy the Eighth Army. Subsequently, in North-West Europe, the field-marshal's desire to capture the whole show not only did great harm to Anglo-American relations but also, it could be argued, resulted in his embarking on the costly and abortive Arnhem adventure in preference to the more mundane, though vital, task of opening up the port of Antwerp.

A significant feature of these wrong decisions was that they involved operations which were inconsistent with the personality of the man who made them.

Of Arnhem, Thompson writes:

To have turned the enemy flank in the north, seizing the bridge-heads on the way, would have demanded daring of high order in conception, in leadership in the field, and in execution. The conception of such a plan was impossible for a man of Montgomery's innate caution ... In fact, Montgomery's decision to mount the operation aimed at the Zuider Zee was as startling as it would have been for an elderly and saintly Bishop suddenly to decide to take up safe-breaking and begin on the Bank of England.[10]

Contemplation of this paradox suggests that Montgomery's 'lapses', his less successful plans, were not just ego-enhancing bids for glory but also reactions against an abiding fear of failure. It is also likely that these errors of judgment, by an otherwise highly competent and professional soldier, were partly a function of another factor underlying military incompetence: interpersonal friction and competition. If, as his biographers suggest, Montgomery irked the Americans by his slowness and caution, coupled with his general mien of 'What a good boy am I', then it is equally certain that the American generals, and Patton in particular, must have grated on *him*, with their manifestly greater dash and 'attack everywhere all the time' philosophy. What more natural than that he should be goaded into going one better—the *first* to cross the Rhine, the *first* to enter Berlin, the *first* to get the plaudits of 'housemaster' (C.I.G.S.) Alanbrooke and 'headmaster' Churchill. It is surely a truism to aver that a proportion of senior military men act out upon the battlefield their 'Billy Liar' schoolboy fantasies and in so doing find themselves in sharp competition with each other. This is not to belittle their performance but rather to emphasize that one reason for their occasionally anomalous behaviour, one potent source of 'noise' in their decision-processes, is the irritating ghost of some long-forgotten Smith minor upon some far-off playing-field.

3. Montgomery's next shortcoming presents something of a paradox. It concerns the matter of communication. For a man who was adept at simplifying the apparently complex, whose ability to extract the essentials from a host of irrelevant factors was second to none, who could communicate his intentions and issue orders to his subordinates with a lucidity that left no room for misinterpretation, and who could write his memoirs with a style that puts most generals to shame, it is extraordinary that he should have been almost incapable of explaining himself to those *above* him. It was not simply a question of them not knowing what he was going to do, but often of not even understanding what he was doing while he was doing it. It was almost as if he took a delight in being misunderstood.

Putting the pieces together, one is driven to the conclusion that Montgomery's generalship was marred by the effects of his unhappy childhood.

Like Lawrence, Churchill and Curzon, Montgomery showed the effects of a poor relationship with his mother. In his case, the combination of too little affection and too much discipline seems to have eventuated in an insatiable need to prove himself, a desire to create the happy family life (symbolized by his staff of young liaison officers) which he had lacked as a child, and a tendency, as far as his superiors were concerned, to keep things to himself. On this latter point, Lewin draws an interesting parallel between Montgomery and T. E. Lawrence, another Irishman who 'endured a mother'. Of *his* parent

Lawrence wrote: 'I have a terror of her knowing anything about my feelings or convictions, or way of life. If she knew they would be damaged, violated, no longer mine.'

In Montgomery's case it was evidently not enough that he remained inscrutable. One gets the impression that he had to act out some much earlier experience by actually contriving situations in which he *would* be misunderstood. He had, it seems, to provoke the inevitable counter-attack.

If he refers back to the chapter on military organizations, the reader will recall that these may be conceptualized as complex devices for legitimizing, handling and controlling aggression. As such they provide admirable vehicles for the acting out of those aggressive fantasies that centre round the parent–child relationship. Much of Montgomery's behaviour becomes explicable in the light of this theory.

There is reason to believe that the lives of many adults represent attempts to find ideal parent-substitutes—people, things or activities which provide what was lacking in the original relationship with father or mother, and which represent an outlet for feelings that could not be expressed at the time of their original occurrence. Thus it was that Hitler, according to Langer's analysis, projected on to Germany the love which he felt towards his mother, and on to Austria the hatred which he bore towards his father.

Montgomery found two such substitutes—first, his wife and marriage, and then, when these were lost to him, the Army. For him the Army became at once the good mother and the good family, something into which he could pour his energy and love and through which he could restore his damaged self-esteem. But it also became a means for expressing his more negative emotions, a place in which he had to find the bad parent whose ire must be aroused and from whom secrets must be kept. Let us examine this theory rather more closely, in the light of certain specific incidents. What follows is in fact an attempt to account for certain recurring features of Montgomery's behaviour in terms of what is known of the psychology of identification.

There may well be other explanations, but it is this one, so it is contended, which most clearly fits *all* the facts. While one hesitates to question a view put forward by his brother Brian, I would suggest that any resemblance between the character of Bernard Montgomery and that of his mother is *primarily* due to the fact that, from a very early age, he identified strongly with this formidable woman, and that anomalies in his behaviour were a consequence of this identification.

Support for this contention comes from the finding that children tend to identify with the more dominant of their parents.[11] In Montgomery's case this would appear to have been his mother.

Now identification entails several components which include adopting the parent's views, and modelling oneself on the parent's behaviour. In Montgomery's case it probably also involved what is technically known as 'identification with the aggressor'. This is a defensive identification, an apparent siding with the enemy—a sort of 'if you can't beat 'em join 'em' response, in which the child attempts to cope with the all-powerful and frustrating parent by adopting as his own the very weapons that were used against him. In Montgomery's case the aggressor with whom he identified was again, almost certainly, his mother. But such identifications are stressful in themselves and this for one very simple reason. The child is made dependent upon someone towards whom he feels basically hostile but from whom he cannot break free because the frustrating person is now, in a sense, inside him— an incorporated part of his own mind. Several results might be expected to follow from this state of affairs. Firstly, the individual may alternate between dependence upon, and rejection of, the person with whom he has identified. Secondly, he may well strive to prove himself as an individual in his own right; hence he will be averse to sharing his achievements with others. Finally, as a reflection of his underlying conflict between dependency and rebelliousness, he will, throughout his life, show periodic regressions of behaviour in which he acts out once again the original childish revolt.

So much for a theoretical analysis, based upon general principles. How does Montgomery's behaviour fit this pattern? The following facts reported by his brother illustrate the main points of contact.

Firstly we have two incidents, minor in themselves, which suggest the continued identification with and dependence upon his mother. The occasion is a summer holiday, when Montgomery took it upon himself to organize the house party 'in a proper military manner'. This was the programme for the day which he posted up for his various friends and relations:

PROGRAMME FOR TODAY

—Date—

Time	Event	Remarks
1000	The men will play squash. Girls will go shopping in the village. Orders by Mother	
1100	Bathing Betty will paint Colin will write sermon	Optional
1200	Girls will pick flowers for the house	
1300	Lunch	Don't be late
1430	All to golf course at Greencastle, less Winsome and Wangy who will prepare tea and convey it to golf course. RV in club house 1700	
1800	Return to New Park[12]	

Though ostensibly a joke, its particular form suggests the internalized voice of the dominant and organizing mother speaking through the mind of her son. He has, as it were, become her Chief of Staff.

The second incident involves a letter which, at the age of thirty-eight, he wrote to his mother. In it he confided that the first girl he had ever proposed to had turned him down. As his brother remarks: ' ... the fact that he had confided in our mother about this whole matter, and had written to her at its ending, points to the growing influence she then had on his life. The days of conflict with her were over, at any rate for the time being.'[13]

The reference to the early conflict, followed by the dependence, followed in turn by the suggestion of yet more conflict to come, hints at the existence of a much deeper and abiding conflict between the need to lean upon and the need to revolt against the internalized parent.

Behind this need to revolt lay the necessity to be a person in his own right, something more than the male embodiment of his mother. This it was perhaps that drove him into the bizarre position of simply not being able to share with another even the smallest fruits of his greatest achievements.

Strangely and very regrettably for all those who know him well, he was seen to reject (perhaps repudiate is not too strong a word) not only his own kith and kin but also some of his best friends. Astonishingly, the latest victim now turned out to be none other than his trusted chief of staff de Guingand. It may be extraordinary to record, but de Guingand had not been allowed to witness the surrender ceremony at Tac. H.Q. 21 Army Group on 4 May 1945; his chief had told him not to attend.[14]

This same high-ranking officer, who had been Montgomery's right-hand man throughout most of the war, suffered three further rebuffs from his erstwhile chief. He was not detailed to take part in the victory parade before the King. He was not given an official seat along the route but had to buy a back seat for himself and his wife. Finally, he was not given the job of V.C.I.G.S. when Montgomery became C.I.G.S., despite the fact that six months earlier the latter had asked him if he would accept the job.

Predictably, Montgomery was no less ruthless with his mother. When this lady requested and received, from Newport council, an invitation to his installation as an Honorary Freeman of the city, the Field-Marshal intervened and insisted that she should be banned from the luncheon room. His brother writes: 'It needs no imagination to appreciate the distress and sorrow that ensued, though in due time our mother fully recovered and was none the worse for it.'[15]

Reading between the lines, one has the impression that Montgomery's mother was herself as tough as old boots and, one might add, just the sort of parent who would be most difficult to dislodge from the big takeover which years previously had occurred in the mind of her son. No doubt she herself increasingly served to exacerbate this conflict in the mind of her boy by her growing desire to bask in his reflected glory.

We have looked at evidence for identification, dependency and self-assertion, but what about those incidents in Montgomery's life which suggest that on occasions he could regress to a much earlier pattern of behaviour — that of the schoolboy who is trying to prove himself?

One of the most revealing occurred while he was a cadet at the R.M.A. Sandhurst. This was the notorious occasion when, as leader of a gang, he cornered a fellow cadet and set fire to his shirt tails. His victim was so badly burned and suffered such agonizing pain that he had to be admitted to hospital. When Montgomery was discovered to have been the instigator and leader of this assault he was threatened with expulsion from the Academy.

Needless to say, when she received news of this happening his mother sprang into action. A phone call to the commandant achieved her first objective—she was invited to Sandhurst for the night. By morning, so it seems, this incredible woman had steam-rollered the head of the R.M.A. into retaining her son. His only penalty for what criminal law now terms 'grievous bodily harm' was six months' loss of promotion. While it could hardly be disputed that Britain has much to thank Mrs Montgomery for, this incident does illustrate several points of pertinence to the theory advanced here. First, the act itself, carried out at the relatively mature age of twenty, is, however entertaining for all but the victim, certainly juvenile in its conception. What is perhaps more remarkable is Montgomery's almost total lack of guilt, shame and compassion. His only regret seems to have been that it cost him six months' seniority. No less significant is his stated reason for the assault: 'He was a dreadful chap.' While the more enlightened reader might well regard this as adding insult to injury, we would only opine that it reveals something of a schoolboy mentality.

In its totality, the episode involves an interesting and perhaps not entirely coincidental repetition of an earlier relationship between the benign father, the all-enveloping and powerful mother, and the troubled child who has to provoke the very parent upon whom he so abjectly depends. Taken by itself, the incident may seem unimportant, but considered alongside other anomalies in the behaviour of Montgomery it does imply, even if it does not prove, the relevance of identification-behaviour.

Of this event Brian Montgomery writes: 'But this time he learnt his lesson and for good.' In one sense this is true—he certainly worked hard from that time on. But in another sense it lacks validity. Throughout his subsequent career, Montgomery continued to show evidence of the unremitting schoolboy in his make-up.

A typical instance was the occasion in 1911 when, as battalion sports officer, he was required to organize a friendly football match between men from his regiment and the sailors from a visiting German battle-ship. For diplomatic reasons he was given strict instructions to field a team of mediocre players. In the event, however, he confronted the Germans with the best footballers he could find. The inevitable result was a crushing and embarrassing defeat for the Germans by forty goals to nil. Montgomery's subsequent excuse was: 'I was not taking any risks with the Germans.'

This can be viewed in two lights. On the one hand it may be regarded as an amusing instance of Nelsonian disobedience coupled with ardent patriotism. But on the other it is the behaviour of a somewhat 'shortsighted' schoolboy who at one stroke manages to rebel against authority, appear a hero in the eyes of his followers, avert his own underlying fear of failure and 'give those rotters a sound thrashing'. He was then twenty-five years old.

In 1944 he had his portrait painted by Augustus John. To some people his subsequent remark to Colonel Daunay—'Who is this chap? He drinks, he's dirty, and I know there are women in the background'—may seem the very embodiment of the military mind; anti-intellectual, anti-sex and anxious about dirt. But in Montgomery's case, judged in the light of his other characteristics, this remark sounds more like the calculated rudeness of a schoolboy whose headmaster has given him a poor report. The poor report was of course the portrait, which Montgomery evidently found unflattering. His vanity regarding photographs of himself lends some countenance to this view.

Then there is the episode which occurred during his term as C.I.G.S. after the last war. Returning with his Military Assistant from New York on the *Mauretania*, Montgomery sought to gain an interview with a famous fellow passenger, Andrzei Gromyko. He was rebuffed, but refusing to take 'no' for an answer hid behind a ventilator and ambushed the Russian diplomat during the latter's morning constitutional.

His brother puts the two sides of this behaviour very well.

Some of my readers may regard the action I have described as a planned antic, or at least a boy-scout episode, but undignified and therefore inappropriate for the professional head of the British Army and his staff officer, particularly when the other party involved was a high-ranking Soviet diplomat. Be that as it may, the sequel will surely show the affair in a different light. For the plan worked in all respects! ... I cannot but see this incident as further evidence, not only of that imp of mischief in my brother, but also of his resolution never to accept defeat.[16]

The last illustration given here is also taken from Brian Montgomery's book, and includes two letters from his brother.

29–5–59   *Isington Mill*

My dear Brian,

You may like to know that Hoyer-Millar is coming here tomorrow, Saturday, at 4.30 to have tea with me. He is coming alone!!

Yrs. ever

BERNARD

Some ten years later, in 1968, he sent me a copy of Hansard with a letter in which he wrote:

Isington Mill

Dear Brian,

You may like to read in Hansard my speech in the Defence Debate in the House of Lords on Wednesday last. The house was full; members from the Commons crowded their place at the Bar of the House, and the steps of the Throne were filled with Privy Councillors!

Yrs. ever

BERNARD

Critics of the Field-Marshal will say that letters such as these are evidence of overwhelming conceit and arrogance; but in reality they serve to show his supreme confidence and judgment in his own ability and opinions.[17]

Here one must beg to differ with the interpretation put on these letters by Brian Montgomery. At a superficial level the letters are certainly those of the braggart schoolboy brimming with self-confidence, but at a deeper level of analysis and in the context of his other characteristics these really rather pathetic and naïve pieces of conceit betray an underlying need to prove himself to others and thereby to himself. People who are really self-confident do not need to boast to their younger brothers, particularly when they are intelligent enough to realize that such boasting detracts from, rather than enhances, their image in the public mind.

Reading over this section on Field-Marshal Montgomery, I foresee trouble with his devotees as well as with his detractors. The former will no doubt be appalled that 'our greatest general since Wellington' should be subjected to a character-analysis which may seem as speculative as it is insulting. His detractors, on the other hand, may view his placement in a list of great generals as totally unwarranted, and the excusing of his less acceptable behaviour, on the grounds of factors over which he had no control, as unjustified exculpation for what was in effect transparent megalomania.

Be that as it may, the fairest conclusion is surely that Montgomery was, in the main, a great and gifted general. Much of his greatness resulted from a confluence of three factors: sheer hard work, a refusal to conform to the dead hand of military tradition, and the possession of a mind as open, clear and sensitive as that of any sharp-eyed, sharp-tongued schoolboy of above-average intelligence.

These three factors resulted from his identification with, and subsequent unsuccessful attempts to free himself from, an invasive and oppressive woman, his mother. To her we owe, in far more than any biological sense, not only the victory at Alamein but also all those quirks of the Field-Marshal which have made him one of the most controversial, criticized and entertaining military commanders in British history.

KITCHENER

Kitchener, who became Secretary of State in the First World War, had, according to Philip Magnus, 'two basic attributes — an unparalleled thoroughness and an unparalleled drive' He was 'an individualist of great conceptions, whose hard and selfish nature was capable, at times, of kindness, sympathy and even affection'.[18] These traits — his excessive drive (Lord Curzon once described Kitchener as 'this molten mass of devouring energy'), his individualism and refusal to conform, the originality of his thinking, and the occasional flashes of underlying warmth and generosity, are hard to reconcile with the notion of authoritarianism.*

By the same token, Kitchener did *not* appear to manifest a need for approval. As Asquith put it: 'He did not pose for posterity; he never

---

* According to K. Macksey, the view that Kitchener's apparent dismissal of tanks as 'pretty mechanical toys' showed a reactionary streak in his nature, is mistaken. The remark was apparently motivated by security reasons, to distract attention from the tank development work then going on. The same writer presents evidence that Kitchener, like Haig, was active in pressing for the construction of tanks.[19]

laid himself out either for contemporary or posthumous applause.' On the contrary, like Montgomery he could be excessively rude, tactless, irritating, and apparently impervious to what others thought of him.

He did not welcome external constraints, and showed scant regard for rules and regulations when these conflicted with his purpose. Nor did he happily accept that sacrosanct feature of authoritarian ideology—the established 'pecking order'—but would 'swing his boot into any system of administration and ... [rend] in pieces any established chain of command. His system was, in reality, the negation of any system and his drive prompted him inexorably to centralize every species of authority in himself. After he had done so, he performed miracles of improvisation and extracted from subordinates whom he trusted and occasionally loved much more than they or any one else believed they had to give.'[20]

Finally, Kitchener betrayed no trace of the ethnocentric feelings which are associated with authoritarianism. On the contrary he showed a marked liking for the society of Egyptians, Turks and Jews.

However, there are reasons for believing that those aspects of his military career which invited unfavourable comment, and eventually brought about his downfall, sprang from an unresolved conflict between innate drive and puritanical conscience. On the one hand we find him 'acquiescing with apparent indifference in the continued infliction of outmoded punishments for trivial offences, such as the lopping off of hands and feet'[21] and on the other showing signs of 'unmistakable discomfort' while watching 'a somewhat improper French Light Opera' in Cairo. 'Interest in the veiled indecencies was as unintelligible to him as the indecencies were themselves intolerable.'[22] By the same token he was obsessive, orderly and mean. 'When his table was laid for guests, Kitchener invariably inspected it himself ... and took immense pains to ensure that no glass or vase, and no knife, fork or spoon was a fraction of an inch out of position.'

Of his meanness Magnus says: ' ... he had a reputation for meanness which was not wholly undeserved ... those who knew him superficially had no conception of the powerful emotional undercurrents which he had schooled himself to repress, and which found an outlet in his passion for art and flowers ... dealers learned to close their shops, and fellow collectors to be suddenly indisposed, whenever it was known that the Commander-in-Chief was engaged upon an artistic prowl.'[23]

The traits revealed here fit in with much that was good and bad in his military performance. Thus his 'unparalleled drive', constrained into channels of administrative thoroughness and military zeal, achieved miracles of organization in Egypt and the Sudan. Even his meanness found valuable expression in the stringent economies which he imposed on military spending. As Magnus remarks: 'Kitchener developed his twin authoritarian attributes ("drive" and "thoroughness") to a unique pitch of obsessional intensity.'[24]

But, of necessity, a price was paid. For all his greatness, Kitchener seems to have been a victim of the repressive forces implanted in him as a child, presumably by his martinet of a father. His aloofness, his unpopularity with many of his fellow officers, his failure to work as part of a team and, most damaging of all, his latter-day indecisiveness and hesitancy in directing the Gallipoli campaign must be ascribed to defects of personality rather than intellect.

> During those critical October days his indecision disgusted his colleagues. His insistence upon the obvious, combined with an elaborate display of stolidity which he used to cover his ineffectiveness in verbal discussion, finally killed the admiration and even the affection which many of his colleagues had previously felt for him.[25]

One final point: it is perhaps significant that, though so alike in their professionalism, their drive, their loneliness, secrecy and their unpopularity, Montgomery and Kitchener differed profoundly in two important ways. Kitchener's childhood had been happy; Montgomery's had not. Kitchener was so self-confident that he lacked vanity, Montgomery so vain that he appeared to have enormous self-confidence. Both in their way were highly competent warriors. Both, when they lapsed, did so as a result of shortcomings in personality rather than intellect.

HAIG

In trying to answer the question whether the recurring features of military incompetence derive from aspects of the authoritarian personality, *even* in a commander who ultimately emerged victorious, one cannot do better than consider the case of Douglas Haig, Commander-in-Chief of the British Armies on the Western Front between 1915 and 1918.

Judging from the war of words which has raged between his

detractors and his devotees, there never was a more controversial military commander!

'Haig, Britain's number one war criminal, expected the Germans to advance in this attack at the same slow pace of his own clumsily planned assaults.'[26]

' ... he seemed to be the most highly equipped thinker in the British Army.'[27]

'Haig perhaps failed to see that a dead man cannot advance, and that to replace him is only to provide another corpse.'[28]

'It is indeed strange that the man whose stubbornness in the offensive had all but ruined us on the Somme should from August 1918 onwards have become the driving force of the Allied armies. Yet this was so and it must stand to his credit, for no man can deny that during the last hundred days of the war he fitted events as a hand fitted a glove.'[29]

'Haig was unimaginative. Maybe he was competent according to his lights, but these were dim; confidence of divine approval appeared to satisfy him. Nothing can excuse the casualties of the Somme and Passchendaele.'[30]

By way of introduction it is proposed to examine his role in just one battle: that of Third Ypres, or, as it sometimes is called, Passchendaele. It is contended that, however great a commander Haig may have been in other ways and at other times, the relationship between the events of Passchendaele and the personality of the man exemplify most of the major points made in this book.

Of the battle Liddell Hart wrote: 'It achieved little except loss—in which again, it repeated the earlier history of this theatre of war. So fruitless in its results, so depressing in its direction was this 1917 offensive that "Passchendaele" has come to be like Walcheren a century before, a synonym for military failure.'[31]

This was written with hindsight. Hence we have to ask if these results were foreseeable and if Haig was aware of the inadvisability of this offensive. The following facts suggest that he was aware but could close his mind to information which did not fit in with his preconceptions or wishes:

1. Lloyd George and Clemenceau tried to restrain him but without success. Even Foch thought the plan absurd.

2. Meteorological advisers warned that weather conditions would be unusually bad, with abnormally high levels of rainfall. A memo to G.H.Q. from Tank H.Q. warned that if an artillery bombardment

destroyed the Belgian land drainage system the battlefield would revert to a swamp. Haig ignored these items of information.

3. Haig's intelligence service knew that the Germans expected the offensive. Haig was evidently undismayed.

4. Haig met opposition to his plans with two rationalizations. He opined that if fighting continued at its present intensity the enemy would run out of men. This forecast exceeded that of even his own optimistic intelligence service. Secondly, he was supported by the Navy in maintaining that *their* continued existence depended upon clearing the Germans from the Belgian coast. As the chief of Haig's Intelligence Staff remarked: 'No one believed this rather amazing view.'

All in all, Haig's optimistic belief that he could defeat the Germans single-handed in Flanders had given rise to a plan that was founded on faith rather than reason.

In total disregard of the evidence regarding land drainage, Haig started the offensive with a bombardment of $4\frac{1}{2}$ million shells (at a cost of £22,000,000). This spectacular release of energy provided four and three-quarter tons of high explosive for every yard of front. It continued for ten days. Predictably, the drains collapsed, the rains came and the ground subsided into a sea of liquid mud.

Into this sodden lunar landscape Haig launched twelve divisions. They advanced in torrential rain. On the left flank they made progress but on the right men and tanks simply disappeared into the mud.

So hopeless was the situation and so appalling the losses that even the impetuous General Gough, one of Haig's army commanders, advised calling off the offensive. He was disregarded. Attacks went on well into October. From being what one war correspondent described as 'a great bloody experiment—a huge gamble' they became 'total and expensive failures'. Their rifles and other weaponry made useless by the mud, the attackers struggled forward only to fall and be replaced by others. And still Haig, buoyed up with lofty optimism, could not let go, but pushed ever more men into the battle, into what had by now become one enormous swamp of rotting flesh.

The following postscripts to the battle probably say all that remains to be said in the present context about 3rd Ypres. The first is by a private soldier, the second by a military historian.

While I and others were taking supplies into the line at Ypres, we waded through mud all the way. It was very necessary to keep following the leader strictly in line, for one false step to the right or left sometimes meant plunging into dangerous and deep mud-pools. One of our men was unfortunate enough to step out of line and fell into one of these mud-holes. Knowing from past experience that quick action was needed if we were to save him from quickly sinking, we got hold of his arms and tried to pull him out. This did not produce much result and we had to be careful not to slip in with him. We finally procured a rope and managed to loop it securely under his armpits. He was now gradually sinking until the mud and water reached almost to his shoulders. We tugged at the rope with the strength of desperation in an effort to save him, but it was useless. He was fast in the mud and beyond human aid. Reluctantly, the party had to leave him to his fate, and that fate was—gradually sinking inch by inch and finally dying of suffocation. The poor fellow now knew he was beyond all aid and begged me to shoot him rather than leave him to die a miserable death by suffocation. I did not want to do this, but thinking of the agonies he would endure if I left him to this horrible death, I decided a quick death would be a merciful ending. I am not afraid to say therefore that I shot this man at his own most urgent request, thus releasing him from a far more agonizing end.[32]

Perhaps the most damning comment on the plan which plunged the British Army in this bath of mud and blood is contained in an incidental revelation of the remorse of one who was largely responsible for it. This highly placed officer from G.H.Q. was on his *first* visit to the battle front—at the end of the four months' battle. Growing increasingly uneasy as the car approached the swamplike edges of the battle area, he eventually burst into tears, crying 'Good God, did we really send men to fight in that?' To which his companion replied that the ground was far worse ahead. If the exclamation was a credit to his heart it revealed on what a foundation of delusion and inexcusable ignorance his indomitable 'offensiveness' had been based.[33]

So much for the battle. The question now arises: Did Haig evince those character-traits that are associated with authoritarianism?

He certainly had most of them. For a start he was conservative, conventional and, in his attitude towards the French, ethnocentric. His diary and dispatches suggest that he was unemotional and totally anti-intraceptive (i.e., not one to reflect upon his own motives). He was manifestly lacking in compassion towards his fellow men. He was a confirmed believer in the direction of events by supernatural powers (according to research a common correlate of authoritarianism), and reserved to the point of being verbally almost inarticulate.

Haig also betrayed that triad of traits which, according to contemporary research, defines the obsessive character and is correlated with authoritarianism. He was obstinate, orderly and mean.

About his obstinacy little further need be said. From beginning to end, his handling of Third Ypres betokened an obstinacy of statuesque proportions.

As for the second trait: in his dress, habits and appearance Haig was immaculate, orderly and quite probably the cleanest man on the Western Front. A contemporary of his at Clifton remembered him particularly for his cleanliness, a remarkable attribute to be recalled of a fellow schoolboy.

And for an example of his love of 'bull' there is this excerpt from a cavalryman's letter: 'He had a personal escort consisting of a full troop of his own regiment. They were easily the smartest thing in France. Not a buckle out of place, stripes of gold for the N.C.O.s, great silver "Skulls and Crossbones" ... '[34] Other writers have commented on his meticulous attention to minute detail, and his habit of planning each day according to a set pattern.

Like many obsessives, in or out of banks, Haig loved the task of handling other people's money but hated parting with his own. Though parliament voted him a grant of £100,000, he never let compassion overrule his personal thriftiness. As Sergeant Secrett records of indigent ex-officers who came to see Haig after the war: 'Never a penny did they get from Lord Haig! That was one thing he would not do; he would never part with money to them. I have seen hundreds of cases where a man had told a most distressing tale, and wound up with a request for a small "grant in aid" only to be disappointed.'[35]

While many rationalizations can be advanced for the withholding of private charity, the following episode, in which Haig's personal servant for twenty-five years seeks a modest rise in wages, is not so easily explained:

All the long years that I was with Earl Haig, though I held the rank of Sergeant it was only acting unpaid rank. I received thirty shillings a month for years until the war increases came. Then, when I came out of the Army with 28 years service and a pension of 19/6 a week the Earl gave me £65 a year. It was impossible to marry on that. I had hoped he would have seen my point, but he was essentially conservative in his arrangements and I believe he thought in his heart I would never break away ... When I told him he scarcely realized it. He hesitated, threw out his heels, took them back, threw them out again, embarrassed, awkward. So was I ... I knew he was on the point of offering something that would have made the break unnecessary, but he could not bring himself to do it.[36]

Further comment seems superfluous.

From research into the nature and aetiology of authoritarianism it was concluded that the condition derives from the impact on the child of status-anxious parents. Subsequent research has confirmed a relationship between sexual repression, militarism, religiosity, aggression and having a restrictive mother. Haig fits this pattern pretty well. His early development presents the picture of an unusually sullen and aggressive child being pushed resolutely forwards and upwards by a strict and puritanical mother. The sad thing is that the strength of her motivation towards his success was more than matched by the extremity of his intellectual backwardness. Under the circumstances it is hardly surprising that, though remaining devoted to his mother, he became somewhat ambivalent towards women, and uneasy, to say the least, about the whole subject of sex. 'The only persons, if one may use such a definition, whom he ever introduced to Oxford, were brothers, and he never (as far as I am aware) entertained any woman except his sister, though I have seen his face set in a silent but obstinate protest, against any loose jokes about women.'[37]

The description of his proposal, and subsequent honeymoon, certainly accords with the theory that Haig was no Don Juan. 'We looked for a quiet seat but not finding one he blurted out "I must propose to you standing". This was very abrupt and I must say unexpected, but I accepted him ... Douglas had only two weeks' leave before we were due to sail for India and during this time we amused ourselves by riding about the countryside.'[38]

Sir George Arthur alludes to Haig's typically authoritarian attitude towards sex.

It is not to peer too intrusively into the arena of a man's life to allude to its austere purity, to suggest that in this respect there are men of high courage who shrink back with something like horror from certain forms of evil, to whom it would be a shame even to speak of those things done in secret ... he looked, of course, for no moral Utopia but no name, nor effort, was subscribed more heartily than his to the famous Memorandum in which officers were urged to encourage in their men a belief in leading a good and healthy life, and in every way—not least by themselves setting an example of self-restraint—to protect them from a grave and devastating evil.[39]

It is interesting to compare this with Montgomery's enlightened and equally famous memorandum on how to prevent V.D. in the Army (see page 277).

As for aggression, it is no exaggeration to say that Haig presents a classic picture of the vicissitudes which attend the aggressive drive in authoritarians. Naturally aggressive and self-willed, Haig, like Montgomery, encountered heavy opposition from his mother. In Haig's case this conflict with mamma appears to have been dealt with by what analysts call introjection and repression. He incorporated his mother as an idealized authority-figure, and her harshly puritanical precepts became his own. With this inner strength he evidently achieved fairly massive repression of his sexuality and that rerouting of aggression which is legitimized, and indeed encouraged, by the teaching of the Scottish Church. These sanctioned outlets include hard work, a belief in the inevitability of punishment for wrongdoing and a preoccupation with the concept of discipline. 'Soldiering was his first and main consideration and perhaps not the least attractive part of it was the discipline of mind and body involved—discipline was so ingrained that neglect of it in any walk of life was intolerable to him.'[40]

As Haig himself wrote: 'Discipline has never had such a vindication in any war as in the present one, and it is their discipline which most distinguishes our new Armies from all similarly created armies of the past. At the outset the lack of deep-seated and instinctive discipline placed our new troops at a disadvantage compared with the methodically trained army.'[41]

Even when teaching his children to ride, Haig remained the awesome disciplinarian laying emphasis upon the fact that wrongdoing (i.e., teasing their ponies) would be punished sooner or later. The guilty would be thrown off.

Predictably, Haig's writings, diaries and dispatches are generally lacking in any emotional fervour. Indeed they have been described by one historian as 'less exciting than the average laundry list'. Though praising the loyalty and discipline of his armies, he rarely mentions the fearful casualties which these armies suffered, and no sign of genuine warmth or compassion leavens the otherwise flatly factual prose. The only time that he shows a ripple of excitement is when discussing enemy losses. Such phrases as 'our forward guns did great execution among his retiring columns', 'magnificent fight' and 'repulsed with great slaughter' carry nuances of a curious relish which are markedly absent from the rest of his writing.

In terms of style, Haig's writing has much in common with the diary of Heinrich Himmler. Both make dull reading of what, in Haig's case anyway, were world-shaking events. Both show that preoccupation with time which characterizes many anal-obsessives: 'Lord Derby called to see me at 9.45 a.m.' 'We took the children to the Pantomime ... The play began at 1.30 p.m. and lasted till 5.30 p.m.' 'I attended a meeting of the war cabinet at 11.30 a.m.' 'At 3 p.m. I attended a conference of the supreme war council at Versailles ... the conference did not break up till 6.30 p.m. ... we got back to Paris by 7.30 p.m.'[42] (For a comparable excerpt from Himmler's diary, see page 312.)

Both diarists seemed to delight in criticism of others. 'He [Lloyd George] is a real bad 'un. The other members of the war cabinet seem afraid of him. Milner is a tired dyspeptic old man. Curzon, a gas-bag. Bonar Law equals Bonar Law. Smuts has good instinct but lacks knowledge.' 'They [Wilson and Rawlinson] are both humbugs.' 'I found Foch most selfish and obstinate.' 'I thought Pershing was very obstinate and stupid.' As to the content of these criticisms, it is a feature of authoritarians that they impute to others their own less pleasant traits. Haig and Himmler shared two other characteristics. To begin with, neither could bear to witness the suffering of others. 'He

[Haig] felt it was his duty to refrain from visiting the casualty clearing stations because these visits made him physically ill.'[43]

The second, and related, common denominator of these two men was their tendency to find a psychosomatic outlet for their surplus aggression. In Himmler's case, as we saw earlier, the favourite outlet was disorders of the gastro-intestinal tract. In Haig's case it was asthma, a psychosomatic complaint which according to some authorities results from repressed aggression. Suffice it to say that Haig suffered one of his worst attacks of this malady before the ill-fated battle of Loos, which, largely thanks to the incompetence of Sir John French, cost the British 50,000 casualties.

The evidence so far suggests that, whatever his virtues as a general, Haig possessed well-marked obsessive and authoritarian traits and that these explain much of the behaviour fastened on by his detractors. It remains to add that Haig appears to have possessed four other characteristics which, as often as not, go along with authoritarianism. He had 'a closed mind', need for social approval, some pathology of achievement-motivation and, to quote A. J. P. Taylor, 'a total lack of imagination'.

It will be recalled that according to Rokeach people differ along a continuum of open-mindedness. At one end are those who are open to fresh ideas, and at the other those who find it hard to accept and act upon information which does not accord with their systems of belief. Haig, like many authoritarians, seems to have belonged to the latter category. Certainly his behaviour before and during Third Ypres appeared to reflect the workings of a mind that was impervious to contrary information. Nor did he make up for his 'closed' mind by having a fertile imagination; for about Haig's lack of this faculty there has been almost complete unanimity of agreement. Lloyd George had no illusions on this score when he wrote: 'I never met anyone in a high position who seemed to me so utterly devoid of imagination.' Wavell described Haig as having 'a one-track mind'; and even Haig's chaplain, Duncan, whose adulation of his chief bordered on the sycophantic, was forced to admit that there may have been grounds for supposing Haig to be unimaginative. In a charitable attempt to make a virtue out of a necessity he goes on to stress that Haig 'recoiled from strategical conceptions that were not in accord with military principles' and that 'his dogged inflexibility prepared the way for final victory'.[44]

Finally, we come to the complex issue of achievement-motivation. The evidence suggests that Haig's burning ambition to succeed overlay a pronounced fear of failure, itself a product of childhood. Since he was devoted to his mother it is reasonable to suppose that her bitter disappointment over his intellectual backwardness caused him considerable anguish. He felt perhaps that he had failed her, and, particularly after her death, worked towards proving himself. In the light of research (described earlier) by Birney and his colleagues he showed several traits associated with pathological achievement-motivation. He was an ardent seeker of social approval, particularly from the King. He disliked publicity and nursed an unreasoning dislike of newspaper reporters. He preferred the company of his inferiors to that of his equals. Though professing a great belief in the value of loyalty, he rarely had a good word to say of his military contemporaries and was quite ruthless in his machinations against his erstwhile chief, Sir John French.

Finally, he showed a predisposition towards persisting in tasks that were so difficult that failure seemed excusable. As Liddell-Hart wrote of Third Ypres, he chose a spot 'most difficult for himself and least vital to the enemy'.

It seems, then, that Haig possessed more than his fair share of traits associated with authoritarianism. He was conventional, conservative, unimaginative and rigid. He had a 'closed' mind, was pathologically ambitious, anti-intraceptive and punitive. He was superstitious, militaristic, obsessive and devoid of real compassion. Finally, and not very surprisingly in view of his other attributes, he was never popular. He commanded respect and adulation but lacked that warmth which elicits affection from one's fellow men.

Now whether or not the Third Battle of Ypres, which was described by A. J. P. Taylor as 'the blindest slaughter of a blind war' and which cost the British over 300,000 casualties for 'trivial gains', exemplifies incompetence is perhaps debatable. People are still divided on this issue.

What *is* certain is that Haig's conceiving of this battle, his conduct of the fight and his apparent inability to let go are consistent with the personality of the man and not attributable, as some would have it, to stupidity.

Besides providing a particularly stark illustration of the relationship between military behaviour and human psychopathology, the case of Haig points up three other issues of some pertinence. First it supports the psycho-analytic notion of 'acting out'. Whether by accident or by design, the events of Third Ypres—the enormous release of destructive energy, the churning up of ground until the overlapping craters coalesced into one great reeking swamp, and the expulsion into this morass of more and yet more 'faecal' bodies—constitute the acting-out of an anal fantasy of impressive proportions. The apparent denial of what would happen as a result of such mundane causal determinants as bad weather and shattered land drains, the obstinate straining until the last soldier had been expelled into the cesspool of Passchendaele, and the gastro-intestinal pain experienced by the progenitor of these excesses (when asked on other occasions to contemplate the results of his aggression) force the conclusion that acting-out was not so much coincidental as deliberately, if unconsciously, motivated.

The second issue, and one of great relevance to the general theory of military incompetence, concerns the relationship between personality and promotion. Whatever his faults and failings, Haig did exceptionally well in his race to the top. His particular set of traits, obstinacy, orderliness, personal ambition and underlying aggression, abetted by Machiavellianism and conformity, were ideally suited to the requirements of a military organization. Far from hindering his ascent, his early intellectual backwardness served to provide that initial wound to his self-esteem which *had* to be redressed.

There remains one issue of somewhat wider psychological interest: the astonishing cleavage of opinion between those who champion and those who condemn Sir Douglas Haig.

Thus in contrast to what Liddell Hart thought about Third Ypres—likening it for depths of military ineptitude to the notorious Walcheren expedition—we find Marshall-Cornwall, in 1973, defending Haig against his critics. This general, who served under Haig as a junior staff officer in 1917, defends Haig's conception and handling of Third Ypres on five main counts:

1. The offensive was forced upon Haig through the need to liberate the Belgian ports, thus striking a blow at the German submarine fleet.

2. He was also moved to take the load off the French, whose armies were debilitated by mutinies following Verdun.

3. He was misled regarding German strength by his over-optimistic chief of intelligence, Brigadier Charteris.

4. Third Ypres was justified by the fact that the 400,000 casualties

sustained by Germany outnumbered the 240,000 suffered by Britain.

5. Haig's prolongation of the battle after October 3rd, 1917, which cost the British a further 100,000 casualties, saved his troops from wintering in flooded marshland under the domination of enemy observation posts on higher ground.

Marshall-Cornwall makes no mention of the evidence which Haig had received regarding rainfall, and the likely effect of a bombardment on the Belgian land drains.[45]

In the light of these arguments it is interesting to consider A. J. P. Taylor on the same events.

Haig manufactured excuses why the Ypres offensive had to be made. He made out that Pétain pleaded for a British offensive in order to divert the Germans from his mutinous army. This was not true. Pétain wanted small actions to keep the Germans busy, not a great offensive which might reduce the British Army to the same state of demoralization as his own. Again Haig recruited Jellicoe to insist that Ostend and Zeebrugge must be taken if the German submarines were to be checked. This, too, was not true. Most German submarines operated from home ports, not from Ostend and Zeebrugge. Haig himself knew that the argument was unsound. He regarded Jellicoe as 'an old woman'; but every argument for the offensive was welcome to him. Haig also made out that this was the last chance for the British to win the war before the Americans arrived. This, too, was an afterthought, and an odd one, when the British claimed that they and the Americans were fighting for the same cause. The truth was simple: Haig had resolved blindly that this was the place where he could win the war. He disregarded the warnings of his own Intelligence staff against the mud. No one else shared his confidence ... criticism only made Haig more obstinate.[46]

On the subject of casualties Taylor has this to say:

The British casualties were something over 300,000; the Germans under 200,000 ... Thirty years later, the British official History turned these figures round: British losses, 250,000; German 400,000. No one believes these farcical calculations ...

Then Haig stopped. The campaign had served its purpose. What purpose? None. The British line stuck out in a sharper and more awkward salient than before the battle began. All the trivial gains were abandoned without a fight in order to shorten the line, when the Germans attacked in the following year.[47]

It is difficult to believe that the loyal junior staff officer and the eminent historian are writing of the same events. Since they cannot both be right several interesting questions are raised. On the one hand we have to ask if loyalty can distort judgment, and on the other if repugnance and compassion at the immensity of the losses which Haig sustained can warp historical objectivity. Thus, when Marshall-Cornwall writes: 'Haig was certainly not the man to cause his troops unnecessary casualties' is this, to many people surprising, opinion the product of misplaced generosity towards an old chief?

Let us look at and try to answer another historian who finds no harm in Haig. In his book *The Western Front 1914–1918*, John Terraine argues that had Haig been appointed Commander-in-Chief of the B.E.F. in December 1917, in time to meet the German Spring Offensive of 1918, and then gone on to achieve the victories which culminated in the Armistice, it is unlikely that there would be any argument about him.[48] His place in history as a great commander would have been assured.

In his final dispatch of March 21st, 1919, Haig wrote: ' ... neither the course of the war itself nor the military lessons to be drawn therefrom can properly be comprehended unless the long succession of battles commenced on the Somme in 1916 and ended in November of last year on the Somme are viewed as forming part of one great and continuous engagement.' Terraine observes: 'It is entirely characteristic that, in asserting his final victory, he should identify himself completely with all that had preceded it: and it is precisely what preceded it that has done most harm to his memory.'[49] He implies that it is to Haig's credit that he did not dissociate himself from the huge losses (420,000 men in four months on the Somme and 300,000 men at Third Ypres), the great suffering and the small apparent gains of preceding years.

But there are perhaps other ways of looking at these data. First, Haig did not become Commander-in-Chief in 1917 but in 1915, and this as a result of some fairly ruthless and, some have said, disloyal machinations against his superior, Sir John French. What might be thought of Haig had he not become Commander-in-Chief until December 1917 is hypothetical and irrelevant.

As for associating himself with the dire events of 1915–17, since they were his doing he really had no option. Moreover, the act was fully consistent with what is known of his character. Only by advancing the thesis that his successes in 1918 were a result of preceding events could he hope to place the latter in a more favourable light. It might be thought that Haig's last dispatch only serves to confirm the theory that here was a man who, though possessing military virtues, was also adroit in covering himself against criticism and improving his image.

There is, moreover, another argument which needs to be considered. Suppose that Haig *had* inherited the situation of December 1917, *without* having been responsible for preceding events, would the successes of 1918 demonstrate his brilliance as a great commander? The answer cannot be an unequivocal yes. For if, as Terraine and Haig himself imply, the successes of 1918 were a natural consequence of the situation existing in December 1917, then there was nothing remarkable about his realizing this natural consequence in the last year of the war. With a Germany half starved, and her army led by a man who was intellectually crippled by physical illness,* he could hardly help but win.

According to Terraine, the consensus of opinion that finds Haig 'insensitive, unreceptive, obstinate and unimaginative' is largely mistaken. He bases his conclusion on four examples of 'grand-scale imagination' shown by Haig between 1916 and the end of the Great War.

The first of these occurred when Haldane, bent upon reform of the Army, sought Haig's advice. Haig opined that what was needed was a citizen army, because the next war when it came would be a protracted struggle involving the resources of the whole country. Not only does Terraine regard this as showing 'a rare vision' but he also suggests that Haig must have been a great thinker, otherwise Haldane would hardly have bothered to consult him. As examples of imagination, on any scale let alone a grand one, these points are not totally convincing. After all, since the Boer War was only four years over, even the least imaginative of men must have realized that the next war

---

* According to L'Etang[50] Ludendorff had been suffering from toxic goitre since 1914. The mental disturbance which results from this condition would have been quite enough to explain that indecision and constant change of plans which doomed to failure the German offensives of 1918.

would be protracted and call for all our resources. As for the notion of a citizen army, this had already been implied by Lord Roberts's call for National Service. Finally, that Haldane should have felt compelled to call on Haig perhaps does no more than reflect upon the intellectual level of the Army at that time. The choice was, after all, a relative one, as Terraine himself makes clear.

> many fine theories [regarding the incompetence of Haig] might go astray; if it should also prove to be the case that no general in any country at that time was able to avoid similar slaughter under certain conditions, while the best achievements of any of them are fully matched by Haig's, then one might find oneself drawn to the more sober conclusion of Sir Winston Churchill that 'He might be, surely he was, unequal to the prodigious scale of events; but no one else was discerned as his equal or his better.'[51]

This 'best of a bad bunch' argument certainly does not militate against the argument that Haig's particular shortcomings were attributable to those features of personality which are endemic to the military scene.

Haig's second important insight, according to Terraine, was when he became convinced 'that the French would not be able to go on shouldering the main burden of fighting Germany in the West for long, and that the British would have to take most of that burden off them'.[52] Having said this, Terraine then allows that 'Lord Kitchener glimpsed it [the same truth] but recoiled from the consequences' and Sir William Robertson 'approached it but never with Haig's clarity'. In other words, it was not such a unique insight, nor did it show a very great gift of imagination. Indeed it could be argued that it was an ugly fact which Haig saw clearly because it fitted in with his own driving ambition towards self-advancement at whatever cost.

Presumably to quell the uncharitable thought that Haig might have welcomed a great burden on the British forces because this would ultimately redound to his credit, Terraine cites evidence which suggested a 'selfless broad-minded recognition' of the fact that we should still have to submit to the general control of operations by the French. But again this picture of self-abnegation in the Allied cause has been disputed. In point of fact Haig, like his Chief of Staff Kiggell, harboured strong prejudices against the French and strongly resisted any question of domination by them until his own reputation was in such jeopardy that he was only too glad to shift responsibility on to someone else's shoulders. With the German breakthrough on March 21st, 1918, Haig 'belatedly appreciated the virtues of the supreme command which he had previously opposed'.[53]

The third occasion upon which Haig is thought to have revealed evidence of his superior imagination was in August 1918 when he told Churchill: ' ... we ought to do our utmost to get a decision this autumn.'

However, since by this time Ludendorff's last great German offensive had already spent itself, thereby wrecking German hopes of final victory, Haig's pronouncement might well be considered more of a foregone conclusion than a brilliant prophecy. His use of the words 'ought to' is perhaps revealing. If he, rather than the French or the Americans, was to be hailed as the final architect of victory then the sooner the war was ended the better. He was not one for sharing prizes with his fellow men.

Of all Terraine's examples of Haig's imaginative brilliance the last is the hardest to swallow—his founding of the British Legion. Of this Terraine says:

> His human quality, the working of the same insight through the channels of compassion and feeling, are revealed in his developing awareness of the nation's duty towards its citizen army when that army was disbanded, and above all to those who had been maimed in the service, and to the descendants of those who had died. By the middle of 1916 he was becoming aware of the scale of this problem; all through the harassing preoccupations of 1917 it was in his mind. In February of that year he addressed the Army Council on this subject, in a long and carefully-thought-out memorandum ... After the armistice ... he gave the remaining years of his life to the formation and guidance of the British Legion ... an everlasting mockery of the notion that he was 'insensitive' to the sufferings and virtues of the men whom he commanded.

In these respects, according to Terraine, 'Haig was out ahead.'[54]

Since the period of Haig's generalship between 1916 and 1917 cost Britain over 700,000 casualties it is hardly surprising that he became 'aware of the scale of this problem', nor that it remained in his mind to the end of the war. Nor is it so surprising that, since the Battle of the

Somme was an 'unredeemed defeat' which should never have been fought, and that of Third Ypres 'the blindest slaughter of a blind war for which Haig bore the greatest responsibility',[55] he should have become aware of a slight dulling of his image in the minds of his compatriots. And, since the peace of mind of many public men depends in large measure upon their public image, it is hardly surprising that he should have set about rehabilitating himself. What better way than by founding the British Legion? It could also be argued that this act was entirely consistent with Haig's record as an astute military politician, bent on personal advantage, and with the other non-mutually-exclusive hypothesis that even he might have felt some guilt at the enormity of the price that had been paid, in other people's lives, for his success. It was perhaps not the first or the last time that people have put flowers, including poppies, on the graves of their victims. Such behaviour is, incidentally, entirely consistent with what is known of the psychology of authoritarianism—a reaction to the release of excessive but rationalized aggression.

However, Terraine also cites several other occasions on which Haig's imagination had an immediate impact on operations. They are:

1. The short bombardment at Neuve Chapelle. In fact this was an operation staged by Sir John French in which shortage of shells *necessitated* a short bombardment. As a result of this fortuitous occurrence the Germans were taken by surprise and British infantry broke through, for the only time in the war. However, as Taylor records: 'The British hesitated to enter the hole they had made. They waited for reinforcements; and by the time these arrived, German reinforcements had arrived also. The gap was closed. The British went on battering to no purpose.'[56]

The relevance of this event to Haig was that he failed to learn the lesson of surprise. In the years which lay ahead hundreds of thousands were doomed to lose their lives because their Commander-in-Chief chose to ignore the serendipitous occurrence of a short bombardment.

2. The dawn attack on the Somme on July 14th, 1916. Despite the lesson of Neuve Chapelle, Haig's tactics continued to follow the rigid and expensive formula of prolonged bombardment, followed by a pause as his guns increased their range, followed by an infantry assault. During the hiatus between the lifting of the barrage and the infantry attack German machine-gunners emerged from their dug-outs and established themselves in the craters which British artillery had so

considerately supplied. Then came the British infantry, each man weighed down with 66 pounds of equipment, moving slowly in extended line—a target impossible to miss. 'The bullets ran across the front in a steady spray. The first British line faltered and fell, a second followed it, a third, and then a fourth, all to no avail. By the early afternoon the survivors were back in their trenches.'[57]

Applied to the first day of the Battle of the Somme, Haig's formula cost 57,000 casualties, of whom 20,000 died—'the heaviest loss ever suffered in a single day by a British army or by any army in the First World War'.

It is against this background that we must consider the events of July 14th, 1916, for it was because of these enormous losses that Rawlinson (not Haig), after another fortnight of comparable carnage, at last decided on an attack under cover of darkness. Haig acquiesced to the plan only after a night of argument with his subordinate commander. In the event the assault was launched at 3.55 a.m. and, because it achieved surprise, succeeded in overrunning two lines of German trenches. Unfortunately the gain could not be exploited because mud hampered the arrival of reinforcements.

Certainly a dawn attack was a refreshing change from what had gone before, but the idea was not Haig's and, even if it had been, would hardly constitute the product of a brilliant imagination.

3. The first British tank attack on September 15th, 1916. If Haig had invented tanks, or even the idea of using armoured fighting vehicles, some credit would be due to him. But he did neither of these things. On the contrary, in the face of considerable opposition he insisted on using the few tanks then available. The tanks were untried and some broke down, leaving one or two to penetrate the German lines. Not only did Haig achieve little by this abortive attack but he lost for ever the surprise of a really heavy tank attack. Certainly, as Sixsmith points out, 'it is doubtful if the evolution of suitable tactics for infantry and artillery working with tanks would have been possible without battle experience'.[58] But as an example of imagination it is not terribly convincing.*

Haig's real impact on the Battle of the Somme lay in his insistence that 'every yard of lost trench must be retaken by counter-attack'. About this the less said the better, for it cost the British nearly half a

---

* On the other hand, it is to Haig's lasting credit that he consistently encouraged the development of tanks.

million casualties and availed them nothing beyond the fact that it 'set the picture by which future generations saw the First World War: brave helpless soldiers; blundering obstinate generals; nothing achieved'.[59]

4. The Battle of Messines. On June 6th, 1917, German defensive positions on the Messines ridge were destroyed by the explosion of one million pounds of high explosive that had been placed in nineteen mines, deep in the hillside. Haig was not responsible for this success. On the contrary, credit for the enterprise, which took two years to prepare, belongs to Sir Herbert Plumer, described by Taylor as 'one of the few sensible British Commanders of the War'.

5. The first mass tank attack at Cambrai, 'the stunning surprise of August 8th, 1918'. Following upon the useless and enormously expensive Battles of the Somme and Third Ypres, Haig allowed tanks to be used at Cambrai. If after four years of mindless destruction this successful though temporary gain can be described as the 'impact' of an imaginative general then the word has lost its meaning.

Terraine's estimate of Haig appeared in 1964. For a more recent and some may think truer summing up there is John Walters's *Aldershot Review*:

From the start of the First World War, Haig, with defeatist zeal, always insisted that it must last for a long time. He believed in the inevitability of 'long fluctuating battles', of a process of 'wearing down' the enemy by the relentless and merciless sacrifice of the lives of his men. 'Germany's resistance', he preached, 'must be worn down by a continuous battery on her frontier.' ... He directed this 'battery' from luxurious château or villa headquarters in cosy isolation from the scene of battle ... from this grand headquarters Haig made big strategic decisions, involving the lives of hundreds of thousands of soldiers, without consultation with the ordinary fighting officers and men who were at the scene of battle ... [thus] more than 90,000 men and youths, the finest of British manhood, perished in the battle, herded to slaughter like cattle in the stockyards of Chicago.[60]

It will be recalled that this somewhat lengthy digression in which we have considered the views of no less than five historians took off from contemplation of a curious phenomenon — the fact that even half a century after the event opinions about Haig still remain extra-ordinarily intense and strikingly divided.

What conclusions, then, can be drawn about the origins of all this heat and cerebration? The first, a rather obvious point, is that figures like Haig command lasting interest simply because they symbolize, and in their role act out, the divided forces within each one of us. It is just because they are neither wholly good nor wholly bad, but perpetrate colossal destruction within the bounds of noble acquiescent conscience, that they fascinate and cannot be left alone. Their increasingly frequent re-emergence in plays, films and books years after the event is also a special instance of the general finding that the longer the time since a major war the greater our interest in large-scale military aggression.

The second, perhaps less obvious, point is that the division of opinion regarding a man like Haig reflects precisely the two sides of his nature — callous butcher versus gentle introverted knight endowed with massive self-control.

The third point, or rather hypothesis, follows naturally. It is simply this: that people who are primarily concerned with preserving the constraining forces in society have a soft spot for authoritarians. Those who aren't do not. This hardly-to-be-wondered-at phenomenon is a microcosmic exemplar of the general thesis that people are drawn towards organizations or individuals who have developed defences against the very anxieties to which they themselves are prey. To demolish a Haig is to demolish the very structure upon which their own peace of mind depends. How else can one explain the disproportionately vitriolic attacks upon those who dare to 'explain' some long-dead individual for whom the attacker cannot surely still entertain a passionate affection?

By the same token it will be interesting, if painful, to witness the rage brought forth by comparing the personalities of Haig and Himmler. On this latter issue, and by way of sparing the feelings of devotees of Haig (not to mention admirers of Himmler), let us be quite clear. We are not saying that Haig was a Himmler, or vice versa. We are not suggesting any necessary identity between two outcomes of authoritarian psychopathology. Clearly in the case of these two men the outcomes were very different. But what we *are* saying is that in terms of aetiology, underlying psychopathology and certain manifest traits there were remarkable similarities. We are also saying that those

ways in which they were so competent—their orderliness, thoroughness, tenacity and singleness of purpose—and those ways in which
they were incompetent stem from their underlying authoritarianism,
not from intellectual brilliance or stupidity. If anything Himmler was
rather more competent in his job than was Haig in his. This is merely
because, vocationally, extreme authoritarianism is more compatible
with the task of the bureaucratized extermination of an 'inferior' race
than it is with waging war against a relatively free, relatively flexible
and equally matched enemy.

Having read so far, the reader may, with some justification, ask why
Haig is included at all in our short list of great but controversial
commanders.

There are two reasons. Firstly, he has been acclaimed as a 'great
captain' by some who are presumably qualified to judge his total
performance. Secondly, it could be argued that given the conditions
of the First World War his authoritarian traits were not without value.
His driving ambition, his obstinacy, his fixity of purpose, his orderly
mind and even his imperviousness to the tragedy of human sacrifice
on a gargantuan scale added up to at least one way of winning a senseless conflict. In his psychopathology he was the very embodiment of
the sort of war which the armies on both sides were compelled to
fight: a war of tightly controlled aggression in which destructive
forces, of hitherto undreamt-of violence, were hurled against the
constraints of mud, wire, protocol and ordered lines of trenches.

He had to be the man he was, to be the perfect figurehead for such
a fight.

From the foregoing account of three controversial commanders three
conclusions may be drawn:

1. By far the greatest of the three, Montgomery, was also the least
authoritarian, and by far the most controversial, Haig, was certainly
the most authoritarian. Kitchener lies somewhere between them on
both counts.

2. Even in the case of Montgomery those aspects of his military
performance for which he has been criticized are clearly attributable
to features of his personality rather than to any intellectual shortcoming
or lack of professional expertise.

3. Whereas Montgomery and Kitchener reached positions of
supreme power despite a relative absence of authoritarian traits and

because of this abundant drive and military excellence, Haig scaled the same heights largely through his authoritarianism. In contrast to Montgomery and Kitchener, his conformity, obedience, anti-intellectualism and thinly veiled but well controlled aggression fitted the needs and ethos of the organization to which he belonged — 'as a hand fitted a glove'.[61]

29

Retreat

Hail, ye indomitable heroes, hail!
Despite of all your generals ye prevail.
LANDOR, 'The Crimean Heroes'

'This difficulty of seeing things correctly, which is one of the greatest sources of friction in war, makes things appear quite different from what was expected.' C. VON CLAUSEWITZ

In the introduction to this book it was suggested that for someone to take offence at an analysis of military incompetence would be as unjustified as a devotee of teeth complaining about a book on dental caries. Now that we are drawing to a close the point needs re-emphasis, for a work devoted to the pathology of a system tends to leave the impression that dysfunction is the rule rather than the exception.

It is not the intention to leave a comparable impression of general-ship, but rather to show that, just as the nature of their job predisposes teeth to decay, so the nature of intra-species aggression predisposes the leaders of armies and navies to certain sorts of error.

Far from diminishing the stature of senior military commanders, the existence of this predisposition makes the performance of the majority of soldiers and sailors doubly creditable. They are, to continue with the dental metaphor, like teeth which survive *despite* the hazards of plaque, gingivitis and calcium deficiency—stalwart fangs indeed!

We have dealt too with the problems of militarism but here again, as the following lines suggest, there is another side to the coin:

What I like about the British Army is that during my lifetime it has almost always fought for good reasons, and almost always

393

in a disciplined fashion. It has done no tarring and feathering that I know of, beaten up no pregnant women that I know of, blown the faces off no shop girls that I know of. And it has not chosen to kill any fathers in front of their wives and children ... I write this with a fairly extensive knowledge of the charges of brutality that have been levelled at the troops in Ulster.[1]

The theory advanced in this book starts from the position that by its very nature military incompetence cannot be attributed to dullness of intellect. There is, it seems, a recurring pattern to military mishaps which defies any explanation in terms of the 'bloody fool' theory.

In its stead it is tentatively suggested that the syndrome occurs through the enormous difficulties of professionalizing the instinctual activity of intra-species aggression. This professionalization entails the growth of militarism, that collection of rules and conventions whereby hostility is controlled and anxiety reduced. That militarism hampers military behaviour is a special instance of the general principle that attempts to minimize the *side*-effects of an event tend to hinder its *main* effect—rather as a silencer reduces the efficiency of an internal combustion engine, or preservatives impair the flavour of jam.

The nature and effects of militarism have more than coincidental similarity to those psychological defences which individuals erect against personal anxieties. These have to do with sex, dirt, aggression, self-esteem and death.

Not very surprisingly, a military career attracts a minority of people with these sorts of anxieties. Within a military organization their neurotic needs are gratified. They, for their part, help to reinforce those very aspects of militarism which are so congenial to their requirements. In return, as it were, for fitting in so well, they may rise to positions of considerable power. Once there, however, they become incapacitated by the very characteristics which hastened their ascent. These traits—need for approval, fear of failure, being deaf to unpalatable information, and the rest—are probably accentuated by their larger responsibilities, and the fact that there is now no longer anyone higher up to whom they might appeal. They are also older, for generalship, like arthritis, is something which besets a man as he approaches old age.

These speculations gained support from a number of phenomena associated with military organizations: 'bull', codes of honour, anti-intellectualism, anti-effeminacy and sensitivity to criticism. They also fitted in with research data on leadership, achievement-motivation and the authoritarian personality.

Finally, a study of highly competent versus incompetent commanders seemed to confirm the belief that ego-weakness and authoritarianism, rather than stupidity, underlie most military ineptitude.

In the light of this conclusion it is perhaps a triusm to aver that while the seeds of military incompetence are sown in very early childhood, they may owe much to such subsequent baneful influence as separation-anxiety, restrictive mothers, status-conscious fathers, monastic schools and the Victorian cult of 'muscular Christianity'.

So much for a theory based on past history. Has it any relevance for the future? Since armies and navies have changed out of all recognition, perhaps the sorts of military incompetence described in these pages are no longer likely to occur. In fact, the evidence suggests this to be a forlorn hope. Some of the same sorts of mistakes occur now as blighted the lives of soldiers one hundred years ago. In Vietnam, in three weeks in 1968, the Tet offensive alone cost the Americans 500 dead and the South Vietnamese 165,000 dead with two million refugees. Why did it happen? One reason was the inability to respond to unexpected military intelligence.

But there are other reasons for a continuance of military incompetence, which, paradoxically, came about with the transition from old-time to modern generalship. Far from diminishing the possibility of error, this transition opens up vistas of potential ineptitude hitherto denied to professionals in violence. It also has results which tend to obscure the precise sources of incompetence.

In the words of a contemporary American: 'One of the chief differences between ourselves and the ancients lies not (unfortunately) in human nature, but rather in the proliferation of our skills, and our institutions, and therefore in the number of niches in which the incompetent can now instal themselves as persons of consequence.'[2]

To be more specific there are three factors which predispose towards errors in modern generalship. Firstly, thanks to Marconi and the thermionic valve, larger armies can now be controlled from much farther away by minds which might still have been selected and trained for an earlier form of warfare and are now prevented by sheer distance from obtaining any real feel for the battle. Secondly, the whole complexity of modern war has meant larger staffs with a consequent multiplication of the sources of distortion in the flow of essential information. Finally, there is the sad irony that the best intentions of modern generals, particularly at the level of commanders-in-chief, may be hazarded by the sheer wealth of technological resources now placed at their disposal.

The first concatenation of these three factors occurred in the First World War, with terrible results. In this war, generals, shielded by distance and their staffs, their hands unstayed by witnessing the outcome of their orders, could send thousands, even millions, of men to their deaths without any feeling of being wasteful, and safe from retribution. In short, they lacked incentives to act otherwise than they did.

And, like an old person with a weak heart, they might be deliberately kept in ignorance of unpalatable facts. For example, one of the reasons for the costly aftermath of the Cambrai offensive in 1917 was that Haig had been grossly misled by his staff as to the German strength. It seems that before the battle his intelligence chief, Brigadier Charteris, was approached by a more junior staff officer with documentary evidence that a German division from Russia and other reinforcements were arriving in the Cambrai area. On Charteris's orders this inconvenient fact was kept from Haig because 'he did not accept the evidence, *and in any case did not wish to weaken the C.-in-C.'s resolution to carry on with the attack.*'[3] (My italics.)

As to why Haig should have retained a man who could perpetrate such deceptions, one can only point out that to preserve loyalties and affectionate ties within the in-group (in this case his staff) at the expense of disasters for an out-group accords with what is known of the authoritarian personality. It requires greater moral courage to fire a congenial subordinate whom one knows personally than to accept the death of an army whom one does not. Haig evidently lacked this particular brand of moral courage.

But all this, product though it was of those factors which distinguish modern from old-fashioned generalship, happened many years ago. Perhaps they were only the teething pains of modern generalship? It seems not.

Any doubts as to whether the three factors of remote control, swollen staffs and a wealth of resources make for incompetence are removed by contemplation of Vietnam. In this most ill-conceived and horrible of wars there was the Commander-in-Chief, Lyndon Johnson, aided by his advisers, dreaming up policies and even selecting targets at a nice safe distance of 12,000 miles. And there was the man on the spot, General Westmoreland, a by no means unintelligent military commander but bemused by the sheer weight of destructive energy and aggressive notions supplied by his President. Together, the Machiavellian mind of the one, coupled with the traditional military mind of the other, produced a pattern of martial lunacy so abject and appalling that it eventually did for both of them.

Like the Boer leaders half a century earlier, the versatile General Giap and *his* commander-in-chief, a little old man with a wispy beard, made the huge professionally trained and over-equipped army of their enemies look utterly ridiculous, and their leaders helplessly irate. Unfettered by traditional militarism, lacking an excess of brute force, and without an obsession with capturing real estate, Ho and Giap relied on poor men's strategies—surprise, deception and the ability to melt away. They relied on the fact that Westmoreland would expend his energies swatting wherever they had last been heard of while they got ready to sting him somewhere else. And as Lyndon and 'Westy' got madder, so that vast tracts of South-East Asia reeled beneath their rage, the North Vietnamese and the Viet Cong flitted round them and through them, puncturing the myth of American supremacy.

This brings up yet another hazard of modern war—government by committee. From the long history of earlier disasters it might well have been concluded that all future military decisions should be left to committees rather than individual commanders, if only to dilute the effects of undesirable personality-traits. Unfortunately such a conclusion would probably be fatal. According to a study by I. L. Janis, four of the worst military disasters in recent American history are directly attributable to the psychological processes which attend group decision-making.

In an analysis of the events which led up to the Bay of Pigs fiasco, the destruction of Pearl Harbor, America's participation in the Korean War, and the escalation of the war in Vietnam, Janis draws attention to the staggering irrationality which can beset the thinking of otherwise highly competent, intelligent, conscientious individuals when they begin acting as a group.

Take the decision to invade Cuba with a group of Cuban exiles. In approving the C.I.A. plan, Kennedy and his key advisers made six assumptions. Each was wrong.

They assumed that no one would guess that the U.S. Government was responsible for the invasion.

In their contempt for the Cuban Air Force they assumed it would be annihilated before the invasion began.

They assumed that the small invasion force led by unpopular ex-officers from the Batista regime would be more than a match for Castro's 'weak' army of 20,000 well-equipped Cuban troops.

They assumed that the invasion would touch off a general revolt behind Castro's lines.

They assumed that even if unsuccessful in their primary objective the exile force could hole up in Cuba and reinforce anti-Castro guerrillas.

In the event each assumption proved a gross miscalculation. Nothing went as planned. Nobody believed the C.I.A. cover story. The ships carrying reserve ammunition for the invasion force failed to arrive— two were sunk and two fled. By the second day the invaders were surrounded by Castro's army, and by the third they were either dead or behind bars. Seven months later the United States recovered what was left of their invasion force for a ransom price to Castro of 53 million dollars. Kennedy was stricken. 'How could I have been so stupid as to let them go ahead?' he asked. As Sorensen wrote: 'His anguish was doubly deepened by the knowledge that the rest of the world was asking the same question.' Arthur Schlesinger Jr noted that 'Kennedy would sometimes refer incredulously to the Bay of Pigs, wondering how a rational and responsible Government could have become involved in so ill-starred an adventure'. Others who had participated in the initial decisions were similarly afflicted. Dulles offered to resign as head of the C.I.A.; McNamara publicly acknowledged his personal responsibility for misguiding the President.

All in all, it had been 'an operation so ill-conceived that among literate people all over the world the name of the invasion site has become the very symbol of perfect failure'.[4]

If the central shared illusion in the Bay of Pigs action was 'the plan can't fail', that which sealed the fate of Pearl Harbor was 'it can't happen here!'

Sunday, December 7th, 1941, had been set aside by Admiral Kimmel (Commander-in-Chief of the Pacific Fleet) for a friendly game of golf with his colleague General Short, ninety-six ships of the American Fleet slept at anchor in the harbour, American planes stood wing-tip to wing-tip on the tarmac, American servicemen were off duty enjoying week-end leave. By the end of the day Pearl Harbor, with its ships, planes and military installations, had been reduced to smoking ruins, 2,000 servicemen had been killed and as many more missing or wounded. By the end of the day Kimmel was offering to resign. Later he was court-martialled, reprimanded and demoted to a position where he was never again required to make decisions of any consequence.

Through our old friends the neglect of intelligence reports and gross underestimation of enemy capabilities, coupled in this instance with an assiduous misinterpretation of warning signals from Washington and amiable dedication to the task of mutual reassurance regarding their invulnerability, Kimmel and his circle of naval and military advisers achieved a state of such supine complacency that they brought upon themselves 'the worst disaster in American history'.

Pearl Harbor, like the Bay of Pigs, confirmed once again that military incompetence is more often a product of personality-characteristics than of intellectual shortcomings. For these American disasters show very clearly that even the *combined* intellects and specialized knowledge of highly intelligent and dedicated men are no proof against decisions so totally unrealistic as subsequently to tax the credulity of even those who had made them. Far from diminishing the chances of ineptitude, the group actually accentuates the effects of those very traits which may lead to incompetence in individual commanders. The symptoms of this process, which Janis terms 'group-think', include:

1. An illusion of invulnerability that becomes shared by most members of the group.

2. Collective attempts to ignore or rationalize away items of information which might otherwise lead the group to reconsider shaky but cherished assumptions.

3. An unquestioned belief in the group's inherent morality, thus enabling members to overlook the ethical consequences of their decision.

4. Stereotyping the enemy as either too evil for negotiation or too stupid and feeble to be a threat.

5. A shared illusion of unanimity in a majority viewpoint, augmented by the false assumption that silence means consent.

6. Self-appointed 'mind-guards' to protect the group from adverse information that might shatter complacency about the effectiveness and morality of their decisions.

Not very surprisingly it has been suggested that those most susceptible to 'group-think' will tend to be people fearful of disapproval and rejection. 'Such people give priority to preserving friendly relationships at the expense of achieving success in the group's work-tasks.'[5] Conversely, the sort of person who, as we have seen, makes the best military commander— the outspoken individualist— clearly cannot give of his best in the group situation. If he fails to hold his tongue, he runs the risk of being ejected by his colleagues.

Yet another factor concerning groups, of particular relevance to the arguments advanced in this book, has been emphasized by Richard Barnet in *The Economy of Death* (cited by Janis). It is the common social and educational backgrounds of the top men in Washington. In as much as such homogeneity will tend to increase the chances of group-think so also will it play a part in exacerbating the problems which confront those larger groups: military organizations. For, as we have seen, few human groupings exceed a corps of officers in homogeneity of educational and social background.

Finally, it is worth noting that the personality-determined malaise of 'group-think' produces once again those four most frequently occurring symptoms of past military incompetence: wastage of manpower, over-confidence, underestimation of the enemy, and the ignoring of intelligence reports. These, it seems, are the enduring hazards of professionalizing violence.

Is there, then, *no* crumb of comfort when we survey the future? Perhaps at least the military mind is fading from the scene? It seems not. For all its demonstrable shortcomings, cultivation of the military mind continues with unabated vigour. In one American training establishment alone two new versions of compulsive 'bull' have been discovered: rectilinear movement, and eating by numbers. In this place where future controllers of nuclear weaponry receive their basic training, cadets must always walk *parallel* to the walls of buildings; diagonal or other directions of locomotion are forbidden. The old belief that a straight line is the shortest distance between two points is evidently a property of a subversive geometry! An interesting feature of eating by numbers, wherein officer cadets are constrained to handle their cutlery with measured uniformity— *up* one-two, *across* one-two, *enter mouth* one-two, *withdraw* one-two, *reload* one-two— is that the drill can never be completed if a junior eater is spoken to in midstream by a senior member of the mess. As a result of this delicately controlled sadism, it is quite possible for a potential future general to die of starvation.

What is particularly sinister about these superficially amusing inanities is that they are embraced and encouraged, not by the authorities, but by the senior cadets. Two conclusions might be drawn. First, they support the thesis (see page 183) of a weak dividing-line between militarism and obsessional neurosis. Secondly, they suggest, as we have argued throughout, that a military career still tends to attract people with peculiarities of personality. We are not alone in this supposition.

While this book was being written, researchers in other parts of the world have been studying what has come to be known as the 'military-industrial personality' — one who is drawn to, and has an emotional investment in, the use of force and the machinery of war to solve world problems. It has been described as follows: 'The militarist is a relatively prejudiced and authoritarian person. He is emotionally dependent, socially conformist and religiously orthodox. His interest in the welfare of others is relatively low. He is extremely distrustful of the new and strange.' Such people are also 'uncreative, unimaginative, narrow-minded, security-seeking, prestige-oriented, parochial, ultra-masculine, anti-intellectual, extraversive, and severely socialized as children'. They are lacking in aesthetic appreciation, complexity of thinking, independence, self-expression and altruism, and relatively high in anxiety. Finally, military professionals are lower in self-esteem than any other occupational group.[6]

Evidently, then, the sorts of people about whom this book has been centred are still with us. That is bad enough. What is worse, judging from a theory based on past events, is their capacity for incompetence in the very trade they seek to ply. This is rather like discovering that ophthalmic surgeons are particularly prone to Parkinson's disease — their prime reason for carrying out eye surgery being to prove that their hands don't tremble!

Nearer home, in Britain, studies of what has been called the 'conservatism syndrome' have reached conclusions that are also consistent with our theory of military incompetence.

The syndrome includes such attitudes as religious dogmatism, ethnocentrism, intolerance of minority groups, punitiveness, anti-hedonism, conformity, conventionality, superstition, resistance to scientific progress, and a liking for *militarism*. In other words, the syndrome is scarcely distinguishable from the Berkeley concept of authoritarianism.[7]

It has been argued that this constellation of attitudes functions as an ego-defence against feelings of inferiority and insecurity. It reflects what Glenn Wilson has called 'a generalized susceptibility to experiencing threat or anxiety in the face of uncertainty'. It works by 'simplifying, ordering, controlling, and rendering more secure, both the *external* world (through perceptual processes ...) and the *internal* world (needs, feelings, desires, etc.). Order is imposed upon inner needs and feelings by subjugating them to rigid and simplistic external codes of conduct (rules, laws, morals, duties, obligations, etc.), thus reducing conflict and averting the anxiety that would accompany awareness of the freedom to choose among alternative modes of action.'[8]

Desirable though they may be in *other* contexts, these defences, which, as we have seen, characterize some military minds and attract those who possess them to the comforting controls of military organizations, are obviously at odds with the requirements of competent generalship. For a start they will be activated by the uncertainties of war – the greater the uncertainties the stronger the defences. Secondly, besides exerting a paralytic effect upon decision processes they will predispose towards a delusional underestimation of the enemy (a 'magical' attempt to minimize the external threat), a failure to seek out, use or act upon military intelligence (denial and perceptual blocking of threatening information), and an implacable resistance to the 'uncertainties' of innovation, novelty and new scientific aids to warfare.

It is indeed ironic that one of the most conservative of professions should be called upon to engage in activities that require the very obverse of conservative mental traits. It is rather like expecting the Pope to run an efficient birth-control clinic. And, when it is considered that the military profession is the single biggest occupational group from which officer cadets are drawn, the findings from a very recent study by Eaves and Eysenck, which point to a *hereditary* factor in conservatism, is hardly cause for jubilation![9]

The potentially dangerous situation implied by the above considerations is exacerbated by another factor – that military organizations might now be called upon to wage three quite different sorts of war: nuclear, conventional, and what Frank Kitson has called 'low-intensity operations' (i.e., defensive measures against subversive elements in a civilian population). One hesitates to pontificate on the psychology of such a crucial issue but, judging again from past events, there are grounds for believing that these types of warfare require management by very different personalities.

Thus the controllers of nuclear weaponry should perhaps be relatively obsessive, rigid, conforming and over-controlled – in short, mildly authoritarian. Since the main problem here is the prevention of accidents, the world cannot afford the excitement of entrusting its hydrogen bombs to impulsive, maverick individualists. For this job we still require the naturally inhibited, totally obedient, 'bullshit'-ridden bureaucrat.

At the other end of the scale, however, for peace-keeping operations like that in Northern Ireland, an authoritarian cast of mind would probably be a crippling disability. For such 'warfare', tact, flexibility, imagination and 'open minds', the very antitheses of authoritarian traits, would seem to be necessary if not sufficient. Perhaps the military should take a lesson from the police who, so one is given to understand, take care to avoid recruiting people with marked authoritarian traits.

One thing is certain: the ways of conventional militarism are ill-suited to 'low-intensity operations'.

At the moment many of these people [officers commanding units dealing with insurgents] deliberately try to present the situation to their subordinates in terms of conventional war. They make rousing speeches about knocking the enemy for a six, and they indulge in frequent redeployments and other activities designed to create the illusion of battle. But quite apart from the tactical disadvantages which accrue, e.g., lack of continuity, they actually manage to aggravate the strains on their subordinates because they are in effect encouraging the development of the characteristics which are unsuited to this particular type of operation, whilst retarding the growth of those which might be useful. In other words they are leading their men away from the real battlefield on to a fictitious one of their own imagining.[10]

It is perhaps *most* difficult to find a suitable prescription for military commanders in conventional warfare. Certainly they should possess two 'virtues' defined by E. H. Erikson. The first is 'purpose – the courage to envisage and pursue valued goals uninhibited by the defeat of infantile fantasies, by guilt and by the foiling fear of punishment'.

The second is wisdom — 'a detached concern with life itself, in the face of death itself'.[11]

Unfortunately the possession of these traits might well deter a man from ever wanting to be a senior military commander.

Afterword

Lest the reader should have doubted my qualifications to write this book, let me reassure him that I have marked authoritarian traits, a weak ego, fear of failure motivation, and no illusions about the fact that I would have made a grossly incompetent general. It takes one to know one!

February 1975   THE AUTHOR

Notes

1. INTRODUCTION

1. Alexis de Tocqueville, *Democracy in America*, p. 280.
2. Denis Judd, *Someone Has Blundered*, p. xx.
3. Hugh Thomas, *The Suez Affair*, p. 183.
4. Russell Braddon, *The Siege*, p. 10.
5. C. von Clausewitz, *On War*, p. 102.

Also consulted

P. Abrams, 'Democracy, technology and the retired British Officer'.
——, 'The late profession of arms'.
S. P. Huntington, *The Soldier and The State*.
Morris Janowitz, *The Professional Soldier*.
——, Introductory chapter in J. A. A. Van Doorn, *Armed Forces and Society*.
J. A. A. Van Doorn, *Armed Forces and Society*.

2. GENERALSHIP

1. See Milton Lettenberg, 'America in Vietnam—statistics of a war'.
2. For more detailed accounts of the nature and application of the theory, see C. E. Shannon and W. Weaver, *A Mathematical Theory of Communication* and P. M. Fitts and M. L. Posner, *Human Performance*.
3. W. Edwards, H. Lindman and L. Phillips, 'Emerging technologies for making decisions'.
4. For the difficulties people have in thinking statistically, see A. Tversky and D. Kahneman, 'Judgement under uncertainty ...'.
5. This episode is described in Stanley Rogers, *Twelve on the Beaufort Scale*, pp. 37–56.

Also consulted
  R. G. S. Bidwell, *Modern Warfare*.
  L. Grinspoon, 'Psychosocial constraints ...'.
  E. Josephson, 'Irrational leadership ...'.
  Z. J. Lipowski, 'Sensory and information inputs overload ...'.

3. THE CRIMEAN WAR

  1. Viscount Montgomery, *A History of Warfare*, p. 426.
  2. David Divine, *The Blunted Sword*, p. 49.
  3. *Observer* colour magazine, April 9th, 1972, p. 19.
  4. Divine, op. cit., p. 50.
  5. W. B. Pemberton, *Battles of the Crimean War*, p. 26.
  6. Ibid., p. 28.
  7. Ibid., p. 45.
  8. Ibid., p. 50.
  9. Ibid., p. 88.
  10. Cecil Woodham Smith, *The Reason Why*, pp. 234-5.
  11. Ibid., p. 229.
  12. Pemberton, op. cit., p. 163.
  13. John Laffin, *Tommy Atkins*, p. 137.
  14. Ibid., loc. cit.
  15. Pemberton, op. cit., p. 206.
  16. Ibid., p. 208.
  17. J. B. Priestley, *Victoria's Heyday*, p. 148.
  18. Christopher Hibbert, *The Destruction of Lord Raglan*, p. 25.

Also consulted
  Correlli Barnett, *Britain and Her Army 1509-1970*.
  Brian Bond, 'The Late-Victorian Army'.
  ——, *The Victorian Army and the Staff College 1854-1914*.
  D. Judd, *Someone Has Blundered*.

4. THE BOER WAR

  1. Rayne Kruger, *Goodbye Dolly Gray*, p. 60.
  2. David Divine, *The Blunted Sword*, p. 56.
  3. Kruger, op. cit., p. 61.
  4. W. B. Pemberton, *Battles of the Boer War*, p. 32.
  5. Kruger, op. cit., p. 98.
  6. Ibid., loc. cit.
  7. Ibid., loc. cit.
  8. *The Times History of the Boer War*, cited by Kruger, op. cit., p. 130.
  9. Pemberton, op. cit., p. 148.

Also consulted

Brian Bond, 'The Late-Victorian Army'.
L. Butler, *Sir Redvers Buller*.
W. Jerrold, *Sir Redvers H. Buller*.
D. Judd, *Someone Has Blundered*.
C. H. Melville, *The Life of General Sir Redvers Buller*.
J. Symons, *Buller's Campaign*.
J. Walters, *Aldershot Review*.

5. INDIAN INTERLUDE

1. P. Scott O'Connor, *The Indian Countryside*, pp. 170–77.
2. Patrick Macrory, *Signal Catastrophe*, p. 244.
3. Sir Gerald Templer, foreword to Macrory, op. cit., p. 7.
4. Vincent Eyre, *The Military Operations at Caubul 1843*, cited in Macrory, op. cit., p. 126.
5. Macrory, op. cit., p. 210.
6. Ibid., p. 284.
7. Ibid., p. 204.
8. Ibid., p. 220.
9. Ibid., p. 258.
10. Ibid., p. 268.

Also consulted

D. Judd, *Someone Has Blundered*.

6. THE FIRST WORLD WAR

1. Charles Fair, *From the Jaws of Victory*, p. 329.
2. A. G. Hales, quoted in B. H. Liddell Hart, *The Tanks*, Vol. 1, p. 13.
3. B. H. Liddell Hart, *History of the First World War*, p. 332.
4. For a discussion of this propensity, see R. G. S. Bidwell, *Gunners at War*, p. 49.
5. A. J. P. Taylor, *The First World War*, p. 84.
6. A. J. Smithers, *The Man Who Disobeyed*, p. 238.

Also consulted

E. K. G. Sixsmith, *British Generalship in the Twentieth Century*.
British Official History, *Military Operations, France and Belgium, 1914–1918*.

7. CAMBRAI

1. Bryan Cooper, *The Ironclads of Cambrai*, p. 28.
2. Ibid., p. 35.

3. Ibid., p. 45.
4. Ibid., p. 117.
5. Ibid., p. 219.

Also consulted
    B. H. Liddell Hart, *The Tanks*, Vol. 1.
    B. H. Liddell Hart, *History of the First World War*.
    R. Woollcombe, *The First Tank Battle: Cambrai 1917*.

8. THE SIEGE OF KUT

1. See also Ronald Millar, *Kut: The Death of an Army*, and A. J. Barker, *Townshend of Kut*.
2. Russell Braddon, *The Siege*, p. 61.
3. Ibid., p. 65.
4. Ibid., pp. 79–80.
5. Ibid., p. 167.
6. Ibid., p. 129.
7. Ibid., pp. 129–30.
8. Ibid., p. 135.
9. Ibid., p. 236.
10. Ibid., p. 289.
11. Ibid., p. 283.
12. Ibid., p. 93.

Also consulted
    General Sir Richard Gale, review article on Townshend of Kut.

9. BETWEEN THE WARS

1. 'The military heresy', review in *The Times Literary Supplement*, June 30th 1972, of Michael Howard, *The Continental Commitment*.
2. B. H. Liddell Hart, *Memoirs*, Vol. 2, p. 50.
3. E. K. G. Sixsmith, *British Generalship in the Twentieth Century*, p. 176.
4. Liddell Hart, op. cit., Vol. 1, p. 241.
5. See John Smyth, *Sandhurst*, p. 230.
6. Liddell Hart, op. cit., Vol. 2, p. 193.
7. David Divine, *The Blunted Sword*, p. 19.
8. Liddell Hart, op. cit., Vol. 1, p. 325–6.
9. See also Donald Macintyre, *U-Boat Killer*.
10. Liddell Hart, op. cit., Vol. 1, p. 329.

Also consulted
    B. H. Liddell Hart, *The Tanks*, Vols. 1 and 2.

10. THE SECOND WORLD WAR

1. B. H. Liddell Hart, *History of the Second World War*, p. 21.
2. Correlli Barnett, *The Desert Generals*, pp. 127, 137.
3. Ibid., p. 157.
4. Liddell Hart, op. cit., p. 284.
5. Barnett, op. cit., p. 99.
6. Ibid., p. 131.

Also consulted
  R. G. S. Bidwell, *Gunners at War*.
  Michael Carver, *Tobruk*.
  E. K. G. Sixsmith, *British Generalship in the Twentieth Century*.

11. SINGAPORE

1. Noel Barber, *Sinister Twilight*, p. 34.
2. Ibid., p. 60.
3. Ibid., p. 65.
4. Ibid., p. 66.
5. Ibid., loc. cit.
6. See Donald Macintyre, *The Naval War Against Hitler*.
7. Barber, op. cit., p. 142.
8. S. Woodburn Kirby, *Singapore: the Chain of Disaster*, p. 254.

Also consulted
  H. Gordon Bennett, *Why Singapore Fell*.
  S. Woodburn Kirby, *The War Against Japan*, Vol. 1.
  E. K. G. Sixsmith, *British Generalship in the Twentieth Century*.
  John Smyth, *Percival and the Tragedy of Singapore*.

12. ARNHEM

1. See Cornelius Ryan, *A Bridge Too Far*, facing p. 454.

Also consulted
  Ronald Lewin, *Montgomery as Military Commander*.
  Viscount Montgomery, *Memoirs*.
  R. W. Thompson, *Montgomery the Field-Marshal*.

13. IS THERE A CASE TO ANSWER?

1. Alfred Vagts, *A History of Militarism*, pp. 25, 27.
2. John Laffin, *Tommy Atkins*, p. 73.
3. See Angus Calder, *The People's War*, p. 223.

14. THE INTELLECTUAL ABILITY OF SENIOR MILITARY COMMANDERS

1. Morris Janowitz, *The Professional Soldier*, p. 105.
2. Viscount Montgomery, *A History of Warfare*, p. 358.
3. Janowitz, op. cit., p. 106.
4. John Smyth, *Sandhurst*, p. 107.
5. Ibid., p. 139.
6. R. S. and C. M. Illingworth, *Lessons from Childhood*, p. 192.
7. Corelli Barnett, *Britain and Her Army 1509–1970*, p. 344.
8. Smyth, op. cit., p. 153.
9. Ibid., p. 231.
10. See S. W. C. Pack, *Britannia at Dartmouth*, p. 185.
11. Janowitz, op. cit., p. 230.
12. *The Times*, May 11th, 1972.
13. Correlli Barnett, *The Desert Generals*, p. 309.
14. D. M. Ingvar and J. Risberg, 'Increase of cerebral blood flow ...'.
15. See Hugh L'Etang, *Pathology of Leadership* and D. B. Bromley, *The Psychology of Human Ageing*.
16. Simon Raven, 'Perish by the Sword', p. 39.
17. N. P. Chaubey, 'Effect of age on expectancy of success ...'.
18. L. Festinger, *Conflict, Decision and Dissonance*, p. 155.
19. See D. C. Glass, 'Theories of consistency ...'.
20. See N. Kogan and M. A. Wallach, 'Risks and deterrents ... '.
21. C. E. Osgood, 'Towards international behaviour ... '.
22. Kogan and Wallach, op. cit., p. 91.
23. M. Deutsch, 'Some considerations relevant to national policy', p. 62.

Also consulted
R. P. Abelson and others, *Theories of Cognitive Consistency*.
L. Festinger, *A Theory of Cognitive Dissonance*.
Hugh L'Etang, 'Some actualities of war'.
P. G. Zimbardo, *The Cognitive Control of Motivation*.

15. MILITARY ORGANIZATIONS

1. Wayland Young, 'Sitting on a fortune ...', p. 19.
2. Albert Einstein, *I Believe*, p. 71.
3. Robert Holt, *Assessing Personality*, pp. 196–7.
4. K. R. L. Hall, 'Aggression in monkey and ape societies', p. 62.
5. I. L. Janis, 'Group identification under conditions of external danger', pp. 227–38.
6. John Laffin, *Americans in Battle*, p. 186.

Also consulted

P. Abrams, 'The late profession of arms'.
——, 'Democracy, technology and the retired British Officer'.
S. P. Huntington, *The Soldier and the State*.
M. Janowitz, *The Professional Soldier*.
C. B. Otley, *Public School and Army*.
——, 'The social origin of British army officers'.
N. Sanford and C. Comstock, *Sanctions for Evil*.
A. Vagts, *A History of Militarism*.
——, 'Generals—old or young'.
J. A. A. Van Doorn, *Armed Forces and Society*.

16. 'BULLSHIT'

1. See Donald Macintyre, *Jutland*.
2. See A. B. Campbell, *Customs and Traditions of the Royal Navy*, p. 117.
3. E. S. Turner, *Gallant Gentlemen*, p. 139.
4. Campbell, op. cit., p. 13.
5. Ibid., loc. cit.
6. A. H. Buss, *Psychopathology*, p. 61.
7. See B. Alapin, personal communication to H. R. Beech (ed.), *Obsessional States*.
8. See I. Sarnoff and S. M. Corwin, 'Castration anxiety and the fear of death'.
9. Richard Hough, *The Fleet that had to Die*, p. 38.
10. Hans Selye, *The Stress of Life*.
11. E. Josephson, 'Irrational leadership ...'.

Also consulted

I. Eibl-Eiblesfeldt, *Love and Hate*.
Konrad Lorenz, *On Aggression*.

17. SOCIALIZATION AND THE ANAL CHARACTER

1. See G. Murphy, *Personality*, p. 747.
2. Paul Kline, *Fact and Fantasy in Freudian Theory*.
3. Halla Beloff, 'The structure and origin of the anal character'. See also D. R. Thomas, 'Conservatism, authoritarianism ...'.
4. H. R. Beech, *Obsessional States*.
5. See M. L. Farber, 'The anal character and political aggression'.
6. Personal communication to the author.
7. A. T. Carr, 'A psychophysiological study of ritual behaviours ... '.
8. P. Janet, *Psychological Healing*, p. 886, and A. J. Lewis, 'Problems of obsessional illness', cited in H. R. Beech, *Obsessional States*, p. 335.

9. C. Rycroft, *Anxiety and Neurosis*, p. 77.
10. H. Orlansky, 'Infant care and personality'. D. M. Vowles, *The Psychobiology of Aggression*.

Also consulted

C. G. Costello, *Symptoms of Psychopathology*.
Sigmund Freud, *Character and Anal Erotism*.

18. CHARACTER AND HONOUR

1. D. Bannister, 'Psychology as an exercise in paradox', p. 21.
2. Karl Demeter, *The German Officer-Corps*, p. 113.
3. John Laffin, *Tommy Atkins*, p. 43.
4. See C. Roetter, *Psychological Warfare*, pp. 136–7.
5. E. Aronson and J. Mills, 'Effect of severity of initiation … '.
6. Cecil Woodham Smith, *The Reason Why*, p. 62.
7. Noel Barber, *Sinister Twilight*, p. 132.
8. H. Gordon Bennett, *Why Singapore Fell*, p. 31.
9. G. M. Carstairs, foreword to E. Stengel, *Suicide and Attempted Suicide*, p. 8.
10. William Makepeace Thackeray, *Henry Esmond*, Book III, chapter 1.
11. See E. S. Turner, *Gallant Gentlemen*.
12. Brian Bond, 'The Late-Victorian Army', p. 616.
13. P. Abrams, 'The late profession of arms'.
14. A. Vagts, *A History of Militarism*.
15. Turner, op. cit., p. 82.
16. Personal communications to the author.
17. Duff Cooper, letter to Basil Liddell Hart, in B. H. Liddell Hart, *Memoirs*, Vol. 1, p. 300.

Also consulted

J. F. C. Flugel, *Man, Morals, and Society*.
J. Huizinga, *The Waning of the Middle Ages*.

19. ANTI-EFFEMINACY

1. See John Laffin, *Tommy Atkins*.
2. S. W. C. Pack, *Britannia at Dartmouth*, p. 62.
3. In B. S. Johnson (ed.), *All Bull*, p. 116.
4. Richard Hough, *Admirals in Collision*.
5. Communication to the author from Michael Joseph Ltd. See also *The Times*, May 8th, 1971.
6. Corroborated by Wing-Commander F. Carroll and Miss Coombes of the Imperial War Museum, who worked with Mrs Pratt: personal communications to the author.

7. Cited in Donald Macintyre, *U-Boat Killer*.
8. Morris Janowitz, *The Professional Soldier*, p. 223.
9. Personal communication by Donald Macintyre to the author.
10. John Laffin, *Americans in Battle*, pp. 2, 6.
11. See Viscount Montgomery, *A History of Warfare*.

20. LEADERS OF MEN

1. See R. White and R. Lippitt, 'Leader behaviour and member reaction ... '.
2. See R. Brown, *Social Psychology*.
3. See R. M. Stogdill, 'Personal factors associated with leadership'.
4. F. L. Greer, 'Leader indulgence and group performance'.
5. See A. J. Smithers, *Sir John Monash*, p. 178.
6. See E. F. Borgatta and others, 'Some findings relevant to the "great man" theory of leadership'.
7. Smithers, op. cit., p. 122.
8. Donald Macintyre, *The Naval War Against Hitler*, p. 35.
9. Elizabeth Longford, *Wellington: The Years of the Sword*, p. 199.
10. Simon Raven, 'Perish by the Sword', p. 43.
11. Ibid., pp. 38–40.
12. Ibid., pp. 45–7.
13. Ibid., p. 48.
14. R. M. Baron, 'Social reinforcement effects ... ', p. 529.
15. See *Encounter*, June 1959.

Also consulted
   F. E. Fiedler, 'Personality and situational determinants ...'.
   D. C. Korten, 'Situational determinants of leadership'.
   J. W. Thibaut and H. H. Kelley, *The Social Psychology of Groups*.

21. MILITARY ACHIEVEMENT

1. Sir Richard Gale, review article on Townshend of Kut.
2. Ronald Lewin, *Montgomery as Military Commander*, p. 33.
3. See J. W. Atkinson, 'The achievement motive and recall ... '.
4. See E. G. French, 'Motivation as a variable ... '.
5. See H. A. Burdick, 'The relationship of attraction ... '.
6. D. C. McClelland, *The Achieving Society*.
7. Lewin, op. cit., p. 29.
8. See R. H. Knapp, 'Achievement and aesthetic preference'.
9. See E. Aronson, 'The need for achievement ... '.
10. In R. W. Thompson, *Montgomery the Field-Marshal*, p. 231.
11. J. W. Atkinson and N. T. Feather (eds.), *A Theory of Achievement Motivation*, p. 16.

12. R. C. Birney and others, *Fear of Failure*.
13. Richard Hough, *Admirals in Collision*, p. 242.
14. W. B. Pemberton, *Battles of the Boer War*, p. 124.
15. For contrasting views of General Percival, see: J. Smyth, *Percival and the Tragedy of Singapore*; S. Woodburn Kirby, *Singapore: the Chain of Disaster*; N. Barber, *Sinister Twilight*; and E. K. G. Sixsmith, *British Generalship in the Twentieth Century*.
16. H. Gordon Bennett, *Why Singapore Fell*, p. 30.
17. B. H. Liddell Hart, *Memoirs*, Vol. 1, p. 81.
18. In Alan Clark, *The Donkeys*, p. 24.
19. Ibid., loc. cit.
20. Ibid., p. 24.
21. Ibid., p. 32.
22. Ibid., p. 23.
23. Ibid., p. 176.
24. Ibid., loc. cit.
25. Ibid., p. 186.
26. See A. H. Buss, *Psychopathology*.
27. See H. Miller and D. W. Baruch, 'A study of hostility in allergic children', and P. G. Mellett, 'Motive and mechanism in asthma'.
28. See J. T. Barendregt, 'A cross-validation study ... '.

Also consulted
   D. P. Crowne and D. Marlowe, *The Approval Motive: Studies in Evaluative Dependence*.
   R. S. and C. M. Illingworth, *Lessons from Childhood*.

22. AUTHORITARIANISM

1. See Roger Brown, *Social Psychology*.
2. See T. W. Adorno and others, *The Authoritarian Personality*.
3. See H. J. Eysenck, *The Structure of Human Personality* and *The Psychology of Politics*.
4. Brown, op. cit., p. 495.
5. D. J. Levinson and P. E. Huffman, 'Traditional family ideology ... '.
6. See P. Adler, *A House is not a Home*, and J. West, *Plainsville, U.S.A.*
7. See R. T. Hare, 'Authoritarianism ... '.
8. Brown, op. cit., p. 504.
9. T. F. Pettigrew and others, 'Binocular resolution and perception of race in South Africa'.
10. See R. Christie and M. Jahoda (eds.), *Studies in the Scope and Method of the 'Authoritarian Personality'*, G. D. Wilson, *The Psychology of Conservatism*, J. J. Ray, 'Militarism, authoritarianism ...', W. Eckhardt and A. G.

Newcombe, 'Comment on Ray's "Militarism, authoritarianism ..." ', D. R. Thomas, 'Conservatism, authoritarianism ...', D. R. Thomas, 'Conservatism and premarital sexual experience', and M. Masling, 'How neurotic ...'.

11. See Milton Rokeach, *The Open and Closed Mind*.
12. W. Haythorn and others, 'The behaviour of authoritarian and equalitarian personalities in groups'.
13. E. P. Hollander, 'Authoritarianism and leadership choice ... '.
14. See Paul Kline, *Fact and Fantasy in Freudian Theory*.
15. J. C. Finney, 'The M.M.P.I. as a measure of character structure ... '.
16. See R. R. Izzet, 'Authoritarianism and attitudes towards the Vietnam War ... '; W. Eckhardt and others, 'Militarism in our culture today'; A. G. Miller and B. J. Morrison, 'Disposition towards military service ... '; W. Eckhardt, 'Psychology of war and peace'; and E. French and R. R. Ernest, 'The relationship between authoritarianism and acceptance ... '.
17. H. H. Anderson and G. L. Anderson, 'Social values of teachers ... '.
18. M. Deutsch, 'Trust, trustworthiness ...'.
19. K. Strong, *Men of Intelligence*, pp. 35, 115.
20. S. Kates and L. S. Klein, 'Authoritarian beliefs ... '.
21. N. Kogan, 'Authoritarianism and repression', *J. Abnorm. Soc. Psychol.*, 53 (1956), pp. 34–7.
22. R. G. S. Bidwell, *Modern Warfare*, pp. 74–8.
23. Strong, op. cit.
24. Peter Kelvin, *The Bases of Social Behaviour*, p. 124.
25. B. R. Sappenfield, 'Stereotypical perception ... '.
26. D. Johnson, *France and the Dreyfus Affair*.
27. M. Janowitz, *The Professional Soldier*, p. 149.
28. Donald Macintyre, *Jutland*, p. 151.
29. See J. Walters, *Aldershot Review*, p. 121.
30. Richard Hough, *Admirals in Collision*, p. 30. For a further account of the disaster see also R. T. V. Gould, *Enigmas*.
31. See Stanley Milgram, *Obedience to Authority*.
32. K. G. B. Dewar, *The Navy from Within*, p. 73.
33. Ibid., p. 104.
34. B. H. Liddell Hart, *Memoirs*, Vol. I, p. 252.
35. See Dewar, op. cit.
36. See L. Gardiner, *Royal Oak Courts Martial*.
37. Dewar, op. cit., p. 18.
38. See Richard Deacon, *A History of the British Secret Service*, pp. 316–17.
39. Elizabeth Longford, *Wellington: The Years of the Sword*, p. 491.
40. See E. A. Ritter, *Shaka Zulu*, p. 245.
41. See W. Eckhardt, 'An essay on compassion'.

42. See Brian Montgomery, *A Field-Marshal in the Family*.
43. J. R. Hale, 'Incitement to violence? ...', pp. 375-6.
44. P. L. Adams, *Obsessive Children — a Socio-Psychiatric Study*, p. 141.
45. Cited in W. Eckhardt, 'The military-industrial personality'. For the relationship between authoritarianism and various cognitive disabilities, see also O. J. Harvey, 'Conceptual systems ...'.

Also consulted

   A. C. Elms and S. Milgram, 'Personality characteristics associated with obedience ... '.

   P. J. Leach, 'A critical study ... '.

   P. B. Warr, *Thought and Personality*.

   R. Wilkinson, *The Broken Rebel*.

23. MOTHERS OF INCOMPETENCE

1. H. A. Witkin, 'Psychological differentiation and forms of pathology'.
2. See J. W. Shafer, 'A specific cognitive deficit ... '.
3. See N. Konstadt and E. Forman, 'Field dependence and external directedness'.
4. See C. I. Hovland and I. L. Janis (eds.), *Personality and Persuasibility*; and H. A. Burdick and A. J. Burnes, 'A test of "strain towards symmetry" theories'.
5. H. I. Kaplan and H. S. Kaplan, 'The psychosomatic concept of obesity'.
6. S. Schacter and others, 'The effects of fear, food deprivation ... '.
7. H. A. Witkin, op. cit., p. 228.
8. R. R. Schopbach and R. A. Matthews, 'The psychological problems in obesity'.
9. H. J. Shorvon and J. S. Richardson, 'Sudden obesity and psychological trauma'.
10. H. Bruch, 'Role of the emotions in hunger and appetite'.
11. John Walters, *Aldershot Review*, p. 153.
12. C. M. Thompson, *Interpersonal Psycho-Analysis*.
13. M. Rokeach and A. Eglash, 'A scale for measuring intellectual conviction'.
14. J. C. Martin and F. R. Westie, 'The tolerant personality'.
15. W. Eckhardt and A. G. Newcombe, 'Militarism, personality and other social attitudes', p. 214.
16. Derek Wright, *The Psychology of Moral Behaviour*, p. 214.
17. R. H. Blum, 'The choice of American heroes ... '.
18. M. L. Farber, 'The anal character and political aggression'.
19. G. M. Lyons, 'The military mind'.
20. Eckhardt and Newcombe, op. cit., p. 215.
21. S. A. Rudin, 'The relationship between rational and irrational authoritarianism'; J. Martin and J. J. Ray, 'Anti-authoritarianism ... '.

24. EDUCATION AND THE CULT OF MUSCULAR CHRISTIANITY

1. See Rupert Wilkinson, *The Prefects*.
2. Ibid., p. 87.
3. Personal communication to the author.
4. Richard Deacon, *A History of the British Secret Service*, p. 123.
5. Ibid., p. 279.
6. P. Rowlands, *Children Apart*.
7. J. W. Masland and L. I. Radway, *Soldiers and Scholars*, p. 391.
8. Ibid., p. 390.
9. Ibid., p. 391.
10. R. V. Jones, R.U.S.I. Journal, August 1947, cited in Masland and Radway, op. cit., p. 389.
11. Masland and Radway, op. cit., p. 387.
12. Correlli Barnett, *The Collapse of British Power*, p. 43.

Also consulted
  D. Newsome, *Godliness and Good Learning*.

25. INDIVIDUAL DIFFERENCES

1. Sigmund Freud, *Civilization, War and Death*, p. 5.
2. J. B. Priestley, *Victoria's Heyday*, p. 148.
3. See C. S. Forester, *The General*.
4. S. P. Huntington, *The Soldier and the State*, pp. 59–60.

26. EXTREMES OF AUTHORITARIANISM

1. H. V. Dicks, *Licensed Mass Murder*.
2. Peter Loewenberg, 'The unsuccessful adolescence of Heinrich Himmler', p. 624.
3. Dicks, op. cit., p. 163.
4. Ibid., p. 114.
5. E. Hanfstaengl, *Unheard Witness*, p. 23.
6. See Loewenberg, op. cit., p. 619.
7. Ibid., p. 624.
8. Lionel Tiger, *Men in Groups*.
9. R. G. S. Bidwell, *Modern Warfare*.
10. M. Janowitz, *The Professional Soldier*, p. 223.
11. W. Jerrold, *Sir Redvers H. Buller*, p. 8.
12. John Walters, *Aldershot Review*, p. 142.
13. Ibid., p. 143.

27. THE WORST AND THE BEST

1. William Langer, *The Mind of Adolf Hitler*.

NOTES

2. G. L. Waite, Afterword to the above.
3. Langer, op. cit., pp. 121, 147.
4. Ibid., p. 124.
5. In W. Richardson and S. Freidin, *The Fatal Decisions*, p. 178.
6. See John Strawson, *Hitler as Military Commander.*
7. Richardson and Freidin, op. cit., p. 151.
8. Strawson, op. cit., p. 227.
9. John Prebble, *Culloden*, p. 122.
10. E. S. Turner, *Gallant Gentlemen*, p. 75.
11. Elizabeth Longford, *Wellington: The Years of the Sword*, pp. 359, 439, 440.
12. Ibid., p. 338.
13. E. A. Ritter, *Shaka Zulu*, p. 62.
14. Spike Milligan, *Adolf Hitler—My Part in His Downfall*, p. 65.
15. Ritter, op. cit., p. 120.
16. Ibid., p. 51.
17. Pieter Geyl, *Napoleon: For and Against*, p. 55.
18. Ibid., p. 9.
19. Quoted in ibid., p. 317.
20. Herbert Fisher, *Napoleon*, p. 85.
21. Frédéric Masson, *Napoleon and the Fair Sex*, p. 58.
22. Ibid., p. 63.
23. Ibid., p. 64.
24. Ibid., p. 94.
25. Ibid., p. 95.
26. Ibid., pp. 113, 123.
27. Fisher, op. cit., p. 54.
28. Masson, op. cit., p. 136.
29. In Oliver Warner, *A Portrait of Lord Nelson*, p. 313.
30. Prince William Henry, cited in ibid., p. 28.
31. Ibid., p. 36.
32. Ibid., p. 37.
33. Ibid., p. 249.
34. Ibid., loc. cit.
35. Ibid., p. 165.
36. Ibid., loc. cit.
37. Richard Hough, *Fisher, First Sea Lord.*
38. Ibid., p. 110.
39. B. H. Liddell Hart, *T. E. Lawrence*, p. 442.
40. Ibid., p. 443.
41. Viscount Montgomery, *A History of Warfare*, p. 489.
42. P. Knightley and C. Simpson, *The Secret Lives of Lawrence of Arabia*, p. 87.
43. G. Evans, *Slim.*

44. Ibid., p. 214.
45. Ibid., p. 215.
46. Ibid., p. 43.
47. Ibid., p. 165.
48. Ibid., p. 134.
49. D. Young, *Rommel*, p. 225.
50. O. P. Chaney Jr, *Zhukov*, p. 419.
51. Ibid., p. 349.
52. Nigel Nicolson, *Alex*, p. 15.
53. B. H. Liddell Hart, *Memoirs*, Vol. 1, p. 271.
54. M. Janowitz, *The Professional Soldier*, p. 151.
55. Yigal Allon, *The Making of Israel's Army*, p. 44.
56. Robert Henriques, *A Hundred Years to Suez*, quoted in Allon, op. cit., p. 12.
57. Allon, op. cit., p. 192.
58. Ibid., p. 45.
59. S. Teveth, *Tanks of Tammuz*, p. 69.
60. See Dennis Chaplin, R.U.S.I. Journal, March 1974.

Also consulted
  J. Bainville, *Napoleon*.
  R. G. S. Bidwell, *Modern Warfare*.
  The Bonaparte Letters.
  Alan Bullock, *Hitler*.
  Dennis Chaplin, 'The Yom Kippur War'.
  J. Connell, *Auchinleck*.
  F. N. Cutlack (ed.), *War Letters of General Monash*.
  Y. Dayan, *A Soldier's Diary*.
  A. Fournier, *Napoleon I*.
  R. N. Gale, *Call to Arms*.
  B. Gardner, *Allenby*.
  David Garnett, *The Essential T. E. Lawrence*.
  Chaim Herzog, *The War of Atonement*.
  D. Howarth, *Trafalgar—the Nelson Touch*.
  A. A. Knopf (ed.), *These are the Generals*.
  P. Loewenberg, 'Hitler's psychodynamics examined'.
  A. T. Mahan, *The Life of Nelson*.
  R. F. Mackay, *Fisher of Kilverstone*.

28. EXCEPTIONS TO THE RULE?

  1. See Brian Montgomery, *A Field-Marshal in the Family*, p. 254.
  2. Ronald Lewin, *Montgomery as Military Commander*.
  3. Viscount Montgomery, *Memoirs*, p. 59.

4. Ibid., p. 72.
5. Ibid., pp. 41–2.
6. Ibid., p. 44.
7. Brian Montgomery, op. cit., p. 289.
8. Lewin, op. cit., p. 271.
9. Ibid., p. 270.
10. R. W. Thompson, *Montgomery the Field-Marshal*, p. 201.
11. See J. Kagan, 'The concept of identification'.
12. Brian Montgomery, op. cit., p. 212.
13. Ibid., p. 196.
14. Ibid., p. 320.
15. Ibid., p. 321.
16. Ibid., p. 327.
17. Ibid., p. 347.
18. Philip Magnus, *Kitchener: Portrait of an Imperialist*, pp. 360–61.
19. K. Macksey, *The Tank Story*, p. 9.
20. Magnus, op. cit., p. 361.
21. Ibid., p. 118.
22. G. Arthur, *Life of Lord Kitchener*, p. 370.
23. Magnus, op. cit., p. 232.
24. Ibid., p. 361.
25. Ibid., p. 340.
26. Alan Sillitoe, *Raw Material*, p. 118.
27. Lord Haldane, cited in John Terraine, *Douglas Haig: the Educated Soldier*, p. xiii.
28. E. K. G. Sixsmith, *British Generalship in the Twentieth Century*, p. 161.
29. J. F. C. Fuller, *Memoirs of an Unconventional Soldier*, p. 341.
30. Viscount Montgomery, *A History of Warfare*, p. 494.
31. B. H. Liddell Hart, *History of the First World War*, p. 327.
32. See Bryan Cooper, *The Ironclads of Cambrai*, p. 219.
33. Liddell Hart, op. cit., p. 336.
34. In George Arthur, *Lord Haig*, p. 79.
35. Sergeant T. Secrett, *Twenty-Five Years with Earl Haig*, pp. 278–9.
36. Ibid., p. 287.
37. *Oxford Magazine*, February 23rd, 1928.
38. Countess Haig, *The Man I Knew*, pp. 34, 37.
39. Arthur, *Lord Haig*, pp. 50–51.
40. Ibid., p. 7.
41. Sir Douglas Haig, *Despatches*, p. 347.
42. Robert Blake (ed.), *The Private Papers of Douglas Haig 1914–1918*, pp. 276–81.
43. Ibid., pp. 9, 252, 290, 302, 307.

44. D. S. Duncan, *Douglas Haig as I Knew Him*, p. 103.
45. See J. Marshall-Cornwall, *Haig as Military Commander*, pp. 228-47.
46. A. J. P. Taylor, *The First World War*, p. 188.
47. Ibid., p. 192.
48. John Terraine, *The Western Front 1914–1918*.
49. Ibid., p. 183.
50. Hugh L'Etang, *Pathology of Leadership*, p. 41.
51. Terraine, *The Western Front*, p. 184.
52. Ibid., p. 186.
53. Taylor, op. cit., p. 218.
54. Terraine, *The Western Front*, p. 189.
55. Taylor, op. cit., p. 194.
56. Ibid., p. 83.
57. Ibid., p. 136.
58. Sixsmith, op. cit., p. 96.
59. Taylor, op. cit., p. 140.
60. John Walters, *Aldershot Review*, pp. 206–7.
61. Fuller, op. cit., p. 341.

Also consulted

C. R. Ballard, *Kitchener*.
V. L. Germains, *The Truth about Kitchener*.
H. G. Grocer, *Lord Kitchener—the Story of his Life*.
J. B. Rye and H. G. Grocer, *Kitchener in His Words*.

29. RETREAT

1. K. Miller, in B. S. Johnson (ed.), *All Bull*, p. 265.
2. Charles Fair, *From the Jaws of Victory*, p. 22.
3. J. Marshall-Cornwall, *Haig as Military Commander*, pp. 250–52.
4. I. L. Janis, *Victims of Groupthink*, p. 49.
5. Ibid., p. 191.
6. M. Rosenberg, *Occupations and Values*.
7. J. J. Ray, 'Conservatism, authoritarianism, and related variables ... '.
8. Glen D. Wilson, 'A dynamic theory of Conservatism'.
9. L. J. Eaves and H. J. Eysenck, 'Genetics and the development of social attitudes'.
10. Frank Kitson, *Low Intensity Operations*, p. 201.
11. E. H. Erikson, *Youth: Change and Challenge*.

Also consulted

F. Fitzgerald, *Fire in the Lake*.
D. C. Watt, 'Lessons of the American defeat in Vietnam'.

Bibliography

ABELSON, R. P., ARONSON, E., MCGUIRE, W. J., NEWCOMB, T. M., ROSENBERG, M. J. and TANNENBAUM, P. H., *Theories of Cognitive Consistency: a Sourcebook* (Chicago: Rand McNally, 1968).

ABRAMS, P., 'Democracy, technology and the retired British Officer', in S. P. Huntington (ed.), *Changing Patterns of Military Politics* (New York: Free Press, 1962),

——, 'The late profession of arms', *Archiv. Europ. Sociol.*, VI, 238–61, 1963.

ADAMS, P. L., *Obsessive Children — a Socio-Psychiatric Study* (London: Butterworth, 1973).

ADLER, P. *A House is not a Home* (London: Heinemann, 1954).

ADORNO, T. W., FRENKEL-BRUNSWIK, E., LEVINSON, D. J. and SANFORD, R. N., *The Authoritarian Personality* (New York: Harper, 1950).

ALAPIN, B., personal communication to H. R. Beech, in Beech (ed.), *Obsessional States* (London: Methuen, 1974).

ALEXANDER, F., *Psychosomatic Medicine. Its Principles and Applications* (London: Allen & Unwin, 1952).

ALLON, Y., *The Making of Israel's Army* (London: Valentine Mitchell, 1970).

ANDERSON, H. H. and ANDERSON, G. L., 'Social values of teachers in Rio de Janeiro, Mexico City and Los Angeles County, California', *J. Soc. Psychol.*, 58, 207–26, 1962.

ANSBACHER, H. L. and ANSBACHER, R. R., 'The individual psychology of Alfred Adler, a systematic presentation' (New York: Basic Books, 1956).

ARONSON, E., 'The need for achievement as measured by graphic expression', in J. W. Atkinson (ed.), *Motives in Fantasy, Action, and Society* (New York: Van Nostrand, 1958).

ARONSON, E. and MILLS, J., 'Effect of severity of initiation on liking for a group', *J. Abnorm. Soc. Psychol.*, 59, 177–81, 1959.

ARTHUR, G., *Life of Lord Kitchener* (London: Macmillan, 1920).

——, *Lord Haig* (London: Heinemann, 1928).

ATKINSON, J. W., 'The achievement motive and recall of interrupted and completed tasks', in D. C. McClelland et al. (eds.), *Studies in Motivation* (New York: Appleton-Century-Crofts, 1955).

ATKINSON, J. W., and FEATHER, N. T. (eds.), *A Theory of Achievement Motivation* (New York: Wiley, 1966).

BAINVILLE, J., *Napoleon*, trans. Hamish Miles (London: Cape, 1932).

BALLARD, C. R., *Kitchener* (London: Newnes, 1936).

BANNISTER, D., 'Psychology as an exercise in paradox', *Bull. Brit. Psychol. Soc.*, 19, 21–7, 1966.

BARBER, N., *Sinister Twilight* (London: Collins, 1968).

BARENDREGT, J. T., 'A cross-validation study of the hypothesis of psychosomatic specificity, with special reference to bronchial asthma', *J. Psychosomat. Res.*, 2, 109–14, 1957.

BARKER, A. J., *Townshend of Kut* (London: Cassell, 1967).

BARNETT, C., *The Desert Generals* (London: Wm. Kimber, 1960).

——, *The Swordbearers* (London: Eyre & Spottiswoode, 1963).

——, *Britain and Her Army 1509–1970* (London: Allen Lane, 1970).

——, *The Collapse of British Power* (London: Methuen, 1972).

BARNETT, R. J., *The Economy of Death* (New York: Atheneum, 1969).

BARON, R. M., 'Social reinforcements effects as a function of social reinforcement history', *Psychol. Rev.*, 73 (6), 527–39, 1966.

BASS, B. M., 'Authoritarianism or acquiescence?', *J. Abnorm. Soc. Psychol.*, 51, 616–23, 1955.

BEECH, H. R., (ed.), *Obsessional States* (London: Methuen, 1974).

BELOFF, H., 'The structure and origin of the anal character', *Genet. Psychol. Monogr.* 55, 161–72, 1957.

BIDWELL, R. G. S., *Gunners at War* (London: Arms & Armour Press, 1970).

——, *Modern Warfare* (London: Allen Lane, 1973).

BIRNEY, R. C., BURDICK, H. and TEEVAN, R. C., *Fear of Failure* (New York: Van Nostrand Reinhold, 1969).

BLAKE, R. (ed.), *The Private Papers of Douglas Haig 1914–1919* (London: Eyre & Spottiswoode, 1952).

BLUM, R. H., 'The choice of American heroes and its relationship to personality structure in an élite', *J. Soc. Psychol.*, 48, 235–46, 1958.

The Bonaparte Letters and Despatches, Secret, Confidential, and Official; From the originals in his private cabinet, Vol. 2 (London: Saunders & Otley, 1846).

BOND, B., 'The Late-Victorian Army', in *History Today*, Vol. XI, 616–24, 1961.

——, *The Victorian Army and the Staff College 1854–1914* (London: Eyre Methuen, 1972).

BORGATTA, E. F., COUCH, A. S. and BALES, R. F., 'Some findings relevant to the great man theory of leadership', *Amer. Sociol. Rev.*, 19, 755–9, 1954.

BOYLE, A., *Trenchard* (London: Collins, 1962).

BRADDON, R., *The Siege* (London: Cape, 1969).

British Official History, Military Operations, France and Belgium, 1914–1918, 14 vols. (London: Macmillan, 1922–39).

BROMLEY, D. B., *The Psychology of Human Ageing* (Harmondsworth: Penguin, 1974).

BROWN, R., *Social Psychology* (New York: The Macmillan Co., 1965).

BRUCH, H., 'The psychosomatic aspects of obesity', *J. Mt. Sinai Hospital*, 20, 1, 1953.

——, 'Role of the emotions in hunger and appetite', *Ann. New York Acad. Sci.*, 63, 68, 1955.

——, *Eating Disorders* (London: Routledge & Kegan Paul, 1974).

BULLOCK, A., *Hitler: A Study in Tyranny* (Harmondsworth: Pelican, 1962).

BURDICK, H. A., 'The relationship of attraction, need achievement, and certainty, to conformity under conditions of simulated group atmosphere', unpublished doct. dissert., University of Michigan, 1955, cited in R. Brown, *Social Psychology*.

BURDICK, H. A. and BURNES, A. J., 'A test of "strain towards symmetry" theories', *J. Abnorm. Soc. Psychol.*, 57, 367–70, 1958.

BUSS, A. H., *Psychopathology* (London: Wiley, 1966).

BUTLER, L., *Sir Redvers Buller* (London: Smith, Elder, 1909).

CALDER, A., *The People's War* (London: Panther, 1971).

CAMPBELL, A. B., *Customs and Traditions of the Royal Navy* (Aldershot: Gale & Polden, 1956).

CARR, A. T., 'A psychophysiological study of ritual behaviours and decision processes in compulsive neurosis', unpublished Ph.D. thesis, University of Birmingham, 1970.

CARR, I., in B. S. Johnson (ed.), *All Bull: the National Serviceman* (London: Quartet Books, 1973).

CARSTAIRS, G. M., Foreword to E. Stengel, *Suicide and Attempted Suicide* (Harmondsworth: Penguin, 1964).

CARVER, M., *Tobruk* (London: Batsford, 1964).

CHANEY, O. P., Jr, *Zhukov* (Newton Abbot: David & Charles, 1972).

CHAPLIN, D., 'The Yom Kippur War', *J. Royal United Services Institute for Defence Studies,* pp. 30–34, March 1974.

CHARTERIS, J., *Field-Marshal Earl Haig* (London: Cassell, 1929).

CHAUBEY, N. P. 'Effect of age on expectancy of success and on risk-taking behaviour', *J. Person. & Soc. Psychol.,* 29, 774–8, 1974.

CHRISTIE, R., 'Authoritarianism re-examined', in R. Christie and M. Jahoda (eds.), *Studies in the Scope and Method of 'The Authoritarian Personality'* (Glencoe, Ill.: Free Press, 1954).

CHRISTIE, R. and JAHODA, M. (eds.), *Studies in the Scope and Method of 'The Authoritarian Personality',* (Glencoe, Ill.: Free Press, 1954).

CLARK, A., *The Donkeys* (London: Hutchinson, 1961).

CLAUSEWITZ, C. VON, *On War,* ed. Anatol Rapoport (Harmondsworth: Pelican, 1968).

CONNELL, J., *Auchinleck* (London: Cassell, 1959).

——, *Wavell* (London: Collins, 1964).

COOPER, B., *The Ironclads of Cambrai* (London: Souvenir Press, 1967).

COSTELLO, C. G., *Symptoms of Psychopathology. A Handbook* (New York: Wiley, 1970).

CRAIK, K. J. W., *The Nature of Explanation* (Cambridge University Press, 1943).

CROWNE, D. P. and MARLOWE, D., *The Approval Motive: Studies in Evaluative Dependence* (New York: Wiley, 1964).

CUTLACK, F. N. (ed.), *War Letters of General Monash* (Sydney: Angus & Robertson, 1934).

DAYAN, Y., *A Soldier's Diary* (Harmondsworth: Penguin, 1969).

DEACON, R., *A History of the British Secret Service* (London: Muller, 1969).

DEMETER, K., *The German Officer-Corps* (London: Weidenfeld, 1965).

DE TOCQUEVILLE, A., *Democracy in America,* trans. H. Reeve (New York: Colonial Press, 1900).

DEUTSCH, M., 'Trust, trustworthiness, and the F Scale', *J. Abnorm. Soc. Psychol.*, 61(1), 138-40, 1960.

——, 'Some considerations relevant to national policy', *J. Soc. Issues*, 17, 3, 57-68, 1961.

DEWAR, K. G. B., *The Navy from Within* (London: Gollancz, 1939).

DICKS, H. V., *Licensed Mass Murder* (London: Heinemann, 1972).

DIVINE, D., *The Blunted Sword* (London: Hutchinson, 1964).

DUNCAN, G. S., *Douglas Haig as I Knew Him* (London: Allen & Unwin, 1966).

EAVES, L. J. and EYSENCK, H. J., 'Genetics and the development of social attitudes', *Nature*, 249, No. 54541, 288-9, May 17, 1974.

ECKHARDT, W., 'Psychology of war and peace', *J. Human Relations*, XVI, (2), 239-49, 1968.

——, 'The military-industrial personality', *J. Contemporary Revolutions*, 3 (4), 74-87, 1971.

——, 'An essay on compassion'. Paper presented at the Centre in Training About Peace and War, Wayne State University, April 1973.

ECKHARDT, W. and LENTZ, T. F., 'Factors of War/Peace attitudes', *Peace Res. Rev.*, 1.5.1967.

ECKHARDT, W., MANNING, M., MORGAN, C., SUBOTNIK, L. and TINKER, L. J., 'Militarism in our culture today', *J. Human Relations*, 15(4), 532-7, 1967.

ECKHARDT, W. and NEWCOMBE, A. G., 'Militarism, Personality, and other social attitudes', *J. Conflict Resolution*, 13, 210-19, 1969.

——, 'Comment on Ray's "Militarism, authoritarianism, neuroticism, and anti-social behaviour" ', *J. Conflict Resolution*, XVI, (3), 353-5, 1972.

EDWARDS, W., LINDMAN, H. and PHILLIPS, L., 'Emerging technologies for making decisions', in T. Newcomb (ed.), *New directions in Psychology*, Vol. 2 (New York: Holt, Rinehart, 1965).

EIBL-EIBLESFELDT, I., *Love and Hate* (London: Methuen, 1971).

EINSTEIN, A., *I Believe* (London: Allen & Unwin, 1961).

ELMS, A. C. and MILGRAM, S., 'Personality characteristics associated with obedience and defiance toward authoritative commands', *J. Exp. Res. Person.*, 1, 282-9, 1966.

ERIKSON, E. H., *Youth: Change and Challenge* (New York, Basic Books, 1963). Also in L. J. Bischop (ed.), *Interpreting Personality Theories* (New York: Harper, 1964).

Encounter, three letters, 12, 87, June 1959.

EVANS, G., *Slim* (London: Batsford, 1969).

EYSENCK, H. J., *The Structure of Human Personality* (London: Methuen, 1970).

——, *The Psychology of Politics* (London: Routledge, 1954).

FAIR, C., *From the Jaws of Victory* (New York: Simon & Schuster, 1971).

FARBER, M. L., 'The anal character and political aggression', *J. Abnorm. Soc. Psychol.*, 51 (3), 486–9, 1955.

FESTINGER, L., *A Theory of Cognitive Dissonance* (Evanston, Ill.: Row Peterson, 1957).

——, *Conflict, Decision and Dissonance* (London: Tavistock, 1964).

FIEDLER, F. E., 'Personality and situational determinants of leadership effectiveness', in D. Cartwright and A. Zander (eds.), *Group Dynamics: Research and Theory* (London: Tavistock, 1968).

FINNEY, J. C., 'The M.M.P.I. as a measure of character structure as revealed by factor analysis', *J. Consult. Psychol.*, 25, 327–36, 1961.

FISHER, H., *Napoleon* (London: Williams & Northgate, 1912).

FITTS, P. M. and POSNER, M. L., *Human Performance* (Belmont, Calif.: Wadsworth Publishing, 1967).

FITZGERALD, F., *Fire in the Lake* (Boston: Little, Brown, 1972).

FLUGEL, J. F. C., *Man, Morals and Society* (London: Duckworth, 1945).

FORESTER, C. S., *The General* (Harmondsworth: Penguin, 1936).

FOURNIER, A., *Napoleon I. A Biography*, trans. Annie Elizabeth Adams (London: Longmans, 1914).

FRENCH, E. G., 'Motivation as a variable in work-partner selection', *J. Abnorm. Soc. Psychol.*, 53, 96–9, 1956.

FRENCH, E. G. and ERNEST, R. R., 'The relationship between authoritarianism and acceptance of a military ideology', *J. Person.*, 24, 181–91, 1955.

FREUD, A., *The Ego and Mechanisms of Defence* (London: Hogarth Press and the Institute of Psychoanalysis, 1948).

FREUD, S., *Character and Anal Erotism*, Vol. 9, 169, 1908.

——, *Civilization, War and Death*, Psycholo-analytical Epitomes No. 4 (London: Hogarth Press, 1939).

——, Introductory Lectures on Psychoanalysis, Vols. 15 & 16, 1916–1917.

——, The standard edition of the complete psychological works (London: Hogarth Press and Institute of Psychoanalysis).

FULLER, J. F. C., *Memoirs of an Unconventional Soldier* (London: Nicholson & Watson, 1931).

GALE, Sir Richard, *Call to Arms* (London: Hutchinson, 1968).

——, Review article in *Royal Central Asian Journal*, LV, Part 3, 320, October 1968.

GARDINER, L., *Royal Oak Courts Martial* (London: William Blackwood, 1965).

——, *The British Admiralty* (Edinburgh: William Blackwood, 1968).

GARDNER, B., *Allenby* (London: Cassell, 1965).

GARNETT, D., *The Essential T. E. Lawrence* (London: Cape, 1951).

GERMAINS, V. L., *The Truth about Kitchener* (London: The Bodley Head, 1925).

GEYL, P., *Napoleon: For and Against* (London: Cape, 1949).

GLASS, D. C., 'Theories of consistency and the study of personality', in E. F. Bargatts and W. W. Lambert (eds.), *Handbook of Personality Theory and Research* (Chicago: Rand McNally, 1968).

GOLIGHTLY, C. and REINEHR, R. C., 'Authoritarianism and choice of a military milieu', *Psychol. Reports*, 26, 854, 1970.

GORDON BENNETT, H., *Why Singapore Fell* (London: Angus & Robertson, 1944).

GOULD, R. T. V., *Enigmas* (London: MacLehose, 1946).

GREER, F. L., 'Leader indulgence and group performance', *Psychol. Monogr.*, 75 (12), 1961.

GROCER, H. G., *Lord Kitchener—the Story of his Life* (London: Pearson, 1916).

GRINSPOON, L., 'Psychosocial constraints on the important decision-maker', *Amer. J. Psychiat.*, 125, 110–18, February 1969.

HAIG, D., *Despatches,* Vols. 1 and 2, ed. J. H. Boraston (London: Dent, 1919).

HAIG, Countess, *The Man I Knew* (Edinburgh: Moray Press, 1935).

HALE, J. R., 'Incitement to violence? English divines on the theme of war 1579–1613', in J. G. Rowe and W. H. Stockdale (eds.), *Florilegium Historiale* (University of Toronto Press, 1971).

HALL, K. R. L., 'Aggression in monkey and ape societies', in J. D. Carthy and F. J. Ebling (eds.), *The Natural History of Aggression* (London: Academic Press, 1964).

HAMILTON, J. O., 'Motivation and risk-taking behaviour', *J. Person. & Soc. Psychol.*, 29 (6), 856–64, 1974.

HAMPDEN-TURNER, C., *Radical Man: the Process of Psycho-Social Development* (Cambridge, Mass.: Schenkman, 1970).

HANFSTAENGL, E. *Unheard Witness* (Philadelphia: Lippincott, 1957).

HARE, R. T., 'Authoritarianism, creativity, success and failure among adolescents', *J. Soc. Psychol.*, 86, 219–26, 1972.

HARVEY, O. J., 'Conceptual systems and attitude change', in P. B. Warr (ed.), *Thought and Personality* (Harmondsworth: Penguin, 1970).

HAYTHORN, W., COUCH, A., HAEFNER, D., LANGHAM, P. and CARTER, L. F., 'The behaviour of authoritarian and equalitarian personalities in groups', *J. Human Relations*, 9, 57–73, 1946.

HENRIQUES, R., *A Hundred Years to Suez* (London: Collins, 1957).

HERZOG, C., *The War of Atonement* (London: Weidenfeld & Nicolson, 1975); extracts entitled 'Why Israel all but lost a war', *Sunday Telegraph*, 6, May 18, 1975.

HIBBERT, C., *The Destruction of Lord Raglan* (London: Longmans, 1961).

HOFFMAN, M. L., 'Some psychodynamic factors in compulsive conformity', *J. Abnorm. Soc. Psychol.*, 48, 383–93, 1953.

HOLLAND, J. L., *The Psychology of Vocational Choice* (Waltham, Mass.: Blaisdell, 1966).

HOLLANDER, E. P., 'Authoritarianism and leadership choice in a military service', *J. Abnorm. Soc. Psychol.*, 49, 365–70, 1954.

HOLT, R. R., *Assessing Personality* (New York: Harcourt Brace, 1971).

HOUGH, R., *The Fleet that had to Die* (London: Hamish Hamilton, 1958).

——, *Admirals in Collision* (London: Hamish Hamilton, 1959).

——, *Fisher, First Sea Lord* (London: Allen & Unwin, 1969).

HOVLAND, C. I. and JANIS, I. L. (eds.), *Personality and Persuasibility* (New Haven: Yale University Press, 1959).

HOWARD, M., *The Continental Commitment* (London: Temple Smith, 1972).

HOWARTH, D., *Trafalgar — the Nelson Touch* (London: Collins, 1969).

HUIZINGA, J., *The Waning of the Middle Ages* (Harmondsworth: Pelican Books, 1972).

HUNTINGTON, S. P., *The Soldier and the State* (Oxford: University Press, 1959).

ILLINGWORTH, R. S. and ILLINGWORTH, C. M., *Lessons from Childhood* (London: E. & S. Livingstone, 1966).

INGVAR, D. M. and RISBERG, J., 'Increase of cerebral blood flow during mental effort in normals and in patients with focal brain disorders', *Exp. Brain Res.*, 3, 195–211, 1967.

IZZET, R. R., 'Authoritarianism and attitudes towards the Vietnam War as reflected in behaviour and self report', *J. Person. & Soc. Psychol.*, 17(2), 145–8, 1971.

JAENSCH, E. R., *Der Gegentypus* (Leipzig: Barth, 1907).

JAMESON, W., *The Fleet that Jack Built* (London: Hart-Davis, 1962).

JANET, P., *Psychological Healing*, Vol. 2 (London: Allen & Unwin, 1925).

JANIS, I. L., 'Group identification under conditions of external danger', *Brit. J. Med. Psychol.*, 36, 227–38, 1963.

——, *Victims of Groupthink* (Boston: Houghton Mifflin, 1972).

JANOWITZ, M., *The Professional Soldier* (New York: Free Press, 1960).

——, Introductory chapter in J. A. A. Van Doorn (ed.), *Armed Forces and Society* (The Hague: Mouton, 1968).

JERROLD, W., *Sir Redvers H. Buller, V.C.* (London: Dalton, 1908).

JOHNSON, B. S. (ed.), *All Bull: the National Serviceman* (London: Quartet Books, 1973).

JOHNSON, D., *France and the Dreyfus Affair* (London: Blandford Press, 1966).

JOSEPHSON, E., 'Irrational leadership in formal organizations', *Soc. Forces*, 31(2), 109–17, December 1952.

JUDD, D., *Someone Has Blundered, Calamities of the British Army in the Victorian Age* (London: Arthur Barker, 1973).

KAGAN, J., 'The concept of identification', *Psychol. Rev.*, 65, 296–305, 1958.

KAPLAN, H. I. and KAPLAN, H. S., 'The psychosomatic concept of obesity', *J. Nerv. Ment. Dis.*, 25(2), 181–201, 1957.

KATES, S. and KLEIN, L. S., 'Authoritarian beliefs and perceptual recognition of emotionally charged words'. Paper presented at 62nd Annual Convention of the American Psychological Association, 1954, and in *Amer. Psychologist*, 9, 8, 403–4, 1954.

KELVIN, P., *The Bases of Social Behaviour: an approach through Order and Value* (London: Holt, Rinehart, 1970).

KITSON, F., *Low Intensity Operations* (London: Faber, 1971).

KLINE, P., *Fact and Fantasy in Freudian Theory* (London: Methuen, 1972).

KNAPP, R. H., 'N achievement and aesthetic preference', in J. W. Atkinson (ed.), *Motives in Fantasy, Action, and Society* (Princeton: Van Nostrand, 1958).

KNIGHTLEY, P. and SIMPSON, C., *The Secret Lives of Lawrence of Arabia* (London: Nelson, 1969).

KNOPF, A. A. (ed.), *These are the Generals* (New York: Knopf, 1943).

KOGAN, N., 'Authoritarianism and repression', *J. Abnorm. Soc. Psychol.*, 52, 34–7, 1956.

KOGAN, N. and WALLACH, M. A., 'Risks and Deterrents: individual determinants and group effects', in M. Schwebel (ed.), *Behavioural Science and Human Survival* (Ben Lomond, Calif.: Science & Behavior Books, 1965).

KONSTADT. N. and FORMAN, E., 'Field dependence and external directness', *J. Person. & Soc. Psychol.*, 1, 490–93, 1965.

KORTEN, D. C., 'Situational determinants of leadership', in D. Cartwright and A. Zander (eds.), *Group Dynamics: Research and Theory* (London: Tavistock, 1968).

KRUGER, R., *Goodbye Dolly Gray* (London: Cassell, 1964).

LAFFIN, J., *Tommy Atkins* (London: Cassell, 1966).

——, *Americans in Battle* (New York: Crown Publishers, 1973).

LANGER, W., *The Mind of Adolf Hitler* (London: Secker & Warburg, 1973).

LEACH, P. J., 'A critical study of the literature concerning rigidity', *Brit. J. Soc. & Clin. Psychol.*, 6, 11–22, 1967.

LEASOR, J., *Follow the Drum* (London: Heinemann, 1972).

LETTENBERG, M., 'America in Vietnam—statistics of a war', *Survival*, XIV (6), 268–74, 1972 (published by the International Institute for Strategic Studies).

L'ETANG, H., *Pathology of Leadership* (London: Heinemann, 1969).

——, 'Some actualities of war', *J. of the Royal United Services Institute for Defence Studies*, 117 (665), 64, March 1972.

——, 'Human factors in naval warfare', *Navy International*, 78(12), 5, Dec. 1973.

LEVINSON, D. J. and HUFFMAN, P. E., 'Traditional family ideology and its relation to personality', *J. Person.*, 23, 251–73, 1955.

LEWIN, R., *Rommel as Military Commander* (London: Batsford, 1968).

——, *Montgomery as Military Commander* (London: Batsford, 1971).

LEWIS, A. J., 'Problems of obsessional illness', *Proc. Royal Soc. Med.*, 29, 225–36, 1936.

LIDDELL HART, B. H., *T. E. Lawrence* (London: Cape, 1934),

——, *The War in Outline* (London: Faber, 1936).

——, *The Tanks*, 2 vols. (London: Cassell, 1959).

——, *Memoirs*, 2 vols. (London: Cassell, 1965).

——, *History of the Second World War* (London: Pan Books, 1970).

——, *History of the First World War* (London: Pan Books, 1972).

LOEWENBERG, P., 'The unsuccessful adolescence of Heinrich Himmler', *Amer. Hist. Rev.*, 76(3), 612–41, 1971.

——, 'The psychohistorical origins of the Nazi youth cohort', *Amer. Hist. Rev.*, 76(5), 1457–1502, 1971.

——, 'Hitler's psychodynamics examined', *Contemp. Psychol.*, 19(2), 89–91, 1974.

LIPOWSKI, Z. J., 'Sensory and information inputs overload: behavioral effects', *Comprehensive Psychiatry*, 16 (3), May–June 1975.

LONGFORD, E., *Wellington: The Years of the Sword* (London: Panther, 1971).

LORENZ, K., *On Aggression* (New York: Bantam Books, 1963).

LOWIS, G. L., *Fabulous Admirals* (London: Putnam's, 1957).

LYONS, G. M., 'The military mind', *Bull. At. Sci.*, 19–22, Nov. 1963.

MACINTYRE, D., *U-Boat Killer* (London: Weidenfeld, 1956).

——, *Jutland* (London: Evans, 1957).

——, *The Battle of the Atlantic* (London: Batsford, 1961).

——, *The Naval War Against Hitler* (London: Batsford, 1971).

MCCLELLAND, D. C., *The Achieving Society* (Princeton: Van Nostrand, 1961).

MCCLELLAND, D. C., ATKINSON, J. W., CLARK, R. A. and LOWELL, E., *The Achievement Motive* (New York: Appleton-Century-Crofts, 1953).

MCKENNA, R. J., 'Some effects of anxiety level and food cues on the eating behaviour of obese and normal subjects', *J. Person. Soc. Psychol.*, 22(3), 311–19, 1972.

MACKAY, R. F., *Fisher of Kilverstone* (Oxford: Clarendon Press, 1973). See also review in *J. of the Royal United Services Institute for Defence Studies*, vol. 119, no. 1, 78, March 1974.

MACKSEY, K., *The Tank Story* (London: B.P.C., 1972).

MACRORY, P., *Signal Catastrophe* (London: Hodder, 1966).

MAGNUS, P., *Kitchener: Portrait of an Imperialist* (London: Arrow Books, 1961).

MAHAN, A. T., *The Life of Nelson* (London: Sampson Low, 1899).

MARSHALL-CORNWALL, J., *Haig as Military Commander* (London: Batsford, 1973).

MARTIN, J. and RAY, J. J., 'Anti-authoritarianism, an indicator of pathology', *Aust. J. Psychol.*, 24, 13–18, 1972.

MARTIN, J. C. and WESTIE, F. R., 'The tolerant personality', *Amer. Soc. Rev.*, 24, 521–28, 1959.

MASLAND, J. W. and RADWAY, L. I., *Soldiers and Scholars* (Princeton University Press, 1967).

MASLING, M., 'How neurotic is the authoritarian?', *J. Abnorm. Soc. Psychol.*, 49, 316–18, 1954.

MASSON, F., *Napoleon and the Fair Sex* (London: Heinemann, 1894).

MELLETT, P. G., 'Motive and mechanism in asthma', paper given to Psychosom. Res. Soc., London, 1970.

MELVILLE, C. H., *The Life of General Sir Redvers Buller*, 2 vols. (London: Edward Arnold, 1923).

MILGRAM, S., *Obedience to Authority* (London: Tavistock, 1974).

MILLAR, R., *Kut: The Death of an Army* (London: Secker, 1969).

MILLER, A. G. and MORRISON, B. J., 'Disposition towards military service: a psychological enquiry', *Amer. Psychologist*, 26, 741–6, 1971.

MILLER, H. and BARUCH, D. W., 'A study of hostility in allergic children', *Amer. J. Orthopsychiat.*, 20, 506–19, 1950.

MILLER, K., in B. S. Johnson (ed.), *All Bull: the National Serviceman* (London: Quartet Books, 1973).

MILLIGAN, S., *Adolf Hitler — My Part in His Downfall* (Harmondsworth: Penguin, 1972).

MONTGOMERY, Viscount, *Memoirs* (London: Collins, 1958).

——, *A History of Warfare* (London: Collins, 1968).

MONTGOMERY, Brian, *A Field-Marshal in the Family* (London: Constable, 1973).

MURPHY, G., *Personality* (New York: Harper, 1947).

NEWSOME, D., *Godliness and Good Learning* (London: John Murray, 1961).

NICOLSON, N., *Alex* (London: Weidenfeld, 1973).

O'CONNOR, P. SCOTT, *The Indian Countryside* (London: Brown, Langham, 1922).

ORLANSKY, H., 'Infant care and personality', *Psychol. Bull.*, 46, 1–48, 1949.

OSGOOD, C. E., 'Towards international behaviour appropriate to a nuclear age', in G. S. Nielsen (ed.), *Psychology and International Affairs* (Copenhagen: Munksgaard, 1962).

OTLEY, C. B., 'Public school and army', *New Society*, 754–7, Nov. 17th, 1966.

——, 'The social origin of British army officers', *Sociol. Rev.*, 18(2), 213–34, 1970.

PACK, S. W. C., *Britannia at Dartmouth* (London: Redman, 1967).

PARTRIDGE, E., *A Dictionary of Slang and Unconventional English* (London: Routledge & Kegan Paul, 1963).

PEMBERTON, W. B., *Battles of the Crimean War* (London: Batsford, 1962).

——, *Battles of the Boer War* (London: Batsford, 1964).

PETER, M. J. and HULL, R., *The Peter Principle* (London: Souvenir Press, 1969).

PETTIGREW, T. F., ALLPORT, G. W. and BARNETT, E. O., 'Binocular resolution and perception of race in South Africa', *Brit. J. Psychol.*, 49(4), 265–78, 1958.

PIAGET, J., *The Moral Judgement of the Child* (London: Routledge & Kegan Paul, 1932).

PREBBLE, J., *Culloden* (London: Secker, 1961).

PRIESTLEY, J. B., *Victoria's Heyday* (Harmondsworth: Penguin, 1974).

RAVEN, S., 'Perish by the Sword', *Encounter*, XII, 37–49, May 1959.

RAY, J. J., 'Militarism, authoritarianism, neuroticism, and antisocial behaviour', *J. Conf. Resol.*, 16, 319–40, 1972.

RAY, J. J., 'Conservatism, authoritarianism, and related variables: a review and empirical study', in Glen D. Wilson (ed.), *The Psychology of Conservatism* (London: Academic Press, 1973).

RICHARDSON, W. and FREIDIN, S., *The Fatal Decisions* (Manchester: World Distributors, 1965).

RITTER, E. A., *Shaka Zulu* (London: Panther, 1958).

ROETTER, C., *Psychological Warfare* (London: Batsford, 1974).

ROGERS, S., *Twelve on the Beaufort Scale* (London: Melrose, 1932).

ROKEACH, M., *The Open and Closed Mind* (New York: Basic Books, 1960).

ROKEACH, M. and EGLASH, A., 'A scale for measuring intellectual conviction', *J. Soc. Psychol.*, 44, 135–41, 1956.

ROSENBERG, M., *Occupations and Values* (Glencoe, Ill.: Free Press, 1957).

ROWLANDS, P., *Children Apart* (London: Dent, 1973).

RUDIN, S. A. I., 'The relationship between rational and irrational authoritarianism', *J. Psychol.*, 52, 179–83, 1961.

RYAN, C., *A Bridge Too Far* (London: Hamish Hamilton, 1974).

RYCROFT, C., *Anxiety and Neurosis* (Harmondsworth: Pelican Books, 1970).

RYE, J. B. and GROCER, H. G., *Kitchener in His Words* (London: Fisher Unwin, 1917).

SAMPSON, D. E., 'Draft resisters at the University of California', Paper to Amer. Orthopsychiat. Ass., April 1st, 1969.

SANFORD, N. and COMSTOCK, C., *Sanctions for Evil* (San Francisco: Jossey-Bass Inc., 1971).

SAPPENFIELD, B. R., 'Stereotypical perception of masculinity-femininity', *J. Psychol.*, 61(2), 177–82, 1965.

SARNOFF, I. and CORWIN, S. M., 'Castration anxiety and the fear of death', *J. Person.*, 27, 374–85, 1959.

SCHACTER, S., GOLDMAN, R. and GORDON, A., 'The effects of fear, food deprivation, and obesity on eating', *J. Person. & Soc. Psychol.*, 10, 91–7, 1968.

SCHOPBACH, R. R. and MATTHEWS, R. A., 'Psychologic factors in the problem of obesity', *Arch. Neurol. Psychiat.*, 54, 157, 1945.

SECRETT, T., *Twenty-Five Years with Earl Haig* (London: Jarrolds, 1929).

SELYE, H., *The Stress of Life* (London: Longman, 1957).

SHAFER, J. W., 'A specific cognitive deficit observed in gonadal aplasia (Turner's Syndrome)', *J. Clin. Psychol.*, 18(4), 403–6, 1962.

SHANNON, C. E. and WEAVER, W., *A Mathematical Theory of Communication* (Urbana: University of Illinois Press, 1949).

SHORVON, H. J. and RICHARDSON, J. S., 'Sudden obesity and psychological trauma', *Brit. Med. J.*, 2, 951–6, 1949.

SILLITOE, A., *Raw Material* (New York: Scribner's, 1972).

SIXSMITH, E. K. G., *British Generalship in the Twentieth Century* (London: Lionel Leventhal, 1970).

SLATER, H. and SLATER, P., 'A heuristic theory of neurosis', *J. Neurol. Psychiat.*, 7, 49–55, 1944.

SMITHERS, A. J., *The Man Who Disobeyed* (London: Leo Cooper, 1970).

——, *Sir John Monash* (London: Leo Cooper, 1972).

SMYTH, J., *Sandhurst* (London: Weidenfeld, 1961).

——, *Percival and the Tragedy of Singapore* (London: Macdonald, 1971).

STOGDILL, R. M., 'Personal factors associated with leadership. A survey of the literature', *J. Psychol.*, 25, 35–71, 1948.

STORR, A., *Human Aggression* (Harmondsworth: Penguin, 1970).

STRAWSON, J., *Hitler as Military Commander* (London: Batsford, 1971).

STRONG, K., *Men of Intelligence* (London: Cassell, 1970).

SYMONS, J., *Buller's Campaign* (London: Cresset Press, 1963).

TAYLOR, A. J. P., *The First World War* (Harmondsworth: Penguin, 1966).

TERRAINE, J., *Douglas Haig: the Educated Soldier* (London: Hutchinson, 1963).

——, *The Western Front 1914–1918* (London: Hutchinson, 1964).

TEVETH, S., *Tanks of Tammuz* (London: Weidenfeld, 1968).

THACKERAY, W. M., *Henry Esmond* (London: Oxford University Press, rev. edn. 1858).

THOMAS, D. R., 'Conservatism, authoritarianism, and child-rearing practices', *Brit. J. Soc. Clin. Psychol.*, 14(1), 97–8, 1975.

——, 'Conservatism and premarital sexual experience', *Brit. J. Clin. Psychol.*, 14(2), 195–6, 1975.

THOMAS, H., *The Suez Affair* (Harmondsworth: Penguin, 1970).

THOMPSON, C. M., *Interpersonal Psycho-Analysis: the selected papers of C. M. Thompson*, ed. M. R. Green (New York: Basic Books, 1964).

THOMPSON, R. W., *Montgomery the Field-Marshal* (London: Allen & Unwin, 1969).

TIGER, L., *Men in Groups* (London: Panther, 1971).

Times Literary Supplement, 'The Military Heresy' (review), Friday, June 30th, 1972.

TREVELYAN, G. M., *British History in the Nineteenth Century and After* (London: Longman, 1938).

TURNER, E. S., *Gallant Gentlemen. A Portrait of the British Officer* (London: Michael Joseph, 1956).

TVERSKY, A. and KAHNEMAN, D., 'Judgement under uncertainty: heuristics and biases', *Science*, 185, 1124–31, Sept. 1974.

VAGTS, A., 'Generals—old or young', *J. Polit.*, 4, 396–406, 1942.

——, *A History of Militarism* (London: Hollis & Carter, 1959).

VAN DOORN, J. A. A., *Armed Forces and Society* (The Hague: Mouton, 1968).

VOWLES, D. M., *The Psychobiology of Aggression* (Edinburgh University Press, 1970).

WAITE, G. L., Afterword to W. Langer, *The Mind of Adolf Hitler* (London: Secker, 1973).

WALTERS, J., *Aldershot Review* (London: Jarrolds, 1970).

WARNER, O., *A Portrait of Lord Nelson* (London: Chatto, 1959).

WARR, P. B., *Thought and Personality* (Harmondsworth: Penguin, 1970).

WATT, D. C., 'Lessons of the American defeat in Vietnam', *J. of the Royal United Services Institute for Defence Studies*, 135–8, June 1973.

WEST, J., *Plainville, U.S.A.* (New York: Columbia University Press, 1945).

WHITE, R. and LIPPITT, R., *Autocracy and Democracy* (New York: Harper & Row, 1960).

——, 'Leader behaviour and member reaction in three "social climates" ', in D. Cartwright and A. Zander (eds.), *Group Dynamics: Research and Theory* (London: Tavistock, 1968).

WILKINSON, R., *The Prefects* (London: Oxford University Press, 1964).

——, *The Broken Rebel* (London: Harper & Row, 1972).

WILSON, G. D. (ed.), *The Psychology of Conservatism* (London: Academic Press, 1973).

——, 'A dynamic theory of Conservatism', in his *The Psychology of Conservatism* (London: Academic Press, 1973).

WINNICOTT, D. W., *Aggression in Relation to Emotional Development* (London: Tavistock, 1958).

WITKIN, H. A., 'Psychological differentiation and forms of pathology', *J. Abnorm. Psychol.*, 70, 317–36, 1965.

WOODBURN KIRBY, S., *History of the Second World War: The War Against Japan*, Vol. 1 (H.M.S.O., 1957).

——, *Singapore: the Chain of Disaster* (London: Cassell, 1971).

WOODHAM SMITH, C., *Florence Nightingale, 1820–1910* (London: Constable, 1955).

——, *The Reason Why* (Harmondsworth: Penguin, 1958).

WOOLLCOMBE, R., *The First Tank Battle: Cambrai 1917* (London: Arthur Barker, 1967).

WORSLEY, J. L., 'The causation and treatment of obsessionality', in L. E. Burns and J. L. Worsley (eds.), *Behaviour Therapy in the 1970s*, Proc. of Sympos. (Bristol: John Wright, 1970).

WRIGHT, D., *The Psychology of Moral Behaviour* (Harmondsworth: Penguin, 1971).

YOUNG, D., *Rommel* (London: Collins, 1952).

YOUNG, W., 'Sitting on a fortune: the prostitute in London', *Encounter*, 12, 19, May 31st, 1959.

ZEITZLER, K., 'Stalingrad', in W. Richardson and S. Freidin, *The Fatal Decisions* (London: World Distributors, 1963).

ZIMBARDO, P. G., *The Cognitive Control of Motivation* (Glenview, Ill.: Scott Foresman, 1969).

Index

Achievement motivation, 238–55, 281, 321, 324, 325, 379–80
Administrative ability, 250, 283
Admirals, 112, 119, 120, 121, 161, 206, 213, 268, 269
Admiralty, 87, 112, 210, 221, 272, 294
Adorno, T. W., et al., 257, 287
Afghan War, First, 71–9
Age, as factor of incompetence, 162, 221
Aggression, 110, 121, 141, 142, 169, 170, 174, 175, 181, 188, 191, 192, 195, 202, 211, 238, 254, 258–60, 278, 311, 315, 316, 319
Aggressive spirit, 155
Aircraft and airpower, 111, 119, 121, 134, 246, 265
Airey, Gen. Richard, 38, 42
Akbar Khan, 72, 77
Akers-Douglas Committee, 160
Alamein, 127, 162, 240
Alanbrooke, Field-Marshal Viscount, 241, 361
Alexander, Field-Marshal Earl, 159, 240, 347
Allenby, Field-Marshal Viscount, 241, 242, 243, 340–41
Alma, battle of, 39, 41, 44, 47
Ambiguity, intolerance of, 257
America, Americans, 121, 125, 175, 210, 258, 278, 299, 300, 348, 369, 382, 395, 397, 400; see also United States
American Civil War, 125, 175
Anal character, 189–95, 319
Anley, Gen., 89
Anti-effeminacy, 207–13
Anti-intellectualism, 112, 161, 162, 168, 286; see also Authoritarianism
Anti-Semitism, 114, 256, 259, 266, 313, 321; see also Authoritarianism

Anxiety, 24, 141, 167, 168, 173, 174, 182, 183, 184, 192, 218, 256, 283, 286
Apia, naval disaster at, 34
Appearance, importance of, 66, 177, 179; see also 'Bull', Dress, Uniforms
Appeasement, 142, 170, 171, 181
Ardennes, 30, 124, 125, 154, 165, 264, 299
Army Estimates, 116–17
Arnhem, 145–8, 154, 165, 299, 360
Artillery, 53, 58, 83, 87, 91, 251
Assault, against strongest point, 69, 81
Asthma, 253, 379
Attack, 194
Auchinleck, Field-Marshal Sir Claude, 126–8, 158, 162, 344, 347
Auckland, Lord, 73, 79
Australia, Australians, 130, 176
Authoritarian family group, 174
Authoritarianism, 202, 255–80, 291; advantages of, 263–4; aggression, 268; anality, 292, 310–12; anti-Semitism, 266, 267; attitudes to militarism, 264; autocracy, 268, 271, 272, 287, 327; belief in supernatural, 274; closed mindedness, 264; conformity, 262, 264; conservatism, 401–2; dishonesty, 265; dislike of experiments, 265; dogma, 265; ethnocentrism, 265; hostility, 274, 275; initiative, 267; intolerance of ambiguity, 257; Israel's army, 350; leadership, 263; military incompetence, 264–79; military intelligence, 264; morality, 267; naval cadets, 263; obedience, 267–9; obsessive traits, 263, 264, 273, 275; rational versus irrational forms of, 287; religion, 277, 278; sex, 258–60, 263, 268, 270; scapegoating, 272; submission to authority, 270; technical innovations, 327; under-

estimation of the enemy, 265–6; venereal disease, 277; vocational choice, 264, 278, 279
and Allenby, 340–41; Fisher, 336–7; Haig, 354, 371–92; Himmler, 309, 311–315; Hitler, 320–21; Kitchener, 354, 369–71; Lawrence, 338; Montgomery, 354, 355–69; Napoleon, 330; Nelson, 334; Rommel, 344–5; Slim, 342, 343; Wellington, 324–6; Wolfe, 323–4; Zhukov, 346–7

Authoritarian organizations, 164, 179, 271, 299

Authoritarian personality, 141, 256–80, 287, 313, 325

Aylmer, Lieut.-Gen. Sir Fenton, 101–3

Balaclava, battle of, 38, 41, 44, 290
Balla Hissar, 73, 74, 76, 77, 155
Barker, Gen. M. G. H., 240
Barrack square, 110, 178, 224
Battle of Britain, 123
Battleships, 119, 120, 130, 246
Bay of Pigs, 45, 148, 397–8
Belmont, battle of, 59
Bennett, Maj.-Gen. Gordon, 138, 141, 155, 203, 206, 247, 248
Boer Army, 53, 81, 397
Boer War, 45, 52–67, 81, 87, 110, 204, 206, 250, 265, 306, 349, 384
Braddock, Gen., 198, 199
Britannia (Royal Naval College), 160, 208
British Expeditionary Force, 81, 83, 84, 140
British Legion, 386–7
British Second Army, 146, 147
Brooke-Popham, Air Chief Marshal Sir Robert, 133, 134, 137, 246, 248
Browning, Lieut.-Gen. Frederick, 147
Brydon, Dr, 78
Buchanan, Sir George, 101–2
'Bull', characteristics and origins of, 176–88; anxiety, 141, 256; authoritarianism, 273; education, 289, 291; Hitler, 320; honour, 198, 200; Israel's army, 349, 352, 353; masculinity, 208; militarism, 169, 286; Montgomery, Viscount, 359; obsessive traits, 191, 291; self-esteem, 110; Shaka-Zulu, 327; vanity, 200; Wellington, 324; see also Appearances, Dress, Uniforms
Buller, Gen. George, 39
Buller, Gen. Sir Redvers: and achievement motivation, 242; aggression, 122, 316, 317; and economy of forces, 155; and field dependency, 283; irresolution and indecision, 61, 126, 221, 246; lack of confidence, 73, 126, 128, 246, 247; leadership, 81, 218; loyalty and affection of troops for, 206, 216; obesity, 316;

317; obstinacy, 222; over-control of aggression, 316, 317; passivity, 211; in the light of the Peter Principle, 220; physical attributes, 55; and publicity, 248; relationship with mother, 283; scapegoating, 93, 245; and Second Boer War, 55–67

Burgess, Guy, 293
Burgos, siege of, 324
Burgoyne, Sir John, 37, 46
Burrard, Gen. Sir Harry, 221
Burton, Sir Richard, 295
Byng, Field-Marshal Viscount, 92, 93, 272

Cambrai, tank offensive, 30, 86–94, 165, 207, 245, 264, 272, 389, 396
Cambridge, Duke of, 37, 250
Campbell, Gen. Sir Colin, 37, 49
Camperdown, H.M.S., 112, 267, 269
Cardigan, Gen. the Earl of, 38, 41, 42, 200, 201, 218
Career choice, 298; see also Vocational choice and Officer selection
Casualties, 21, 44, 47, 50, 53, 57, 58, 59, 60, 64, 65, 70, 71, 76–9, 82, 84, 85, 91, 92, 95, 99, 102, 105, 126, 130, 137, 144, 148, 154, 155, 299, 326, 379, 380, 381, 382, 386, 388, 395, 399
Caution, vs. boldness, 221
Cavalry, 87, 91, 116, 117, 125, 246, 251; see also Horses
Cavan, Field-Marshal Lord, 112
Censorship, 134, 205, 248
Chaplains, 277, 278
Character, 195–207, 225, 297
Charles, Sir Ronald, 164
Charteris, Brig.-Gen. John, 381, 396
Chatham, Lord, 153
Chetwode, Sir Philip, 117, 121
Childhood, 256, 259, 260, 268, 272, 279, 282, 285, 287, 296, 329, 353, 361, 371, 395
Chivalry, 197, 199n, 220
Churchill, Winston, 65, 87–8, 113, 128, 142, 294, 296, 361, 385, 386
C.I.A., 300, 397, 398
Civilians, 133, 134, 135, 139, 202
Cleanliness, 177–9, 182, 184, 188, 189, 191; see also Bull
'Closed mind', 166, 262, 276, 281, 282, 321, 379
Cognitive dissonance, 164–6
Colenso, battle of, 59, 61, 284
Colley, Gen. Sir George Pomeroy, 158
Commanders, great, 19, 323–48, 354–92; see also under individual names
Committee, government by, 397–400
Communication, lines of, 96, 98, 101

Communications, 27–35, 65, 138, 146; breakdowns in, 91, 215, 361
Compassion, 76, 107, 108, 276, 317
Computers, 32–4
Compulsiveness, 178, 179, 183, 184, 192
Conformity, 34, 162, 182, 185, 239, 244, 245
Conscience, 15, 196, 197
Conservatism, 30, 82, 115, 152, 172, 179; 259, 271, 300, 401–2; hereditary factors in, 402; and authoritarianism, 402
Conventionalism, 258
Convoys, 210–11
Cork and Orrery, Earl of, 160
Corps, XXX, at Arnhem, 146
Costs, 19, 21, 66, 95, 130
Cotton, Sir Willoughby, 72
Courage, 55, 66, 107, 109, 137, 145, 198, 212, 245, 246
Courtesy, 39, 51, 79, 144, 155, 268
Courts martial, 171, 244, 272, 273
Crimean War, 26, 31, 36–51, 52, 61, 66, 102, 131, 162, 172, 204, 218, 292, 306, 359; winter campaign, 44–7
Criticism, dislike of and sensitivity to, 108, 115, 140, 204–7, 245, 253, 254, 286, 322
Cuba, 397–8
Culloden, battle of, 323
Curzon, Lord, 361

Death, fear of, 184–7, 197, 256
Deception, 120, 135, 179; lack of, 153
Decision-making, 61, 128, 164, 167, 168, 222, 274
Decision process, 27–35
Decisions and group decisions, 80, 132, 166, 275, 315, 397–400
Defences: military, 138, 139, 140, 143; psychological, 173, 174, 191, 194, 203, 207
Defensive positions, building of, 55
Defensive responses: military, 141, 167; psychological, 209
Delusions, 74, 100, 133
de Mole, E. L., 86–7
Denial, mechanism of, 65, 168
Dependency, feelings of, 239, 253, 281
Deverell, Field-Marshal Sir Cyril, 115
Dirt, anxieties about, 89; see also 'Bull', Anal character, Obsessive traits
Discipline, 53, 58, 66, 127, 160, 182, 197, 198, 256, 264, 272, 329, 353
Dishonesty, 101, 102, 103, 107, 114, 134, 135, 155, 239
Dissonance Theory, 165
Dogma and dogmatism, 136, 137, 144, 169, 179, 262, 286, 318
Domville, Admiral Sir Barry, 293
Dorman-Smith, Maj.-Gen., 161–2

Dress, 77, 179, 200; see also Appearances, 'Bull', Uniforms
Dreyfus case, 267
Drill, 59, 160, 171, 178
Duelling, 188, 197, 198, 200
Duff, Sir Beauchamp, 95, 96, 102, 106
Dulles, John F., 398
Duncan, Rev. G. S. (Haig's chaplain), 379
Dundonald, Lord, 63
Dunkirk, 123, 125, 126, 240, 322

Economy in manpower, 47, 62, 97, 152, 155, 275
Edmonds, Gen. J. E., 117
Education, 20, 118, 159, 161, 288–301, 400
Effeminacy, 141, 207, 285, 313
Ego and ego weakness, 115, 116, 166, 183, 196, 207, 239, 255, 309, 310, 318, 360
Eichmann, Adolf, 106, 270
Einstein, Albert, attitude to military organizations, 173
Elles, Sir Hugh, 164
Elphinstone, Maj.-Gen. William George, 73–9, 122, 126, 128, 154, 155, 216, 242
Emotion, 32, 168, 197, 244, 311
Enemy, underestimation of, 47, 80
Erskine, Gen. Sir William, 347
Ethnic prejudice, 257, 258
Ethnocentrism, 259, 265, 321, 322, 325, 385
Eyre, Gen. Sir Vincent, 46–7, 72
Expeditionary Force D, 95, 102
Exploiting military gains, 91

'F', authoritarianism scale, 258–9, 262
Fear, 197, 198
Fear of failure, 34, 102, 222, 239, 244, 246, 247, 253, 254, 256, 318
Featherstonehaugh, Gen., 59
Field-dependency, 281–4; and aggression, 284; authoritarianism, 281; extraction of information, 282; genetics, 282; obesity, 283
First Airborne Division, 146, 147
Fisher, Admiral 'Jackie', 268, 271, 272, 336–7, 348
Flesquières, 90–91
Foch, Marshal F., 101, 372, 378
'Fragging', 175
French, Field-Marshal Sir John, 80, 83–5, 122, 152, 206, 250, 251, 252, 383, 387
Frontal assaults, 81, 107, 127, 153
Frustration, 260
Fuller, Maj.-Gen. J. F. C., 81, 89, 90, 112, 113, 114, 117, 158, 162, 163, 207, 235

Gallabat, evacuation of, 342
Gallipoli, 81, 145, 221
Gas warfare, 84, 154
Gatacre, Maj.-Gen. Sir William, 59, 60

General Adaptation Syndrome, 187

Generals, 83, 161, 205–6; and generalship, 23, 27–35, 88, 143, 161, 205 6, 213, 249, 393, 394; training, 158; and age, 162, 163

German Army, 146, 220, 236, 265; General Staff, 113, 179; Officer Corps, 197

German Navy, 259

Germany, 112, 113, 118, 122

Giap, Gen. Vo Nguyen, 397

Gordon, Gen. Charles, 317

Gordon-Finlayson, Gen., 118

Göring, Hermann, 109

Gort, Field-Marshal Viscount, 113, 115, 118, 140, 240

Gough, Gen. Sir Hubert, 373

Gough, Brig.-Gen. John, 251

'Great Captains', 219, 323–48; see also under Alexander, Allenby, Auchinleck, Fisher, Guderian, Lawrence, T. E., MacArthur, Monash, Montgomery, Napoleon, Nelson, O'Connor, Rommel, Shaka–Zulu, Slim, Wellington, Wolfe, Yamamoto, Zhukov

Group-think, symptoms of, 399–400

Guderian, Gen. Heinz, 113, 347

Guilt, absense of, 106, 197

Haig, Field-Marshal Lord: academic record, 158; and achievement motivation, 242, 243, 249–53, 379; aggression, 377; authoritarianism, 354, 371–92; and British Legion, 386–7; and 'bull', 375; and Charteris, Brig.-Gen., 381, 396; childhood, 380; 'closed mindedness', 379; discipline, 377; ethnocentrism, 385; similarities to Himmler, 378, 390, 391; imagination, lack of, 379; intelligence, 250; loyalty, 380; attitude to maching-guns, 82; disregard of military intelligence, 373, 382; mother relationship, 283, 376, 380; obsessive traits, 375, 379; opinion of others, 251, 252, 378; obstinacy, 88, 375; preoccupation with time, 378; psychosomatics, 379; and publicity, 248, 380; scapegoating, 93, 252, 272; self-advancement, 155; self-esteem, 381; and sex, 376, 377; and Sir John French, 380; wish for social approval, 380; and tanks, 88, 89, 388; writings, 378; and Ypres, 3rd battle of, 372–4, 381, 387
others' views on, 80, 219, 235, 355, 372, 381–9

Haldane, Lord, 250, 252, 384, 385

Halgar-Ultra bullet, 117

Hardinge, Lord, 95, 102

Harper, Lieut.-Gen. Sir George, 90

Harris, Air Marshal Sir Arthur, 294

Hart, Maj.-Gen., 59

Hawke, H.M.S., 273

Hawley, Maj.-Gen. Henry, 323–4

Health, see Welfare

Hess, Rudolph, 274

Himmler, 309, 311–15; authoritarianism, 320; compared to Gen. Sir Redvers Buller, 317; similarities with Haig, 378; over-control of aggression, 317, 322; relationship with Rommel, 344, 345

History, and psychology, 17–18

Hitler, 125, 274, 292, 313, 317, 318–23, 344, 345, 362; and authoritarianism, 274, 318–19; response to captured plan, 124; and censorship and communiqués, 134; in inter-war years, 112–14, 116, 164; leadership, 215, 220; military incompetence, 318–29; monorchism, 319; relationship with parents, 318–19, 362

Hobart, Maj.-Gen. Sir Percy, 114, 118, 207

Ho Chi Minh, 397

Honour, 194–207, 292

Hore-Belisha, Rt. Hon. Lord, 113, 114, 140, 245, 267

Horses, 92, 111, 115–19, 125, 347, 351; see also Cavalry

Hostility, 254; see also Aggression

Hughes, Admiral Sir Richard, 335

Hughes, Thomas, 290

Imperial Defence Paper (1926), 110–11

Indecision, 61, 73, 75, 76, 126, 137, 153, 221, 246, 254, 319

Indian Mutiny, 68

Inferiority complex, 202, 244

Inflexibility, 266; see also Rigidity

Information, 23, 27–35; withholding of, 133, 135, 153

Initiation rites, 199, 200

Initiative, in authoritarian organizations, 40, 145, 183, 209, 221, 222, 227, 268

Inkerman, battle of, 44, 47

Instinct, and instinctual behaviour, 24, 141, 169, 170, 180, 197, 256

Intellect, 196, 197; intellectual processes, 166; intellectual level/ability, 20, 157–68, 250; intellectual pursuits, dislike of, 51, 112, 286

Intelligence (mental ability) and intellectual ability, 162, 168; in officer selection, 212; and promotion, 22

Intelligence (military), 299–300; and American disasters, 299–300, 399–400; and American War Colleges, 299–300; and Arnhem, 145, 147, 148; and authoritarianism, 264, 265; and Boer War, 65; and cognitive dissonance, 165,

Intelligence (military)—*cont.*
  166; and Crimean War, 50, 293; and
  First World War, 92; and Gazala, 127;
  and Haig, 373, 382; ignoring of, and
  inadequate supply of, 50, 65, 80, 82, 92,
  99, 135, 147, 148, 152, 165, 166, 264,
  265, 299, 373, 382, 399–400; and
  General Nixon, 99; and Second World
  War, 124, 299; Victorian attitudes
  towards, 292–5
Inter-Service rivalry, 222, 343
Intolerance of ambiguity, 257
Ironside, Field-Marshal Lord, 115, 123,
  140, 162, 163, 248
Irrationality, 167, 177, 198, 199, 202
Israel 113, 209, 350
Israeli Army, 349–53; and obedience, 350;
  and 'bull', 349, 351, 352

Japanese, 130, 131–43, 199, 212, 222;
  Air Force, 137, 246
Jealousy, 83, 108
Jellicoe, Admiral Lord, 348, 382
Jews, 257, 259, 267, 314
Jingoism, 20
Jodl, Gen. Alfred, 322
Joffre, Marshal J. J. C., 80, 155
John, Augustus, 367
Johnson, President Lyndon Baines, 396
Jutland, battle of, 122n, 179, 268

Kabul, retreat from, 71–9, 126, 154, 245
Kavanagh, A. McMorrough, 295
Keitel, Field-Marshal W., 322
Kennedy, President John F., 397–8
Khalil Pasha, 99
Kiggell, Lieut.-Gen. Sir Lancelot, 385
Kimmel, Admiral H. E., 398, 399
King's Regulations, 107
Kingsley, Charles, 290
Kirke Committee, 115, 207, 245
Kitchener, Field-Marshal Earl, 55, 242,
  272, 273, 349, 354, 355, 359, 369–71,
  385, 391
  authoritarianism, 369; childhood, 371;
  conformity, 369; dislike of rules and
  regulations, 370; lack of ethnocentricity,
  370; need for approval, 369; obsessive
  traits, 370; and sex, 370; and short-
  comings, 370–71; and tanks, 369
Kluck, Gen. A. von, 265
Knox, Gen. Sir Harry, 271
Korean War, 397
Kut, siege of, 95–109, 131, 148, 154, 166,
  206, 240, 245, 265, 268

Ladysmith, siege of, 60
Lambton, Maj.-Gen. Sir W., 252
Lawrence, Sir George, 76, 79

Lawrence, T. E., 15, 157, 219, 241, 242,
  243, 245, 275, 337–40, 354, 361
Leadership, 34, 45, 66, 74, 80, 128, 143,
  212, 213, 214, 290, 297, 315
Lee, Gen. Robert E., 161
Liddell Hart, Sir Basil, 113, 114, 117, 121,
  123
Light Brigade, charge of, 38, 41
Loos, battle of, 82, 85, 254
Low-intensity operations, 402–3
Lowry, Rear-Admiral R. S., 271
Loyalty, 94, 183, 226, 290, 309
Lucan, Earl of, 37, 41, 42, 218
Ludendorff, Gen. E., 22, 384, 386
Lumley, Maj.-Gen., 79
Lyttleton, Gen., 62

MacArthur, Gen. Douglas, 235, 347
Machine-guns, 52, 82, 86, 265
McNaghten, Sir William, 73
McNamara, Robert, 398
Mafeking, 60
Magersfontein, battle of, 58, 61
My-Lai massacre, 174, 270
Making, Brig., 116
Malay Peninsula, and tanks, 132
Malaya Command, 138
Malta, 121
Maltby, Air Vice Marshal Sir Paul, 137
Marlborough, Duke of, 204–5
Market-Garden, Operation, 26, 145–8
Markham, Rear Admiral A. H., 209, 245,
  268
Marshall-Cornwall, Gen. Sir James, 381–3
Masculinity, doubts about, 141, 184, 185,
  203, 208, 211, 279, 282, 285, 313, 319
Massey Committee, 169
Maternal relationships: and field-depen-
  dency, 282–4; *see also* under individual
  commanders
Maude, George A., 37
Medical services, 38, 102, 105
Mellis, Maj.-Gen., 107
Memory, 32, 162, 178, 241, 286
Mental activity and cerebral blood flow,
  162
Mesopotamia, 95, 96
Messervy, Gen. Sir Francis, 126–7
Messines, battle of, 389
Methuen, Gen. Lord, 57, 58, 60
Militarism, 10, 21, 110, 125, 169–75, 176,
  180, 191, 194, 197, 200, 203, 236, 286,
  287, 289, 295, 300, 301, 315, 348, 349,
  394
Military achievement, 238–55
Military behaviour, 17
Military commanders, 28
Military commanders: competent vs.
  incompetent, 18

Military disasters, 17, 18, 148, 150; costs of, 19

Military Establishment, 112, 114, 162

Military hierarchy, 171

Military incompetence, summaries of factors in, 152, 306, 394–5, 399–400, 402

Military-industrial personality, 401

Military mind, 235, 256, 279, 307, 400, 401

Military organizations, 21, 23, 128, 169–75, 197, 201, 203, 208, 211, 218, 219, 222, 238, 239, 246, 254, 255, 256, 270, 272, 285, 289, 305, 306, 319

Military parades, 54, 141

Military profession, attraction to, 157, 173, 279, 319, 402

Milne, Field-Marshal Sir George, 115

Modder River, battle of, 57

Monash, Gen. Sir John, 81, 220, 275, 348

Montgomery, Field-Marshal Viscount: achievement motivation, 239–42, 245, 356; and Americans, 360, 361; and Arnhem, 145–8, 165, 360; attention to detail, 273; attitude to First World War, 358; and authoritarianism, 355–92; and 'bull', 359; and casualties, 358; childhood of, 358, 361, 362; and church parades, 359; comparison with Kitchener and Haig, 391; views on Crimean War, 36; dislike of sharing achievements with others, 365; ego needs, 360; views on First World War generals, 358; and health and welfare of troops, 275, 276–8, 327–8, 355–7, 359; humour, 355, 357; initiative, 357; intellectual ability, 157, 158; and leadership, 219, 226; marriage, 358, 362; and North African Campaign, 360; obsessive traits, 359; prevention of venereal disease, 277, 278, 355, 356; and publicity, 248; readiness to accept unpopularity, 359; rebellious behaviour, 367; regard for human life, 358; relationship with mother, 362–9; religiosity, 359; self-confidence, 347; and sex, 330, 356–7; shortcomings, 359–61; on unimportance of physical attributes, 212; unpopularity, 254

Montgomery, Lieut.-Col. Brian, 357, 359, 362, 366, 368

Montgomery-Massingberd, Field-Marshal Sir Archibald, 113, 115, 116, 164, 207, 245, 267

Moral courage, 55, 61, 76, 82, 212, 396

Moral cowardice, 254

Morale, 102, 134, 139, 140, 141, 142, 143, 178, 198, 216, 219, 220, 221, 236, 248, 274, 275

Morality, 196, 225, 267, 285, 399

Motivation, 32, 171, 173, 221

Murray, Maj.-Gen. Sir Archibald, 251

Murray, Gen. Sir James Wolf, 97, 98, 100

Muscular Christianity, 290, 295, 301, 313, 395

Mutiny, 171, 216, 273

Napoleon, 88, 100, 154, 212, 241, 242, 272, 275, 294, 329–34

Narpat Sing, 68

Narvik, 221

Naval commanders, 28, 34, 35–6, 112

Naval Discipline Act, 272

Naval officers, 87, 88, 222

Navy, and navies, 19, 119, 121, 123, 136, 141, 179, 183, 186, 210–11, 263, 264, 271, 273, 289, 293, 294, 336

Nazis, 108, 109, 257, 274, 293, 310

Needs, biological and social, 32, 173

Need to achieve, 238

Nelson, 181, 212, 219, 241, 276, 282, 294, 323, 325, 334–6, 354

Neuve Chapelle, battle of, 252, 387

Nightingale, Florence, 46, 102, 209, 359

Nixon, Gen. Sir John, 95, 96, 98, 99, 102, 106, 155

'Noise', 28, 31, 38

Nolan, Capt. L. E., 42, 43, 44, 93

Northern Ireland, 403

North-West Frontier, 117

Norway expedition, 221

Norwegian campaign, 265

Obedience, 82, 160, 171, 172, 178, 183, 192, 194, 209, 220, 224, 239, 267–9, 270, 280, 282, 290, 291, 297, 309, 310, 320, 350

Obesity, 283, 284, 316

Obsession, 62, 183, 191, 238, 255, 291, 359, 370

Obsessional neurosis, 167, 183, 192, 194, 401

Obsessive traits, 184, 289, 352, 379

Obstinacy, 62, 91, 96, 128, 136, 144, 153, 222, 321

O'Connor, Gen. Sir Richard, 162, 347

Officers' mess, 295, 313; conversation in, 185, 188

Officer selection, 172, 173, 212, 228, 236, 246, 248

Officer training, 223, 238; comments on, 161

Official Secrets Acts, 294

Orderliness, 177, 178, 182, 184, 187, 191, 192, 194, 280, 283

Orders, 134, 171

Over-confidence, 45, 47, 80, 124, 127

Over-control of aggression, 121, 315, 322

Passchendaele, 82, 155, 358, 372

Passivity, 144, 155, 166, 246, 259, 281

Patton, Gen. George S., 118, 235, 361
Patullo, Maj. J. B., 44
Pay-offs, 21, 166
Pearl Harbor, 121, 299, 397–9
Peninsular War, 181
Perception, 32, 280
Percival, Lieut.-Gen. A. E., 131, 135, 138, 139–44, 155, 159, 245, 247, 248
Personality, 94, 167, 169, 188, 195, 211, 212, 235, 255, 258, 259, 268, 282, 287, 288, 310, 329, 371, 380
Pétain, Marshal, 382
Peter Principle, 220
Phillips, Admiral Sir Tom, 136, 245
Physical attributes, 159, 171, 212, 213, 225, 246, 266
Pile, Lieut.-Gen. Sir Frederick, 114
Plumer, Field-Marshal Viscount, 81, 249, 389
Poland, invasion of, 125
Polaris submarine, 119
Political decisions, 21
Political pressures, 119
Politicians, 114
Pollock, J. C., 78
Pontification, 163, 164, 168, 286
Pornography, 295
Portal, Lord, Marshal of the Royal Air Force, 293
Positive transfer, lack of, 52, 81
Pottinger, Maj. E., 76
Pratt, Margaret, 209
Prejudice, 34, 88, 257, 259, 260, 282; and binocular rivalry, 261n
Prince George, H.M.S., 271
Prince of Wales, H.M.S., 136
Prisoners-of-war, 95, 105, 106, 107, 199
Procrastination, 74, 75, 144, 153, 322
Progressive thinking, 119, 161
Projection, 260–62
Promotion, 239, 243, 254
Prostitutes, 19, 170, 180, 295
Psycho-analytic theory, 189, 190, 196, 218, 298
Psychopathology, 10, 23, 247, 287
Psychopathy, 107, 109, 215, 220
Psychosomatics, 49, 253, 283, 312, 379
Publicity, resistance to and dislike of, 51, 206, 248
Public schools, 223, 289
Pulling of punches, 65, 66, 127
Punctuality, 182
Pyke, Geoffrey, 115

Raglan, Lord, 36, 39, 40, 42, 48, 56, 72, 93, 122, 126, 155, 218, 242, 245
Randomness, combatting of, 186, 187
Rationalization, 134
Rawlinson, Gen. Lord, 252, 378, 388

Reactionary motives, 87, 111, 112, 115
Reconnaissance, 62, 69, 74, 119, 153, 293, 325, 326
Recruitment, 20, 21, 111
Redan, battles of, 47–9, 50
Red tape, 45
Reductionism, 18
Rees, Goronwy, 241
Religion, 359
Repulse, H.M.S., 136
Resistance to technological progress, 52, 53, 86, 88, 111, 113, 116, 119, 127, 246, 265
Responsibility, evasion of, 62, 76
Rigidity, 76, 124, 128, 144, 296, 300, 318
Rimington, Brig.-Gen., 100
Risk-taking, 145, 147, 148, 166–8, 217, 218, 240, 244
Ritchie, Gen. Sir Neil, 126–9
Ritual, 141, 169, 178, 182, 184, 187, 191, 192, 273
Roberts, Field-Marshal Earl, 54, 58, 349
Robertson, Field-Marshal Sir William, 238, 252, 385
Rod-and-frame test, 281
Rommel, Field-Marshal, 126, 127, 128, 239, 344–5
Rooyah, fort, 68–70, 154
Royal Air Force, 119, 120, 123, 132, 137
Royal Commission on Awards to Inventors, 86
Royal Military Academy, see Sandhurst
Royal Military College, 159
Royal Oak courts martial, 272, 273
Royal United Services Institute, 112, 271
Russia, and Russians, 45, 49, 51, 53, 113, 220, 235

Sale, Geoffrey, 161
Salute, military, 181, 238, 224
Samoa, 34–6
Sandhurst, 57, 158, 159, 161, 224, 225, 228, 234, 249, 365, 366
Scapegoats, 43, 44, 61, 66, 78, 92, 93, 153, 205, 245, 272, 286, 322, 324
Scarlett, James, 38, 42
School of Infantry, Warminster, 224, 225, 228
Sebastopol, 45, 47
Secretiveness, 254
Secret Service, 294; see also Intelligence, military, Special Operations Excutive, Spies
Secrett, Sergeant T., 375–6
Selection of subordinates, 248, 249, 266, 267; see also Officer selection
Self-confidence, 128, 241, 324
Self-deception, 119, 120

Self-esteem, 31, 94, 108, 110, 166, 167, 202, 204, 207, 245, 246, 250, 254, 297, 319
Self-interest, 96
Separation anxiety, 296, 297, 298
Sepoys, 78
Sex, 141, 180, 184, 203, 211, 258, 259, 260, 262, 269, 277, 278, 282, 284, 285, 290, 295, 310, 311, 315, 321, 325, 329
SHAEF, 147
Shah Soojah, 71, 78
Shaka-Zulu, 241, 276, 326-9
Shelton, Brig. John, 73
Sign stimuli, 180
Simpson, Gen. J., 49, 122
Sims, Admiral William, 210
Simson, Brig. I., 138-40, 142
Singapore, 61, 95, 125, 130-44, 154, 155, 201, 206, 221, 243, 245, 265
Slim, Field-Marshal Viscount, 219, 241, 242, 303, 341-4, 354
Smith-Dorrien, Gen. Sir Horace, 81, 83, 206
Smuts, Field-Marshal Rt Hon. J. C., 93, 272, 378
Snobbishness, 201-3, 311
Social approval, 31, 94, 109, 239, 244, 245, 246, 317
Social class, 400
Social reinforcement standard, 232, 233
Socialization, 189-95, 311
Somme offensive, 82, 88, 89, 115, 245, 372, 387
Soyer, Alexis, 46
Special Operations Executive, 293, 294
Spies, dislike of, 51, 292-4
Spion Kop, 61, 63, 64, 65, 221, 245,
Sport, 116, 160, 290
S.S. war criminals, 309, 311, 313, 345
Staff College, 54, 158, 236, 247, 249, 250, 343
Stalin, Joseph, 346
Status anxiety, 259, 312
Stephenson, William, 294
Stereotypes, 259, 266, 399
Stilwell, Gen. J. W. 161, 343
Stormberg Junction, 56, 59, 61
Strachan, Sir Richard, 154
Stress, 167, 168, 186, 187, 212, 216, 217, 218, 237, 276, 283
Stupidity, 34, 35, 83, 94, 96, 101, 107, 134, 145, 157, 159, 167
Submissiveness, 320
Suez, 38, 45, 131
Superstition, 153, 258, 259, 321, 329
Suvla, 221
Swinton, Maj.-Gen. Sir Ernest, 89

Tal, Gen., 350
Talbot-Coke, Gen., 63, 65

Tank Brigade, 116
Tank Corps, 89, 90, 92, 111
Tanks, 82, 86-92, 114, 116, 117, 119, 124, 125, 127, 132, 136, 138, 246, 265, 388; Kitchener's attitude to, 369n
Technology, 19, 52, 53, 66, 82, 110, 117, 119, 127, 145, 146, 148, 246, 265, 325
Tet offensive, 395
Thinking, 32, 162, 286
Threatening gestures, 181
Time, concern with, 184, 312
Tinker, Mr, M. P., 116
Tobruk, 124, 125, 128
Toilet training, 190, 311, 312
Totemism, 83
Townshend, Maj.-Gen. C. V. F., 95-109, 155, 165, 166, 167, 216, 217, 218, 240, 243, 245, 275
Tradition, 160, 169, 265
Training: neglect of, 54, 66, 135; of officers, 223
Trenchard, Viscount, 132
Tryon, Admiral Sir George, 112, 209, 268, 269
Tsu-Shima, battle of, 186
Turkish army, 96, 98, 99, 105
Turks, 38, 95, 105, 243

Uncertainty, reduction of, 185, 194
Unconventionality, 254
Underestimation of enemy, 47, 54, 61, 80, 127, 134, 148, 153, 235, 266, 322, 325, 353, 402
Uniforms, 46, 54, 177, 313; see also Appearance, 'Bull', Dress
United States, 21, 27, 157, 163, 211, 259, 267, 283, 348, 398; see also America

Venereal disease, 277, 278, 352, 377
Verdun, 83, 155
Victoria Cross, 209
Victoria, H.M.S., 112, 245, 267, 269
Victorian Army, 52, 172, 200, 205, 215
Victorian attitudes, 292, 294, 296, 300, 353
Victorian educational system, 299; see also Education and Muscular Christianity
Victorian family, 138, 172
Vietcong, 397
Vietnam, 21, 27, 265, 278, 305, 395, 396, 397
Vilification of the human, 258
Virility, 141, 200, 211, 238, 319
Vocational selection, 254, 266, 289
Vocations, comparisons between, 170
Von Donop, Gen., 87
Von Rundstedt, Field-Marshal G., 322, 356
Voyeurism, 295

War cabinet, 93

War Colleges, 299
War Ministry, 114
War Office, 53, 56, 86, 88, 110, 113, 116, 118, 135, 163
War and sport, 44
Warfare, modern, 20, 123, 265
Wastage of manpower, 80, 105, 124, 152, 153, 327
Wauchope, Maj.-Gen. Andy, 58
Welfare, 45, 53, 76, 78, 104, 108, 275, 324, 327, 328, 359
Wellington, Duke of, 36, 100, 123, 219, 221, 226, 239, 241, 272, 273, 275, 276, 277, 282, 294, 324-6
Western Front, 80, 86
White, Gen. Sir George, 60
Wilson, Field-Marshal Sir Henry, 158, 251, 378
Wingate, Maj.-Gen. O. C., 343
Wolfe, Gen. Sir James, 212, 275, 323-4
Wolseley, Field-Marshal Lord, 49, 50, 56, 208, 247, 316

Wood, Field-Marshal Sir Evelyn, 53, 268n
Woollcombe, Lieut.-Gen. Sir C. L., 91
World War: First, 22, 45, 54, 80-110, 115, 119, 151, 172, 179, 207, 210, 250, 284, 290, 306, 320, 391, 396; Second, 21, 45, 87, 118, 123, 161, 173, 221, 264, 294, 306
Wounded, care of, 78, 98, 102, 103, 264

Yamamoto, Admiral I., 343
Yom Kippur War, 353
Ypres, 83, 84, 92
Ypres, 3rd battle of, 82, 89, 245, 372-4, 379, 380, 387

Zeitzler, Col.-Gen. Kurt, 321, 322
Zhukov, Marshal Georgi, 239, 241, 272, 273, 346-7
Zululand Wars, 56